Chandan Singh   Gurpreet Singh Lehal
Jyotsna Sengupta   Dharam Veer Sharma
Vishal Goyal (Eds.)

Information Systems
for Indian Languages

International Conference, ICISIL 2011
Patiala, India, March 9-11, 2011
Proceedings

 Springer

Volume Editors

Chandan Singh
Department of Computer Science, Punjabi University, Patiala, India
E-mail: chandan.csp@gmail.com

Gurpreet Singh Lehal
Department of Computer Science, Punjabi University, Patiala, India
E-mail: gslehal@gmail.com

Jyotsna Sengupta
Department of Computer Science, Punjabi University, Patiala, India
E-mail: jyotsna.sengupta@gmail.com

Dharam Veer Sharma
Department of Computer Science, Punjabi University, Patiala, India
E-mail: dveer72@gmail.com

Vishal Goyal
Department of Computer Science, Punjabi University, Patiala, India
E-mail: vishal.pup@gmail.com

ISSN 1865-0929   e-ISSN 1865-0937
ISBN 978-3-642-19402-3   e-ISBN 978-3-642-19403-0
DOI 10.1007/978-3-642-19403-0
Springer Heidelberg Dordrecht London New York

Library of Congress Control Number: 2011921553

CR Subject Classification (1998): I.2.7, F.4.2-3, I.2, H.3, I.7

Typesetting: Camera-ready by author, data conversion by Scientific Publishing Services, Chennai, India

Printed on acid-free paper

Springer is part of Springer Science+Business Media (www.springer.com)

Communications
in Computer and Information Science 139

Preface

The International Conference on Information Systems for Indian Languages (ICISIL 2011) has provided a first-of-its-kind platform to scientists, engineers, professionals, researchers and practitioners from India and abroad working on automation of Indian languages. The primary goal of the conference is to present state-of-the-art techniques and promote collaborative research and developmental activities in natural language processing of Indian languages and related fields. This pioneering effort has won many accolades and brought together the community, otherwise so diverse.

In the technical session, seven topics were identified relating to the theme of the conferences. In all, 126 submissions were received. Each paper was sent to four reviewers from within India and abroad. Each submission was reviewed by at least three reviewers. Based on the reviewers' evaluation, 58 papers were selected, 27 for oral presentation and 31 for poster. Some other good papers were left out due to paucity of time at the conference.

A special demo session was also arranged to demonstrate the systems developed for Indian languages. The demo session generated a good response, and 25 demos of working systems for text-to-speech, optical character recognition, M translation, and Web corpora for Indian languages were received. Finally, 14 systems were selected and their abstracts are published in the conference proceedings.

Very special thanks are due to all the reviewers who extended their maximum co-operation in finishing the job of reviewing and selecting the papers. We wish to express our thanks to Angarai Ganesan Ramakrishnan for organizing a special tutorial on "Online Handwritten Recognition". We deeply acknowledge the financial assistance provided by UGC and DST.

Many deserve special thanks for contributing to the cause of the conference, including all the faculty and administrative staff members of the Department of Computer Science, Punjabi University, Patiala, as well as the entire staff of the Advanced Centre for Technical Development of Punjabi Language, Literature and Culture.

We aim to make this conference a regular event and hope you will enjoy reading the proceedings ICISIL 2011! We heartily thank everybody who contributed in making ICISIL 2011 a grand success.

Chandan Singh
Gurpreet Singh Lehal
Jyotsna Sengupta
Dharam Veer Sharma
Vishal Goyal

Organization

ICISIL 2011 Committee

Patron

Jaspal Singh, Vice-Chancellor, Punjabi University, Patiala, India

General Chair

Santanu Chaudhury, IIT, Delhi, India

Program Chair

Gurpreet Singh Lehal

Organizing Chairs

Dharam Veer Sharma
Sukhjeet Kaur Ranade
Amandeep Kaur
Vishal Goyal

Editorial Chairs (LNCS Volume Editors)

Chandan Singh
Gurpreet Singh Lehal
Jyotsna Sengupta
Dharam Veer Sharma
Vishal Goyal

Liaison Chairs

Gagandeep Kaur
Vishal Goyal
Arun Sharma

Finance Chairs

Neeraj Sharma
Arun Bansal
Maninder Singh

Advisory Board

Adam Kilgarriff	Lexical Computing Ltd., UK
Amba Kulkarni	University of Hyderabad, India
Aravind Joshi	Univeristy of Pennsylvania, USA
B.B. Chaudhuri	ISI, Kolkatta, India
Gerald Penn	University of Toronto, Canada
Hema Murthy	IIT, Chennai, India
K.K. Bhardwaj	JNU, India
K.K. Biswas	IIT Delhi, India
Karmeshu	JNU, India
Kavi Narayana Murthy	University of Hyderabad, India
Mahesh Kulkarni	CDAC, Pune, India
Martin Kay	University of Saarland, Germany
Naftali Tishby	The Herbew University, Israel
Owen Rambow	Columbia University, USA
Sarmad Hussain	NUCES, Pakistan
Srinivas Bangalore	AT&T Labs, USA
Srirangaraj Setlur	CEDAR, USA
Venu Govindaraju	University at Buffalo, USA

Reviewers

Adam Kilgarriff	Lexical Computing Ltd., UK
Aditi Sharan	Jawahar Lal Nehru University, India
Amba P. Kulkarni	University of Hyderabad, India
Asif Ekbal	University of Trento, Italy
Avinash Singh	Machine Intelligence & Learning Labs, India
Besacier	University of Grenoble, France
Dharamveer Sharma	Punjabi University, India
Dhiren B. Patel	Gujarat Vidyapith, India
Dorai Swamy	Eeaswari Engineering College, India
Ekta Walia	M. M. University, India
Eric Castelli	Hanoi University of Science and Technology, Vietnam
Girish Nath Jha	Jawahar Lal Nehru University, India
Gurpreet Singh Josan	Rayat & Bahra Institute of Engineering & BioTechnology, India
Gurpreet Singh Lehal	Punjabi University, India
Hardeep Singh	Guru Nanak Dev University, India
Jugal Kalita	University of Colorado, USA
K.G. Sulochana	C-DAC Trivandrum, India
Kandarpa Kumar Sarma	Gauhati University, India
Karunesh Arora	C-DAC Noida, India
M.B. Rajarshi	University of Pune, India

Mohammad Abid Khan	University of Peshawar, Pakistan
Monojit Choudhury	Microsoft Research Lab, India
Kiran Pala	IIIT Hyderabad, India
Muhammad Ghulam Abbas Malik	CIIT Lahore, Pakistan
Muhammad Humayoun	University of Savoie, France
Manish Jindal	Panjab University, India
Neeta Shah	Gujarat Informatics Ltd., India
Parminder Singh	Guru Nanak Dev Engineering College, India
Prateek Bhatia	Thapar University, India
Rajesh Kumar Aggarwal	NIT Kurukshetra, India
Renu Dhir	NIT Jalandhar, India
Ritesh Srivastava	IMSEC Ghaziabad, India
S.P. Kishore	IIIT Hyderabad, India
Sarabjit Singh	Panjab University, India
Sarmad Hussain	CRULP, Pakistan
Sivaji Bandyopadhyay	Jadavpur University, India
Sopan Kolte	K. J. Somaiya Polytechnic, India
Srinivas Bangalore	AT&T Labs, USA
Suryakant Gangashetty	IIIT Hyderabad, India
Swapnil Belhe	C-DAC Pune, India
Tushar Patnaik	C-DAC Noida, India
Vishal Goyal	Punjabi University, India
Yogesh Chabba	Guru Jambeshwar University, India
Zak Shafran	Center for Spoken Language Understanding, USA

Table of Contents

Oral

Poster

Demo Abstracts

A Novel Method to Segment Online Gurmukhi Script

Manoj K. Sachan[1], Gurpreet Singh Lehal[2], and Vijender Kumar Jain[1]

[1] Sant Longowal Institute of Engineering & Technology, Longowal, India
[2] Punjabi University Patiala, India
manojsachan@gmail.com

Abstract. Segmentation of handwritten script in general and Gurmukhi Script in particular is a critical task due to the type of shapes of character and large variation in writing style of different users. The data captured for online Gurmukhi Script consists of x,y coordinates of pen position on the tablet, pressure of the pen on the tablet, time of each point and pen down status of the pen. A novel method to segment the Gurmukhi script based on pressure, pen down status and time together is presented in the paper. In case of some characters getting over segmented due to the shape of character and user's style of writing, a method to merge the substrokes is presented. The proposed algorithms have been applied on a set of 2150 words and have given very good results.

Keywords: Stroke, Substroke, Merging, Segmentation, Gurmukhi, Devanagari.

1   Introduction

The process of online handwriting recognition consists of steps such as preprocessing of the user handwriting, segmentation of script into meaningful units or shapes, recognition of shapes, and post processing to refine the results of recognition [1, 2]. Gurmukhi Script like Devanagari script consists of large number of strokes with high variation. The number and type of strokes constituting the character may vary from writer to writer. For example as shown in fig 1a and 1b, the character ਲ (lalla) is written with three strokes and two strokes by two different writers respectively.

ਲ							
	Stroke1	Stroke2	Stroke3			Stroke1	Stroke2

Fig. 1. a) ਲ(lalla) = stroke1+stroke2+stroke3 b) ਲ(lalla) = stroke1+stroke2

Similarly, the headline (shirorekha) which is an important feature of Gurmukhi/Devanagari script is generally drawn from left to right as one stroke as shown in fig 2b, but in some cases it may be drawn in parts. For example, the word ਧੱਕਾ (yakka) is written using two headline stroke as shown in fig 2a and with one headline stroke as shown in fig 2b.

C. Singh et al. (Eds.): ICISIL 2011, CCIS 139, pp. 1–8, 2011.

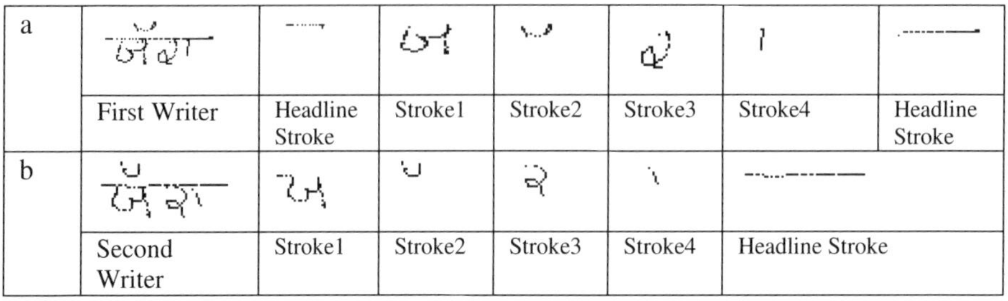

a	ਗੁਰਾ		ੳਸ਼		ੳ	1	
	First Writer	Headline Stroke	Stroke1	Stroke2	Stroke3	Stroke4	Headline Stroke
b	ਚੇਸ਼ ਗੁ	ੳਸ਼	ੳ	੨	\		
	Second Writer	Stroke1	Stroke2	Stroke3	Stroke4	Headline Stroke	

Fig. 2. Words with single or multiple headlines

Thus, there is high variation in the number and type of strokes constituting the Gurmukhi word. In the past, Niranjan Joshi et al. [10] have used syntactic and structural approaches to recognize and segment shirorekha and vowels from isolated Devanagari characters. Anuj Sharma et.al [14] have used point based method to segment isolated Gurmukhi character. In this method, each stroke is observed for number of points and if number of points exceed 300 (an empirically observed value) and the direction of stroke at that point is less than 90 degree, the stroke is segmented at that point. We have taken complete word instead of isolated character in online Gurmukhi Script for segmentation. In subsequent sections, the techniques of segmenting online Gurmukhi Script are described. [3, 4, 12].

2   Characteristics of Gurmukhi

The script of Gurmukhi is cursive and has 41 consonants, 12 vowels and 2 half characters which lie at the feet of consonants .The consonants, vowels and half characters of Gurmukhi are shown in Fig 3.

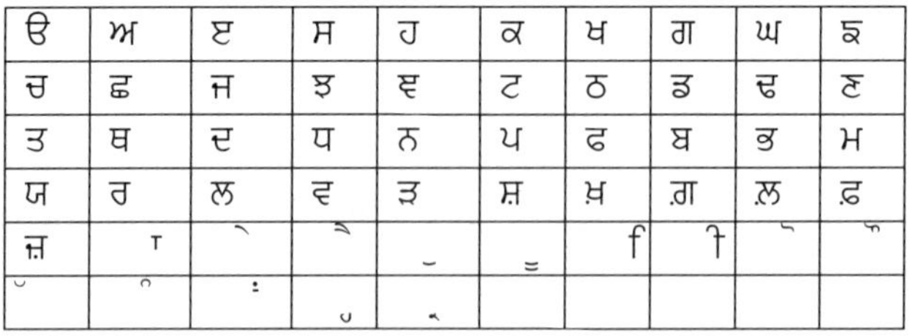

ੳ	ਅ	ੲ	ਸ	ਹ	ਕ	ਖ	ਗ	ਘ	ਙ
ਚ	ਛ	ਜ	ਝ	ਞ	ਟ	ਠ	ਡ	ਢ	ਣ
ਤ	ਥ	ਦ	ਧ	ਨ	ਪ	ਫ	ਬ	ਭ	ਮ
ਯ	ਰ	ਲ	ਵ	ੜ	ਸ਼	ਖ਼	ਗ਼	ਲ਼	ਫ਼
ੜ	T	੍	੖	੍	ੲ	f	ੀ	੝	੝
੝	ੰ	:		ੁ	੍				

Fig. 3. Gurmukhi character set

Most of the characters have a horizontal line (shirorekha) at the upper part. The characters of word are connected mostly by this line called the head line (shirorekha). Hence there is no vertical inter-character gap in the letters of the word. A word in the Gurmukhi can be partitioned into three horizontal zones. The upper zone denotes the region above the head line where vowels reside, while the middle zone represents the area below the head line where the consonants and some parts of vowels are present. The middle zone is the busiest zone. The lower zone represents the area below the middle zone where some vowels and some half characters lie at the foot of consonants. The three zones are shown in Fig 4. [3, 4, 12].

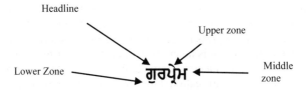

Fig. 4. Three zones in Gurmukhi Script

3 Data Capture and Preprocessing

The data is collected through a Wacom Tablet and Digitizer Pen. As the pen touches the tablet and moves on the surface it sends data packets. The data captured in each packet includes pressure of the pen, position of the pen in terms of x, y coordinates, button which captures the touch of pentip on the surface and time which captures system time of capturing each packet. Fig 5 shows some of the samples of the data captured. In general, the Gurmukhi script is written left to right. Each word is composed of a sequence of strokes. Each stroke consists of the trace of pentip from pendown to penup. From segmentation point of view, the pre-processing steps followed are rescaling and duplicate point removal.

Rescaling: In rescaling, the x,y coordinates are rescaled to origin. The pressure and time are also recalculated w.r.t to their minimum values. The button value remains unchanged.

Duplicate Point Removal: Each stroke contains a large number of duplicate points that occur either in the beginning of the stroke or when the stylus or digitizer pen is stationary on the surface of the tablet. In duplicate point removal, the point (x_i, y_i) is compared with point (x_{i+1}, y_{i+1}). If $(x_i == x_{i+1})$ and $(y_i == y_{i+1})$ then point (x_{i+1}, y_{i+1}) is removed from the stroke.[11]. The pressure and button component at (x_i, y_i) remains unchanged. The time component is averaged and is stored at point (x_i, y_i).

a	日日日日	b	日日日日
c	日日日日	d	日日日日

Fig. 5. Samples of words captured

4 Proposed Segmentation Algorithm

The segmentation technique is based on the idea to separate the strokes based on the way user write them on the tablet. The segmetation algorithm consists of two phases namely Extraction of Strokes and Merging of Strokes.

4.1 Extraction of Strokes

In this step, the strokes in the word as written by the user are extracted. The packet captured by the digitizer tablet consists of information such as button, pressure and time apart from x-y position.When the user places the digitizer pen on the tablet surface, the pressure increases and when the pen is taken off the tablet surface it approaches to zero. Thus, a pressure gap is created in every stroke. Also a time gap is created when the digitizer pen is takenoff the tablet and put on the tablet. Each tip of the digitizer pen have some identification number. Thus, when the tip of digitizer pen touches the surface its identification number is recorded with every packet. The information in button field is the identification number of the pentip. Therefore the strokes in the word can be extratcted on the basis of pressure, time and button. In our algorithm, a combination of all parameters are used for extracting the strokes. The algorithm for extraction of strokes named as ESPPBT (Extraction of strokes based on position,pressure,button and time) is described below.

```
Algorithm ESPPBT
    varibles
    x,y // x,y coordinates of pen
    button    // pentip identification
    pressure // pressure of pentip
    time_diff_cur // current time - prev time
    time_diff_next // next time - current time
    pressure_diff  //diff in pressure
    pressure_th // pressure threshold
    time_i //time at i^th point,
    time_{i+1} //time at i+1^th point,
    number_of_points // total number of points,
    input[] //array of all input packets,
    strokes[][]  //stroke data,
   packet_data//packet of all input parameters from tablet
   point_id // index into stroke data,
    stroke_id //index of strokes
1.Repeat for  i = 0 to number_of_points
        if button == 0
          {  x← -1; y← -1;
            pressure_diff← pressure_i - pressure_{i-1};
            time_diff_cur←time_i-time_{i-1};
            time_diff_next←time_{i+1}-time_i    }
2. Repeat for  i = 0 to number_of_points
        If(x! = -1)and(y! = -1)and(button! = 0)
        and(pressure_diff>pressure_th)and
        (time_diff_current<time_th)and
        (time_diff_next < time_th)
        {
        strokes[stroke_id][point_id].packet_data
        = input[point_id].packet_data;
        point_id←point_id+1;
        }
        else   stroke_id ← stroke_id + 1;
```

The words shown in fig 5a - 5d have been segmented by algorithm ESPPBT and the results are shown in fig 6a – 6d respectively. The results of segmentation show that the strokes are extracted as they are written by the writer on the tablet. All the strokes show left to right order. The fig 6a shows that the strokes forming the ਅੰਬ (pronounced as 'amb') is clearly segmented into characters ਅ, ਬ and ੋ respectively. The fig 6b to fig 6d show that the characters of words are segmented into substrokes or subparts because the writer has written the characters in that way. For example, characters ੲ, ਜ and ੲ as shown in fig 6b and character ਜ as shown in fig 6c is segmented into two substrokes. However, character ਲ਼ as shown in fig 6d, is segmented into three substrokes.

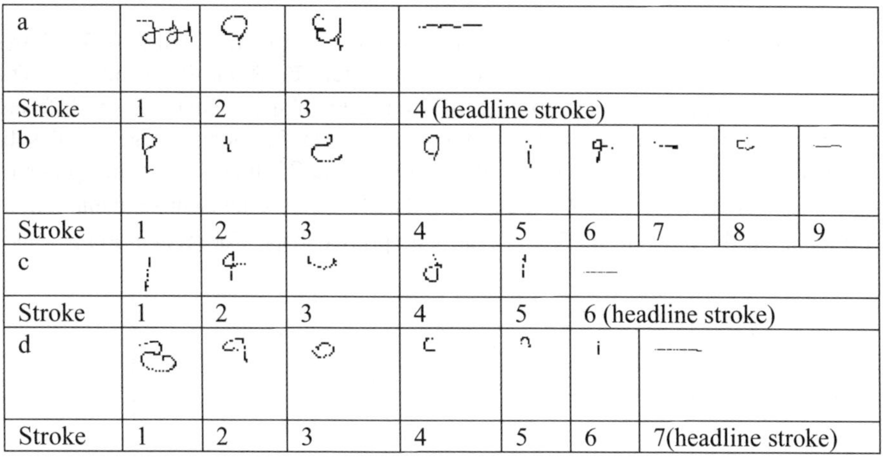

a	ਤਮ	੧	੪	⎯⎯⎯					
Stroke	1	2	3	4 (headline stroke)					
b	੧	੧	੮	੧	੸	੯·	·⎯	੮	⎯
Stroke	1	2	3	4	5	6	7	8	9
c	੶	੯··	ਖ਼	ੰ	੸	⎯⎯			
Stroke	1	2	3	4	5	6 (headline stroke)			
d	੭	੮੧	੦	੮	੭	੸	⎯⎯⎯		
Stroke	1	2	3	4	5	6	7(headline stroke)		

Fig. 6. Extraction of Strokes

The results of segmentation also show that headline is extracted as a different stroke as can be seen from fig 6a to fig 6d. Thus, the characters which were segmented into substrokes as shown in fig 6b to fig 6d requires merging of substrokes in order to form proper character shapes.

4.2 Merging of Substrokes

In merging of substrokes, the two substrokes are compared to find whether the substrokes lie below the headline and overlap horizontally. If the substrokes overlap then these can be merged. The algorithm for merging of strokes (MOS) is described below.

```
Algorithm MOS
    varibles
        x,y //position coordinates of each point in a stroke
        strxmin,strxmax //xmin and xmax for each stroke
```

```
strymin,strymax //ymin and ymax for each stroke
strqueue //queue of all the strokes of a word
    1. for each stroke find strxmin,strxmax,
       strymin, strymax ;
    2. Store all the strokes in strqueue
    3. Find the headline stroke
4. Repeat while (strqueue !=empty) thru 4a-4b
           4a. Pick stroke_i and stroke_{i+1};
           4b. if stroke_i and stroke_{i+1} are  below
               Headline stroke and overlap
               horizontally then
               merge stroke_{i+1} and  stroke_i
```

The results of mergence are shown in fig 7. Comparison of fig 7a and fig 7b shows that subparts or substrokes which form the characters ੲ, ਜ and ੲ are merged. The merging of substrokes or subparts are illustrated by line drawing among different parts in fig 7. Similarly, comparison of fig 7c and fig 7d shows mergence of sub-strokes of character ਜ , comparison of fig 7e and fig 7f shows shows mergence of substrokes forming character ਲ . The fig 7b,7d and 7f shows the clear segmentation of Gurmukhi words shown in fig 5b to fig 5d into proper shapes for recognition.

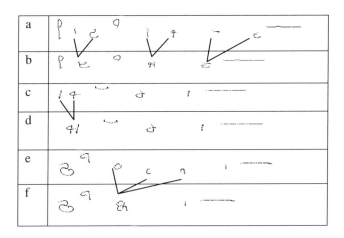

Fig. 7. Mergence of sub-strokes

5   Results, Discussions and Future Scope

The results of ESPPBT produces substrokes constituting the individual Gurmukhi character. The parameters such as pressure, time and button are used only to segment the strokes from Gurmukhi word. The number of substrokes constituting individual Gurmukhi character varies from writer to writer. It is difficult to fix the substrokes for individual Gurmukhi character. Therefore, the substrokes constituting the Gurmukhi character are merged using MOS to form the proper Gurmukhi character shape. The correctness of the MOS algorithm is checked manually by comparing the outcome with the desired shape of Gurmukhi character. The final outcome from MOS algorithm is size normalized to 16x16. The sized normalized character is used for feature extraction and classification. The proposed algorithm for extraction of strokes and merging of substrokes were applied on 2150 words collected from 50 users. The extraction of strokes algorithm (ESPPBT) gives 100% accuracy. However the merging of substrokes algorithm (MOS) gives accuracy varying between 80% to 100% with the average accuracy being 86.4%. The MOS algorithm gives inaccurate results in the following cases:

- If the words are written slanted or the headline drawn is slanted then it results in incorrect finding of headline stroke   which results in incorrect merging of substrokes. For example, in fig 8a, subcharacter '-' of character ਥ (thatha) is detected as headline stroke due to which the subcharacter '-' is not considered for merging with stroke ਪ.
- In some cases, the stroke forming the vowels such as aunkar '' or dulainkar '' is detected as headline, resulting in incorrect merging of substrokes. For example in fig 8b vowel aunkar is detected as headline as the desired headline stroke is slanted.
- If the stroke to be merged is cutting above the headline at some point then it results in incorrect merging. For example in fig 8c the substroke or subcharacter of character ੲ is cutting the headline.

Thus, MOS algorithm gives inaccurate results mainly due to the incorrect finding of headline stroke or improper alignment of headline stroke. In case, the user writes the strokes in proper left to right direction then the performance of both algorithms touches 100%. The performance of MOS algorithm can be improved by adding smoothing operations on strokes and by incorporating fuzzy rules for merging.

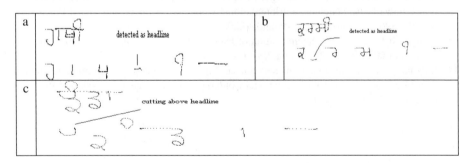

Fig. 8. Cases of failure of MOS algorithm

References

1. Tappert, C.C., Suen, C.Y., Wakahara, T.: The State of the Art in Online Handwriting Recognition. IEEE Trans. on Pattern Analysis and Machine Intelligence 12(8), 787–808 (1990)

2. Plamondon, R., Srihari, S.N.: On-Line and Off-Line Handwriting Recognition A Comprehensive Survey. IEEE Trans. on Pattern Analysis and Machine Intelligence 22(1), 63–84 (2000)
3. Lehal, G.S., Singh, C.: A Gurmukhi Script Recognition System. In: International Conference on Pattern Recognition, pp. 557–560. IEEE Press, Los Alamitos (2000)
4. Connell, S.D., Sinha, R.M.K., Jain, A.K.: Recognition of Unconstrained On-Line Devanagari Characters. In: International Conference on Pattern Recognition, pp. 368–371. IEEE Press, Los Alamitos (2000)
5. Connell, S.D., Jain, A.K.: Template- based online character recognition. Pattern Recognition 34, 1–14 (2001)
6. Deepu, V., Sriganesh, M., Ramakrishnan, A.G.: Principal Component Analysis for Online Handwritten Character Recognition. In: 17th International Conference on Pattern Recognition. IEEE Press, Cambridge (2004)
7. Aparna, K.H., Subramanian, V., Kasrajan, M., Prakash, G.V., Chakravarthy, V.S.: Online Handwritten Recognition for Tamil. In: 9th International Workshop on Frontiers in Handwriting Recognition, pp. 438–443. IEEE Press, Los Alamitos (2004)
8. Joshi, N., Sita, G., Ramakrishnan, A.G., Madhavanath, S.: Comparison of Elastic Matching Algorithm for Online Tamil Handwritten Character Recognition. In: 9th International Workshop on Frontiers in Handwriting Recognition, pp. 444–449. IEEE Press, Los Alamitos (2004)
9. Namboodiri, M., Jain, A.K.: Online Handwritten Script Recognition. IEEE Trans. on Pattern Analysis and Machine Intelligence 26(1), 124–130 (2004)
10. Joshi, N., Sita, G., Ramakrishnan, A.G., Madhavanath, S., Deepu, V.: Machine Recognition of Online Handwritten Devanagari Characters. In: 8th International Conference on Document Analysis and Research, vol. 2, pp. 1156–1160. IEEE Press, Los Alamitos (2005)
11. Huang, B.Q., Zhang, Y.B., Kechadi, M.T.: Preprocessing Techniques for Online Handwriting Recognition. In: 7th International Conference on Intelligent Systems Design and Applications, pp. 793–798. IEEE Press, Los Alamitos (2007)
12. Sharma, A., Kumar, R., Sharma, R.K.: Online Handwritten Gurmukhi Character Recognition Using Elastic Matching. In: International Congress on Image and Signal Processing, pp. 391–396. IEEE Press, Los Alamitos (2008)
13. Sharma, R.K., Singh, A.: Segmentation of Handwritten Text in Gurmukhi Script. International Journal of Image Processing 2(3), 12–17 (2008)
14. Sharma, A.: Rearrangement of Recognized Strokes in Online Handwritten Gurmukhi Word Recognition. In: 10th International Conference on Document Analysis and Recognition, pp. 1241–1245 (2009)

Automatic Speech Segmentation and Multi Level Labeling Tool

R. Ravindra Kumar, K.G. Sulochana, and Jose Stephen

Centre for Development of Advance Computing, Vellayambalam,Trivandrum, Kerala
{ravi,sulochana,jose_stephen}@cdactvm.in

Abstract. An accurate, properly labeled speech corpus is very important for speech research. However, manual segmentation and labeling is very laborious and error prone. This paper describes an automatic tool for segmenting and labeling of Malayalam speech data. The tool is based on Hidden Markov Model (HMM). HMM Tool Kit is used for training, segmentation and labeling the data. Special care was taken in the preparation of pronunciation dictionary so that it will cover most of the possible pronunciation variations. Syllabification rule is applied in the phone label for generating syllable label also.. Segmentation and labeling experiment was done on the speech corpus collected for building text-to-speech system. The performance of the tool is reasonably good as it shows only 19ms average deviation compared to manual labels.

Keywords: Annotated Speech Corpus, Speech segmentation, Speech labeling, automatic segmentation tool, Hidden Markov Model, HMM Tool Kit.

1 Introduction

Annotated speech corpus is the prime raw material in spoken language research [5]. It is an essential component for both Automatic Speech Recognition (ASR) and Text-to-Speech Systems (TTS). Segmenting and labeling of speech corpus is the main bottleneck in the creation of annotated speech corpus. It involves defining a speech segment and assigning suitable label to it. The manual segmentation and labeling are laborious, time consuming and error prone. These difficulties make the development of segmentation and labeling tool, either automatic or semi automatic a necessity.

The core technology behind either automatic or semi automatic labeling is speech recognition. In the segmentation process the speech recognition is aided by the transcription of corresponding speech. Even though there are many approaches available for speech recognition, the sole technique that gains the acceptance of the researchers is Hidden Markov Model (HMM). HMM based technique is a statistical based model of speech recognition and requires training before using it for the segmentation and labeling. The inputs for the training are speech, its transcription and the pronunciation dictionary. The pronunciation dictionary maps each word in the transcription to its corresponding pronunciation. The functional block diagram of the tool is given in figure 1.

C. Singh et al. (Eds.): ICISIL 2011, CCIS 139, pp. 9–14, 2011.

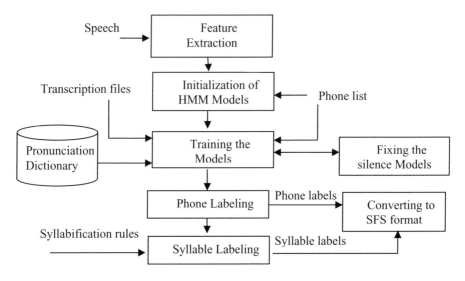

Fig. 1. The functional block diagram of the system

2   System – Overview

2.1   Feature Extraction

Feature extraction is the process of transforming the speech waveform into a set of feature vectors. The system is configured to extract features based on Mel Frequency Cepstral Coefficients (MFCCs) [2]. Speech will be converted to a sequence of feature vectors of length 39. This number, 39, is computed from the length of the parameterized static vector ($MFCC\_0 = 13$) plus the delta coefficients (+13) plus the acceleration coefficients (+13) [2].   System is configured with 10ms frame rate with 25ms Hamming window.

2.2   Initialization of HMM Models

HMM initialization process involves the creation of HMM definition file, finding and storing of global mean and variance. The HMM definition file contains

- Type of observation vector
- Number and width of each data stream
- Number of states
- Transition matrix
- Mixture component weights or discrete probabilities
- Means and covariances

There will be HMM definition for each phone in the phone list. Three state left-to-right HMM with no skip is used to represent each phone. The command "HCompV" scans the data, and computes the global mean and variance for the whole corpus and outputs to a file [2].

2.3 Training the Models

The inputs to the training module are parameterized speech files, transcription, pronunciation dictionary and phone list.

Creation of transcription files. Three levels of transcription of the input speech are essential for the functioning of the tool ie sentence, word and phone level

The sentence level transcription is kept in the following format

text0001 yuvatiye mar:r:evit:eyoo vechch kolappet:uttiya ......
text0002 yuur:oopp r:ashhya jeitaakkal:~ khattar:in!~r:e ......

The word level transcription was done by using a script file and phone level transcription is done using HLEd command in HTK tool kit [2].

Creation of pronunciation dictionary. The pronunciation dictionary maps the orthographic representation in the transcription file to its corresponding pronunciations. For a language like Malayalam which is phonetic, creation of pronunciation dictionary is considered to be comparatively easy. But the existence of words pronounced differently from orthographical representation and common usage of foreign words having valid multiple pronunciations makes the creation of pronunciation dictionary difficult.

Malayalam in general has three types of words [1]

Type-1- Pronunciation in correspondence to the orthographic representation
Type-2- Pronunciation different from respective orthographic representation
Type-3- Pronunciation different from respective orthographic representation and having multiple valid pronunciations

Eg:-   utpannam   u t p a n# n# a m
          utpannam(2)   u l p a n# n# a m

For accurate labeling, all the pronunciation variations for the words in the transcript file should also be available in the pronunciation dictionary. A hybrid method combining both the rule and statistical method is applied to generate the pronunciation variations [1].

The pronunciation dictionary is then converted to HTK format by using the HTK tool kit which take the phone list, created pronunciation dictionary and the sorted unique word list as input and creates dictionary in HTK format.

Training process. Training constitutes the re-estimation of initial parameters such as means, covariances and transition matrix. The training process is iterative and uses the Baum-Welsh algorithm for re-estimating the parameters and stores it to a file. After the three iterations the silence model is fixed and then the training iteration will be continued up to ten so that parameters get stabilized.

2.4 Fixing the Silence Models

There are two types of silence in the input speech data, one at the beginning and end of sentences which is long (sil) and other between the phrases which is short (sp). The fixing of silence models involves the following processes [2].

- Making the long silence model more robust to various impulsive noises in the speech data by adding an extra transition between states.
- Creating one state short pause model which has direct transition from entry to exit node. This model has its emitting state tied to the centre of the silence model

2.5 Phone Labeling

The phone models created so far can be used to realign the training data and create new transcriptions. It is implemented using the Viterbi Algorithm, an ingenuous method for finding the most likely sequence in a probability distribution. It requires the "pronunciation'" for the silence model to be present in the dictionary.

The following is the format of the labeled output

```
#!MLF!#
"mfc/text0001.lab"  # the name of the file
# Format :-   Start duration <space> End duration<space> phone name<space>
Word [optional]
0 6400000 SIL SIL
6400000 7400000 y yu
7400000 7900000 u
12200000 12200000 sp
```

2.6 Syllable Labeling

The phone labels from the previous module are converted to syllable label by applying the syllabification rule [4]. The phone segments are merged and segment durations are recalculated.

The labeled output is as follows

```
0 6400000 SIL SIL
6400000 7900000 yu  yu
12200000 12200000 sp
```

2.7 Converting to SFS Format

Speech Filing System (SFS) is used as a visualizing tool for the labeled file. It can also be used for manual correction of the labels and train them for getting more accurate results.

The following steps are to be executed for viewing both speech and labeled file in SFS [3]

1. Creates a SFS file- hed -n <name of SFS file>
2. Loads the label file- anload -H <name of label file> <name of transcription file> <name of the created SFS file>

3. Link the corresponding speech file to the created SFS file- slink -isp -f
    <sampling rate> <path speech file> <name of SFS file>

After executing these steps open the SFS file and check both label and speech file and
click view for viewing both labeled and speech file in the same window as given in
Fig 2.

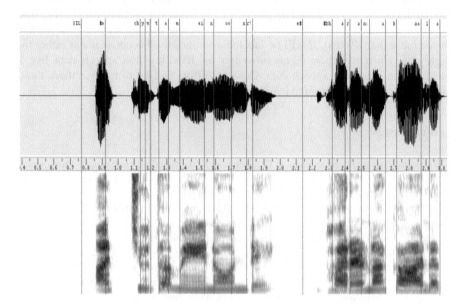

Fig. 2. Screenshot of labeled file

3 Performance Results

The tool has been tested to label Malayalam speech (~ 700 sentences). The difference
between estimated begin-end points and the actual begin-end points was computed
and summed to call it as deviation. The segments are broadly classified as consonants
(other than nasals), vowels and nasals. The deviations were computed for all the segments and an average deviation noted in milliseconds is used as measure of performance. The results are tabularized in Table 1. The speed of labeling is also found to be
reasonably good as it took only 25% less time compared to EHMM based labeling
tool provided in festival.

Table 1. The results of testing

Segment type	Avg. Deviation (ms)
Nasals	22
Consonants (Other than nasals)	15
Vowels	20

4 Conclusion

The comparison of labels generated by the tool with the manual one shows only an average deviation of 19 ms.. The labeling speed is also found to be quite satisfactory with around 5second per sentence. But the average deviation of nasals is found to be more compared to others. The syllable labels are derived from phone label and therefore the deviation will be added up and hence the error will be more for syllable labeling. Another limitation of the tool is with the labeling of long silence (i. e. with the silence more than 40 ms). It will be labeled as a part of adjacent segment rather than single one. Currently there is no provision for handling non speech data like lip sound, cough etc in the speech data. Our next effort will be to remove these limitations and improve the system to handle any speech data.

References

1. Ravindra Kumar, R., Sulochana, K.G., Stephen, J.: Automatic Generation of Pronunciation Lexicon for Malayalam-A Hybrid Approach: LDC-IL, Creation of Multilingual Speech Resources Academic and Technical Issues, 1–5 (2010)
2. Young, S., Evermann, G., et al.: The HTK Book, pp. 23–47. Cambridge University Engineering Department (2006), http://htk.eng.cam.ac.uk/docs/docs.shtml
3. How To: Use HTK Hidden Markov modeling toolkit with SFS, http://www.phon.ucl.ac.uk/resource/sfs/howto/htk.htm
4. Prabodhachandran Nayar, V.R.: Swanavijnanam, Kerala Bhasha Institute, 115–123 (1980)
5. Annotated Speech Corpora Development in Indian languages – A Technical Report on Process and Standards, Speech processing lab, C-DAC Kolkata, 2–3

Computational Aspect of Verb Classification in Malayalam

R. Ravindra Kumar, K.G. Sulochana, and V. Jayan

Centre for Development of Advanced Computing (C-DAC),
Thiruvananthapuram, Kerala, India
{ravi,sulochana,jayan}@cdactvm.in

Abstract. In applications like Morphological Analyzer, Machine Aided Translation (MAT), Spell checker, etc. the verb synthesis and or generation are prime tasks. For paradigm approach verb classification is needed. There exist many verb classifications in Malayalam. Suranad Kunjan Pillai's classification contains sixteen classes, Wickremasinghe and Menon proposed eight, Sekhar and Glazov have twelve, Asher and Prabodhchandran Nair have four and Valentine have two.[1] All descriptions focus on past tense forms, because the much simpler forms present and future tense forms are easily predictable. In regard to verbs an entirely new item of work had to be undertaken. The verbs in the language present a multiplicity of conjugational forms which may perplex anyone who is not thoroughly familiar with them.[3] This paper is focused on the classification of verbs based on the past forms and the morphophonemic changes in the verb roots. This classification is basically done for the rule based MAT System and can be used in the similar NLP applications.

Keywords: Verb morphology, Morphophonemic changes, verb classes, causative suffix.

1 Introduction

The first step in natural language processing (NLP) is to recognize the words in a sentence. We have to look into the way they are created, placement of morphemes in a word, combinations of the morphemes or words and the rules associated with the formation of a semantic category... The analysis of words will provide the syntactic and semantic information. There must be an accord between the language rules and computation. An effective linguistic rule customized for computational purpose can make a paradigm shift in the field of NLP. Malayalam is a morphologically rich language. Even a complete sentence can be combined together to form a word. Decomposition of such sentence is highly complicated. A verb can have up to ten suffix attachments. [5] This makes the analysis of verb a complex task. Notable works on Malayalam verb classification are by Prof. A.R.Rajaraja Varma and Prof. Suranad Kunjan Pillai (SKP). We have derived the 53 classes from the 16 classes of SKPs verb classification. Roman notations used for Malayalam representation is given in appendix 1.

C. Singh et al. (Eds.): ICISIL 2011, CCIS 139, pp. 15–22, 2011.
© Springer-Verlag Berlin Heidelberg 2011

2   Verb Classification of Prof. A.R. Rajaraja Varma (ARR)

ARR was an Indian poet, grammarian and Professor of Oriental Languages at Maharaja's College (Present University College) Trivandrum (1910-18). He wrote widely in Sanskrit and Malayalam. He is known as **Kerala Panini** for his contributions to Malayalam Grammar. He had listed and classified Malayalam verbs into 38 different classes in his monumental work Keralapaniniyam. The classification has been analyzed in detail and found not suitable for computational purpose. For computational purposes, the root word belonging to each class mentioned by ARR has to be rearranged to a new classification based on root word ending and its behavior in various morphological processes. The available classification is a listing of verbs on the basis of *karita* and *akarita* forms and root word ending [2]. For Example, root word ending in Alveolar 'l' akarita (akal, wal, iyal, etc.), ending in Alveolar 'l' karita (El, wOl, nil, etc), etc. This classification can be treated only as a listing of root verbs considering word ending and not suitable for use in language computing as such. An extensive study is needed for the reclassification in this case. At the same time the root words listed by ARR was used for verifying SKP's list of root words. For the verb "iruwwi" SKP has given two root words "iru/iZ_". We verified this with the ARR's root forms and have taken the apt root 'iZ_' for classification.

3   Verb Classifications by Prof. Suranad Kunjan Pillai (SKP)

SKP is a historian, researcher, lexicographer, poet, essayist, literary critic, orator, socio-cultural leader, grammarian, educationist, and scholar of Malayalam language. His verb classification has been published in the first volume of Malayalam lexicon which is based on the tense suffixes especially the past form of the verb. This contains a total of 2881 root verbs. When a tense suffix or causative suffix is added to a root verb certain morphophonemic changes will occur. SKP has considered this change in a pedagogical view. [1] The classification has been analyzed thoroughly to find the compatibility for various computational purposes.

In the preface to the volume he mentioned the parameters used for the classification. The morphophonemic change which occurs in the root verb while adding a tense suffix has been taken care of in the classification. He paid much attention to classify Sanskrit derived Malayalam verbs, which is the largest group of Malayalam verbs.. Before each class he has demonstrated the inflection process of the members in the class except the causative forms. Due to these features the classification is found suitable for computation. But it can not be accepted as such for computation. Root words which are belonging to a particular class contain various word ending. Those word endings will have different morphophonemic changes. SKP's classification has two major groups of verbs evolved from the realization of past tense. One group i.e. classes from 13-16 the verb ending is 'i' and for the other group the verb ending is 'wu' and its variants. The variants are wu, ttu, nnu, ZZu, ccu, FFu, and NNu. This classification will allow the prediction of a past tense form from a statement of the phonology of the stem. Considering this a reclassification was done with minimum deviation from the base classification.

Table 1. SKP's 16 classes of Verbs

Class No	Class Name	No of words	Past Form	Past tense formation
			Category -'wu'	
1	uYuka	13	uYuwu	u-wu->wu
2	uNNuka	7	uNtu	N_+wu->Ntu
3	ituka	14	ittu	t+wu->ttu, Z+wu->ZZu
4	atayuka	202	ataFFu	Y+n+wu->FFu
5	karaLuka	27	karaNtu	L_+n+wu ->Ntu
6	akaYuka	22	akaNNu	Y+n+wu->NNu, nnu
7	akaluka	102	akannu	l+n+wu->nnu, r+n+wu->nnu
8	ataZ_kkuka	155	ataZ_wwu	kk+wu->wwu
9	El_kkuka	10	EZZu	l+kk+wu->ZZu, L_+kk+wu->ttu
10	ayaykkuka	422	ayaccu	y+kk+wu->ccu
11	aMSikkuka	1013	aMSiccu	i+kk+wu->ccu
12	anal_kkuka	30	anannu	kk+n+wu ->nnu
			Category- 'i'	
13	akaZUka	530	akaZi	i
14	ataffuka	127	ataffi	n+k+i->ffi
15	atakkuka	132	ataccu	kk+i->kki
16	ataZ_wwuka	75	ataZ_wwi	ww+i->wwi

Morphotactics for class 1

1) Three tense forms →Root Word (RW) + unnu, RW + uM, RW + u+ wu.
2) Causative (Single) → RW + uvi + kk + unnu/uM or RW + uvi + ccu.
   (Double)→RW+ uvi+ppi+kk+ unnu/uM or RW + uvi+ppi+ccu.
3) Verbal Noun → RW + al_

Here SKP's classification is chosen as base because each class has a unique word ending for a verb in the past tense forms. That will make the computation simpler and easier. After going through all the classes of SKP's verb classification we found that for computational applications 16 classes will not be sufficient as the root word endings are different in same class. Analysis will be difficult with this classification. If we use the rules derived for 'uY' for 'nO' in class 2, the rule will fail and will not give proper verb formation. Considering the computational aspect and without much sacrificing the syntactic rules we have derived 53 verb classes from the 16 classification. These 53 classifications are based on the addition or deletion of characters of the root verb in the formation of different tense forms. Some of the SKP's verb classes were merged together for computational purpose...

The bases of classifications are as follows:

1. Make the computational processing simpler
2. Minimum number of classes for handling all verb forms
3. Each class must take care of transitive, intransitive, and their causative and noun form of the verb formed by a root word.
4. Minimum deviation from SKP's verb classification while grouping of root words.
5. No alterations in the language rules

In some cases same root words are occurring in different classes of SKP's verb classification. For example consider the root word 'vIY'. This is placed in Classes 16 and 6. The past form "vIYwwi" and "vINu" are the transitive and intransitive forms respectively of the verb "vIY"... Thus they are included in the same class... See table: 2 for example.

Table 2. Same paradigm for different SKP classes

Sl No	Paradigm Class	Past form	Past tense formation	SKP class
1	8	vIYwwi	Y+i->Ywwi	16
		vINu	Y+wu->Nu	6
2	27	varuwwi	r+i ->wwi	13
		vannu	r+wu->nnu	7

Consider the SKP Class 2 verbs. The root words are ending in three different forms like 'vE' for 'venwu','nO' for 'nonwu' and 'uN_' for 'uNtu'. Generation of all the verb forms is not possible using the same rule. So SKP Class 2 is subdivided into three paradigm classes. SKP Class 10 has 37 different root word endings. This class is divided into 11 sub classes. Verb formation for all the verbs with the 37 different root word ending are contained in the 11 sub classes. Similarly past form verb ending for SKP Class 13 is 'i' and almost all the root words are different. This class is also categorized into 11 sub classes and is able to generate all the verb formation of SKP Class 13 root words.

In applications like Morphological Synthesizer the selected paradigm class should work for both transitive and intransitive forms. For the same root word the classes are different for some verbs in the SKP's classification. This will make the synthesis ambiguous. To avoid that such verb classes can be placed in the same category.

In applications like Machine Translation we use morphological synthesizers. If we are adopting rule based system, we need a dictionary that contains root word and its paradigm number. The paradigm number will give the inflection details of the verb according to its Tense, Aspect and Modality. While going through a set of verb roots like 'nil' (ninnu) and 'cEr' (cEZ_nnu) it is found that the root word is different for its transitive and intransitive form. In such case the root word itself is changed to accommodate it in the same class. This is a violation of language rule. But applications like machine translation should give more importance to the apt meaning of the word. We need a single root word that is able to generate the tense, aspect and modality of the verb. So the violation of language rule is inevitable here.

Table 3. Different root word for words with same meaning in transitive and intransitive form

Sl No	Transitive Form		Intransitive Form	
	Root Word	Past	Root Word	Past
1	nil	niZ_wwi	nil_	ninnu
2	urutt	urutti	uruL_	uruNtu

Some compound verbs behave differently considering its final root word. The verb like kAwwirunnu (kAww+irunnu) is not having the similar properties of verb 'irunnu'. It does not have an intransitive form as in verb 'irunnu'. Such verb forms are listed in separate classes.

Table 4. Compound Verbs

Sl No	Class Name	Root	Past tense formation	SKP class
1	52	kAwwiZ_	Z_+wu→nnu	12
		--	--	
2	53	vEZ_pet	t+i->uwwi	16
		vEZ_pet	t+wu->ttu	3

The illustration of generation of the verb form based on the paradigm number approach is given below:

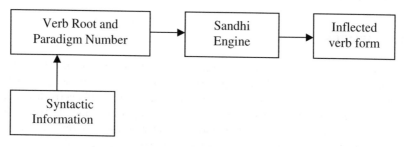

Fig. 1. Morph Synthesizer

Morph Synthesizer gets the root word along with its paradigm number. >From the paradigm info the suffixes needed for inflection is extracted using the syntactic information. Consider the sentence "Raman crossed the river". The Malayalam translation will be "rAman_ puYa muZiccu katannu".. The Malayalam equivalent of "crossed" will be "muZiccu katannu". The root word will be "muZiccu kata". The assigned paradigm is 22 and the Synthesiser will identify the word "crossed" as intransitive verb. Using this info it will identify the suffix "nnu" and combine it with the verb root to generate the past tense form "muZiccu katannu".

An example for the verb formation structure in a morph synthesiser for a verb with paradigm number 22, using C language is given below:

```
#include "struct.h"
struct verb_paradigms verb_parad_list[1090] =
{
    0, "wwunnu",         //katawwunnu /*trans*/
    0, "wwuM",           //katawwuM
    0, "wwi",            //katawwi
    0, "kkunnu",         //katakkunnu /*Intrans*/
    0, "kkuM",           //katakkuM
    0, "nnu",            //katannu
```

```
0, "wwikkunnu",          //katawwikkunnu /*Causative*/
0, "wwikkuM",            //katawwikkuM
0, "wwiccu",             //katawwiccu
0, "wwippikkunnu",       //katawwippikkunnu /*Causative*/
0, "wwippikkuM",         //katawwippikkuM
0, "wwippiccu",          //katawwippiccu
0, "wwal_",              //katawwal_ /*verbal noun*/
```

'0' indicates the deletion of characters from the root word while inflection/derivation, the string in double quotes is the suffix to be added. In the example above we can observe how the verb forms are generated for the root word 'kata' in transitive, intransitive, their causative forms and the verbal noun formation.

Some Sanskrit verbs in SKP Class 11 will also have some deviation from usual verb formation in the transitive form. There will be no transitive form for certain root verb like 'Ananx'. If the sentence has an *object* then the verb will appear in its causative form 'Ananxippiccu'.

4 Conclusions

The 53 verb classes can be used to realize most of the verb roots in Malayalam for NLP applications. The verbs belonging to two different groups are merged together in some classes. We cannot say exactly that this is a subset of SKP's verb classification. Many of the verb classes are added to handle the exceptions in the classifications. In some cases the root word of transitive and intransitive form of a verb with the same meaning is different. Violation of language rules were necessitated here. The compound verb will have different formations compared to that of the final root. We have analyzed the rules for about 6700 verbs and they are giving proper verb formation. This includes commonly occurring compound verbs that are combined with common nouns like "nqwwaM ceywu" for the English word "danced". This classification is found to give the intended verb formation for the 6700 verbs tested. Even if it is derived based on the SKP's classification, this classification can be treated as a general one and can be used for all NLP applications of similar nature.

References

1. Asher, R.E., Kumari, T.C.: Malayalam - Descriptive grammars, pp. 317–319. Routledge, New York (2007)
2. Rajaraja Varma, A.R.: Keralapaniniyam, pp. 177–269. DC Books (2000)
3. Pillai, S.K.: Malayalam Lexicon, Appendix(1-105), Introduction xviii, vol. 1. The University of Kerala (2000)
4. Sinha, R.M.K., Thakur, A.: Synthesizing verb form in English to Hindi Translation, Case of mapping Infinitive and Gerund in English to Hindi. In: Proceedings of International Symposium on Machine Translation NLP and TSS, pp. 52–55 (2005)
5. Madhavan, P.: A PARSER'S PARADOX: Reflections on Malayalam Verb Morphology. In: Proceedings of the First National Symposium on Modeling and Shallow Parsing of Indian Languages, pp. 1–7 (2007)

Appendix 1: Roman Notations for Malayalam

Unicode Glyph	Unicode Value in Hex	ASCII Character	ASCII Value in Hex
അ	0D05	a	61
ആ	0D06	A	41
ഇ	0D07	i	69
ഈ	0D08	I	49
ഉ	0D09	u	75
ഊ	0D0A	U	55
ഋ	0D0B	q	71
എ	0D0E	e	65
ഏ	0D0F	E	45
ഐ	0D10	Q	51
ഒ	0D12	o	6F
ഓ	0D13	O	4F
ഔ	0D14	V	56
അം	0D02	M	4D
അഃ	0D03	H	48
ക	0D15	k	6B
ഖ	0D16	K	4B
ഗ	0D17	g	67
ഘ	0D18	G	47
ങ	0D19	f	66
ച	0D1A	c	63
ഛ	0D1B	C	43
ജ	0D1C	j	6A
ഝ	0D1D	J	4A
ഞ	0D1E	F	46
ട	0D1F	t	74
ഠ	0D20	T	54
ഡ	0D21	d	64
ഢ	0D22	D	44
ണ	0D23	N	4E
ത	0D24	w	77
ഥ	0D25	W	57
ദ	0D26	x	78

ധ	0D27	X	58
ന	0D28	n	6E
പ	0D2A	p	70
ഫ	0D2B	P	50
ബ	0D2C	b	62
ഭ	0D2D	B	42
മ	0D2E	m	6D
യ	0D2F	y	59
ര	0D30	r	72
ല	0D32	l	6C
വ	0D35	v	76
ശ	0D36	S	53
ഷ	0D37	R	52
സ	0D38	s	73
ഹ	0D39	h	68
ള	0D33	L	4C
ഴ	0D34	Y	59
റ	0D31	Z	5A
ൻ	0D28 + 0D4D + 200D	n_	6E + 5F
ൾ	0D33 + 0D4D + 200D	L_	4C + 5F
ൺ	0D23 + 0D4D + 200D	N_	4E + 5F
ർ	0D30 + 0D4D + 200D	Z_	5A + 5F
ൽ	0D32 + 0D4D + 200D	l_	6C + 5F
ാ	0D3E	A	41
ി	0D3F	i	69
ീ	0D40	I	49
ു	0D41	u	75
ൂ	0D42	U	55
ൃ	0D43	q	71
െ	0D46	e	65
േ	0D47	E	45
ൈ	0D48	Q	51
ൊ	0D4A	o	6F
ോ	0D4B	O	4F
ൌ	0D4C	V	56

Period Prediction System for Tamil Epigraphical Scripts Based on Support Vector Machine

P. Subashini[1], M. Krishnaveni[2], and N. Sridevi[3]

[1] Associate Professor, [2] Research Assistant, [3] Research Scholar,
Department of Computer Science, Avinashilingam Deemed University for Women,
Coimbatore -43

Abstract. Tamil is one of the ancient languages of the world with records in the language dating back over two millennia. Epigraphical scripts are the inscription written on various materials and the study of it is vital in knowing the civilized past and hence classification of character belonging to various periods is imperative before using the character bank of the particular period. Therefore a system is proposed for prediction of the period and it is being done by examining a few character referred to as test characters in Tamil language. These test characters are sampled from the script automatically and matched with the characters available for different periods using machine intelligence. The proposed system here has various modules like binarization, thinning, segmentation, feature extraction and finally classification and period prediction using Support Vector Machine. Its performance is most successful in differentiating between four centuries of character. The performance of the system is measured using the four parameters such as prediction rate, Correction rate, Error rate and Time taken to predict the centuries. The system achieves overall accuracy of 90.45%.

Keywords: Prediction, Support Vector Machine, Tamil language, Epigraphical Scripts, Feature Extraction.

1 Introduction

For more than thirty years, researchers have been working on Tamil handwritten character recognition and prediction of the period of an epigraphical script. Epigraphical scripts are the inscription written on various materials and the study of theses inscription is vital in knowing the civilized past. To decipher the script belonging to different periods it is necessary to use the character pertaining to those periods. Hence classification of character belonging to various periods is imperative before using the character bank of the particular period [1]. Several scholars have developed techniques for the recognition of characters in Indian languages such as Devanagari, Bangla, and Telegu whose characters are fairly complex and large in numbers. In this research, an experiment is concerned for period prediction of Tamil Epigraphical scripts using SVM. In the following discussion, section 2 gives an overview of the proposed system, section 3 discuss about the preprocessing

C. Singh et al. (Eds.): ICISIL 2011, CCIS 139, pp. 23–30, 2011.

techniques which are adopted in the system, section 4 deals with segmentation of line and character, section 5 discuss about the feature extraction, section 6 deals with classification and period prediction and in section 7 and 8 the performance of the proposed system and conclusion are presented respectively.

2 System Overview

The underlying objective of this system is to make the interface of a computer more natural for a human being. This prediction method is designed to ease the manual barrier by helping the computer to understand human handwritten characters through an automated system. The design mainly aims in implementation of character period prediction system in which the computer will be able to understand a few simple commands, and identify the century of these characters. *Figure 1 shows the overall system for prediction and figure 2 shows the design for the proposed methodology. Figure 3 shows the sample image of different century characters.*

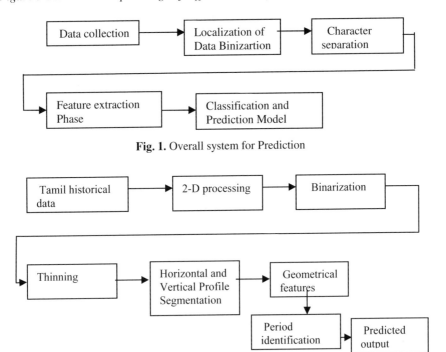

Fig. 1. Overall system for Prediction

Fig. 2. Design of the proposed system

Fig. 3. Sample images of 6th century character

3 Preprocessing

The aim of preprocessing is to process the images in raw form and obtain images suitable for prediction. Since the documents are scanned, the image consists of various gray levels. Hence, thresholding becomes an important part of any handwritten character prediction system.

3.1 Binarization

Before performing binarization, this system is in need of noise removal processing. Hence Median filter technique is adopted for removing noise from the script image. The equ (1) for Median filter is

$$w_i = \begin{cases} 1 & if \ i = \dfrac{\Omega - 1}{2} \\ 0 & otherwise \end{cases} \tag{1}$$

Then the filtered image is been binarized using Otsu's method. The process is estimated using the following equ (2)

$$\sigma_w^2(t) = w_1(t)\sigma_1^2(t) + \omega_2(t)\sigma_2^2(t) \tag{2}$$

Weights ω_i are the probabilities of the two classes separated by a threshold t and σ_i^2 variances of these classes. This is done since the binary document images allow the use of fast binary arithmetic during processing, and also require less space to store and this process is achieved by calculating the optimal threshold value. [2] Figure 3 shows the preprocess phase of the proposed methodology.

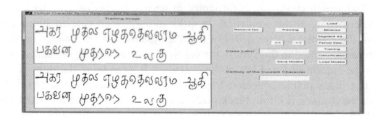

Fig. 4. Binarization process

3.2 Thinning

Binarized image is therefore skeletoinzed using standard thinning algorithm. Thinning is a morphological operation that is used to remove selected foreground pixels from binary images. Algorithm iteratively deletes pixels inside the shape to shrink it without shortening it or breaking it apart.[3] The algorithm makes two passes. The results after the two passes decide if a pixel is to be removed to get the skeleton of the object of interest. the equ (3) for thinning is

$$N(P_1) = P_2 + P_3 + \ldots + P_9 \tag{3}$$

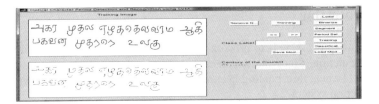

Fig. 5. Thinning process

4   Efficient Segmentation Method for the Proposed Methodology

In this process it includes two tasks such as line segmentation and character segmentation. This segmentation process use both Vertical and horizontal projection profile techniques. By using horizontal projection technique the system segments the lines form the document image. [4] The lines in the paragraphs are scanned for horizontal space intersection with respect to the background. Histogram of the image is used to detect the width of the horizontal lines. Then the lines are scanned vertically for vertical space intersection. The vertical projection profile is obtained by summing pixel values along the horizontal axis for each y value using equ (4),

$$\text{Profile}(y) = \sum f(x,y)$$

$$1 <= x <= M$$

(4)

Here histograms are used to detect the width of the words. Then the words are decomposed into characters using character width computation. Characters are segmented using Vertical projection technique. The letters could be segmented using the vertical projection. However, this method fails for the scripts having no space below two text line which could cause difficulty in period prediction and also it may increase the error rate of the prediction. In order to over this a confirmation of top and bottom for the character is needed. For this the system uses an optional confirmation algorithm implemented in the system for segmentation and the algorithm is: A. start at the top of the current line and left of the character. B. scan up to the right of the character (a).if a black pixels is detected register y as the confirmed top. (b).if not continue to the next pixel. (c).if no black pixels are found increment y and reset x to scan the next horizontal line To increase the efficiency of the system a two pass character segmentation algorithm is proposed for Tamil period prediction system.

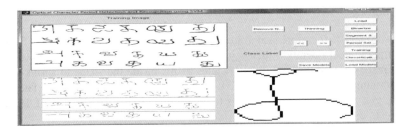

Fig. 6. Line and character segmentation

5 Feature Extraction Phase

In this phase, all the segmented images are then scaled into a common height and width using bilinear interpolation technique. Here each image is divided into equal number of horizontal and vertical strips. The key idea is to perform linear interpolation first in one direction, and then again in the other direction. Linear interpolation is yield in the x-direction using equ (5)

$$f\left(R_1\right) \approx \frac{x_2 - x}{x_2 - x_1} f\left(Q_{11}\right) + \frac{x - x_1}{x_2 - x_1} f\left(Q_{21}\right) \tag{5}$$

Linear interpolation in the y-direction is calculated using equ (6).

$$f\left(P\right) \approx \frac{y_2 - y}{y_2 - y_1} f\left(R_1\right) + \frac{y - y_1}{y_2 - y_1} f\left(R_2\right) \tag{6}$$

From the above equation f(x, y) is estimated in the desired form. The characters are defined by the following attributes such as Height, Weight, Horizontal Projection, Vertical Projection, Hcenter, Vcenter, HPSkewness and VPSkewness [5]. The following table 1 represents the features of Tamil handwritten characters from various periods.

Table 1. Features of Tamil characters

6th century characters	Height	Weight	HCenter	VCenter	HPSkewness	VPSkewness
	65	57	34.1125	28.3079	-4.9061e-004	-5.1122e-004
	53	48	27.1397	25.3691	-0.0029	-0.4082
	54	53	1.6667	1.6667	-0.4082	9.3628e-004
	57	47	27.8016	27.8357	-0.0016	5.4699e-004
	52	69	29.1780	24.7507	-0.0026	5.3371e-004
	67	67	26.5670	35.1809	-2.0049e-004	2.2908e-004
	79	32	33.1747	34.1570	-0.0010	3.5151e-004
	52	33	39.7919	16.5904	5.6234e-005	7.7817e-004
	57	55	27.3419	17.2038	2.6469e-004	-5.7609e-004
	64	62	28.4698	27.7714	7.8819e-004	7.2982e-004

6 Classification and Period Prediction

Classification is done using the features extracted in the previous step, which corresponding to the each character attribute. Here the system uses a Support Vector Machine (SVM) for classification. SVMs are set of related supervised learning method used for classification. It is based on the concept of decision planes that define decision boundaries [6]. To construct an optimal hyperplane, SVM provides an iterative training algorithm which will minimize an error function. For this, training involves the minimization of the error function using equ (7) as below:

$$\frac{1}{2} w^T w + C \sum_{i=1}^{N} \xi_i \tag{7}$$

Subject to the constraints

$$y_i\left(w^T\phi\left(x_i\right)+b\right)\geq\rho-\xi_i \text{ and } \xi_i\geq 0, i=1,......N;\rho\geq 0 \qquad (8)$$

where C is the capacity constant, w is the vector of coefficients, b a constant and ξi are parameters for handling nonseparable data (inputs). The index i label the N training cases. Note that $y\pm1$ represents the class labels and x_i is the independent variables. The kernel ϕ is used to transform data from the input (independent) to the feature space. The RBF kernel type is selected here to improve the efficiency of Support Vector Machines in this system. [7]

$$\phi=\exp\left(-\gamma\left|x_i-x_j\right|^2\right) \qquad (9)$$

This is mainly because of their localized and finite responses across the entire range of the real x-axis. Here the training model takes the input file, target file and trains the network. The last phase prediction is so done based on the precision of the classification module. Multiclass SVM turned out to be a very efficient method in process of prediction here. The accuracy of the algorithm depended on two parameter settings (RBF Kernel parameter σ and regularization parameter C).

Fig. 7. Classification and Period predictions

7 Performance of the System

The accuracy rate yielded by the SVM classifier was quite commendable.

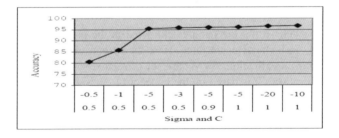

Fig. 8. SVM classifier Accuracy

The Sigma and C value is –5 or 0.5 the accuracy is achieved 95 %. The values of the parameter is 1 the accuracy 97 %. Figure 7 shows the improving performance/accuracy of the system with changing of the parameters. In this system the Sigma and C value is static. Here the Sigma and C value is 1. The System was tested using 4 centuries of Tamil character and 11 most similar characters respectively. The accuracy rate was calculated using 10 fold cross validation technique. Table2 represent performance of the SVM based on 4 parameters and Table 3 depicts the Classification and Period prediction accuracy yielded by the various classifiers.

8 Conclusion

The period prediction of epigraphical script is the area of research where it is possible for the user to the computer and has it understand or recognize what the character is. A calculative approach is proposed here for predicting the Tamil scripts using SVM. To perform this, proposed methodology uses many tests, like finding whether the character is on which century. This research work has successfully managed to identify the centuries of characters. The proposed system successfully differentiated between four different centuries of character. The system achieved overall accuracy is 88.45%.

Table 2. Performance of the System

Parameters	SVM Classification & Period prediction
Prediction rate	97.7%
Correction rate	97.7%
Error rate	2.2%
Time	0.031000s

Table 3. Classifier performances and Period prediction

Classification & Centuries	SVM		Accuracy with %
6th centuries 11 characters	CR	10	90.9%
	ER	1	
10th centuries 11 characters	CR	10	90.9%
	ER	1	
11th centuries 11 characters	CR	9	81.1%
	ER	2	
13th centuries 11 characters	CR	10	90.9%
	ER	1	

References

1. Mantas, J.: An Overview of Character Recognition Methodologies. Pattern Recognition 19(12), 1749–1770 (1986)
2. Aparna, K.H., Chakravarthy, V.S.: A complete OCR system Development of Tamil Magazine Documents. In: Tamil Internet 2003, Chennai, Tamilnadu, India, pp. 45–51 (2003)

3. Sutha, J., Ramaraj, N.: Neural Network Based Offline Tamil Handwritten Character Recognition System. In: IEEE International Conference on Computational Intelligence and Multimedia Applications, pp. 446–450 (2007)
4. Chaudhuri, B.B., Bera, S.: Handwritten Text Line Identification in Indian Scripts. In: 2009 10th International Conference on Document Analysis and Recognition, pp. 636–640 (2009)
5. Subashini, P., Krishnaveni, M.: Prediction of Period for Tamil handwritten Epigraphical scripts using Support Vector Machine. In: Proceedings of International Conference on Mathematices and Computer Science, ICMCS 2010, Loyola College, Chennai, February 5-6, pp. 222–226 (2010) ISBN: 978-81-908234-2-5
6. Campbell, C.: An Introduction to kernel Methods, Radial Basis Function Networks: Design and Applications. Springer, Berlin (2000)
7. Christianini, N., Shawe-Taylor, J.: An introduction to support vector machines: and other kernel-based learning methods. Cambridge University Press, Cambridge (2000)

Name Entity Recognition Systems for Hindi Using CRF Approach

Rajesh Sharma[1] and Vishal Goyal[2]

[1] Lecturer, Kanya Maha Vidyalaya, Jalandhar City, Punjab
rajeshsharma1234@gmail.com
[2] Seniour Lecturer, Department of Computer Science, Punjabi University, Patiala, Punjab
vishal.pup@gmail.com

Abstract. This paper describes the named Entity Recognition (NER) System for Hindi using CRF approach. In this paper, our experiments with various feature combinations for Hindi NER have been explained. The training set has been manually annotated with a Named Entity (NE) tagset of 12 tags. The performance of the system has shown improvements by using the part of speech (POS) information of the current and surrounding words, name list, location name list, organization list, person prefix gazetteers list etc. It has been observed that using prefix and suffix feature helped a lot in improving the results. We have achieved Precision, Recall and F-score of 72.78%, 65.82% and 70.45% respectively for the current NER Hindi system. We have used CRF++ toolkit for training and testing data.

Keywords: Named Entity, Named Entity Recognizer, Conditional Random Field, CRF++ toolkit, Tagset, IOB Tagging, Word Prefix, Word Suffix, Context Word Feature.

1 Introduction

Named Entity Recognition (NER) is the task of identifying and classifying tokens in a text document into predefined set of classes such as person, organization, location and miscellaneous. It is an important tool in almost all Natural Language Processing (NLP). The ability to determine the Named Entity in a text has been established as an important task for several natural language processing areas, including Information Retrieval, Machine Translation, Information Extraction and Language understanding. NER emerged as one of the sub-tasks of the DARPA-Sponsored Message Understanding Conference (MUCs).The current trend in NER is to use the Machine-Learning approach, which is more attractive in that it is trainable and adoptable and the maintenance of a machine learning system is much cheaper than that a rule based one.

2 Related Work for Indian Languages

Parmod Kumar Gupta et al. (2009) presents the hypothesis by making experiment on NER system in Hindi language by using CRF approach. Recall, Precision and

C. Singh et al. (Eds.): ICISIL 2011, CCIS 139, pp. 31–35, 2011.
© Springer-Verlag Berlin Heidelberg 2011

F-score are claimed to be 66.7%, 69.5% and 58% respectively [4]. **Asif Ekbal et al. (2008)** reports about the development of a NER system for Bengali and Hindi language by using CRF approach. Recall, Precision and F-score are claimed to be 51.63%, 59.60% and 55.36% for Bengali and 71.05%, 23.54% and 35.37% respectively for Hindi language [5]. **Parveen Kumar et al. (2008)** reports about the development of a NER system for 5 languages (Hindi, Bengali, Oriya, Telugu, Urdu) by using Hidden Markov Language. They have used POS-Tag and Chunk information. They obtained a decent F-measure of 39.77%, 46.84%, 45.84%, 46.58% and 44.78% respectively for all 5 languages [2]. **Sujan Kumar Saha et al. (2008)** presents the hypothesis by making experiments on NER system in Hindi language by using the Maximum Entropy and Transliteration approach. They have reported Maximal F-score of 55.36%, nested F-score of 61.46% and lexical F-score of 59.39% for Hindi language respectively [1]. **Praneeth M Shishtla et al. (2008)** presents the hypothesis by making experiments on NER system in Telugu language by adopting CRF approach. Recall, Precision and F-score are claimed to be 64.70%, 34.57% and 44.91% respectively [3].

3   Experiment Setup

3.1   Named Entity Tagset

We have used the tagset[1] released in IJCNLP-08 workshop on NER for South and South East Asian (SSEA) language.[2] We have concentrated on recognizing 12 Named Entities. For our experiments we have used the C++ based OpenNLP CRF++4 package for segmenting/labeling sequential data.

3.2   Tagging Scheme

The corpus is tagged using the IOB (Immediately Outside Beginning) tagging scheme. POS tagging of the data has been done manually.

3.3   Named Entity Feature

The named entity feature contains 29 columns. The first column contains the current word, second word POS feature contains one column, the third column is IOB format which contains one column, the fourth column contains the word length and next column contains the prefixes and suffixes. It contains window size 5 to 7. The next is gazetteer list, which contains 12 columns. We have considered different combination from the following set for inspecting the best feature set for NER task:
Following are the details of the set of the feature that are applied to the NER tasks

o   **Context Word Feature.** The previous and next words of a particular word might be used as a feature. In our work we have experimented on word window of length 5 and 7.

---

[1] http://ltrc.iiit.ac.in/ner-ssea-08/index.cgi?topic=3
[2] http://ltrc.iiit.ac.in/ner-ssea-08/

o **Word Prefix.** The prefix information of a word is helpful to identify NEs. A fixed length prefix of the current and the surrounding words might be treated as feature. In our work, the length of prefixes is upto 4. We have experimented on the word by categorizing them on the basis of word prefix. So, the fourth, third, second and first letter is used as a prefix.

o **Word Suffix.** Word suffix information is helpful to identify NEs. A fixed length word of the current and surrounding words might be treated as feature. In our work, the length of suffixes is upto 7. So, the last seventh, sixth, fifth, fourth, third, second and first letter of the word is used as a suffix in our experiments.

o **POS Information.** The first categorization of named entity depends upon POS tag. POS tagger is very helpful in tagging the data. POS feature is done manually for training data only in our experiments. In the experiment we have used all the 8 POS tags.

o **Gazetteer Lists.** The lists of gazetteers have been used for preparing the training data. 12 different lists which were prepared with the help of telephone directory and various web sites have been used. These lists are not exhaustive. Following are the description of these lists.

• **Person Name.** This list contains near about 8000 entries for the first name of the person, middle name and the last names. This feature is set to 1 for the current word otherwise it set to 0.

• **Location Name.** This list contains near about 1000 entries for the location names like cities of India, different state names, district names, country names etc and the feature is set to 1 for the current word.

• **Organization Name.** This list contains near about 900 entries for the organization names like political parties, college names etc and the feature is set to 1 for the current word.

• **Person Prefix.** This list consists on 120 entries. This feature is set to 1 for the current and next 2 words.

• **Abbreviation Name.** This list contains 600 abbreviations. This feature set to 1 for the current word. There are two types of entries in the lists. In the first list there is no any space between two characters and another list contains a dot between every character.

• **Month Name and Day Name.** This list contains the name of all the twelve different months of both English and Hindi calendars also it contains the day names. This list total contains 31 entries. This feature is set to 1 for the current word.

3.4 Training and Test Set Collection

For the evaluation of Hindi NER, we have developed training and test data which has been manually tagged. This corpus has been developed from the Online Hindi newspapers like http://aajtak.intoday.in/, http://www.bhaskar.com/ etc. and we used http://www.google.com/transliterate/ for transliterating the data. We have also refined the training and test data sets in order to remove the sentences which do not contain any named entity. After this the refinement the training set consists of 21,000 words and test data contains 12000 words. There are total 1200 NEs in training set and 395 NEs in test set data.

4 Evaluation and Results

We have used the window [-3,-2,-1,0,+1,+2,+3] for seven words, then we have used other features to find out the optimal feature set. The feature sets and the corresponding F-scores values are mentioned below. After applying all these features we have achieved an overall result Precision, Recall and F-score of 72.78%, 65.82% and 70.45% respectively for the current NER Hindi system. We conducted experiments on a testing data of 21,000 words. All the results are shown in Table 1.

Table 1. Feature set by Overall Code

Feature set	F-Score (%)
Pw,cw,nw	34.12
Pw2,pw,cw,nw,nw2	42.67
Pw3,pw2,pw,cw,nw,nw2,nw3	42.44
Pw2,pw,cw,nw,nw2,pt,pp,cp, 1<|prefix|<5,1<suffix|<5	76.34
Pw2,pw,cw,nw,pt,cp,np1<|prefix|<5,1<suffix|<5	77.67
Pw2,pw,cw,nw,nw2,pt,cp,np,1<prefix|<5,1<suffix<5,1 name list	82.13
Pw2,pw,cw,nw,nw2,pt,cp,np,0<prefix|<4,0<suffix<4,1 name list	80.80
Pw2,pw,cw,nw,nw2,pt,cp,np,1<prefix|<5,1<suffix<5, name list, location list, person prefix list	81.12
Pw2,pw,cw,nw,nw2,pt,cp,np,1<prefix|<5,1<suffix<5, name list, location list, person prefix list, title object list, number list, time list	78.02
Cw: current word, pw: previous word, nw: next word, pw2: previous to previous word, nw2: next to next word, cp: POS tag of current word, pp: POS tag of previous word, np: POS tag of next word, pt: NE tag of previous word |prefix| length of the prefix of the current word, |suffix| length of the suffix of the current word. Cwi,pwi,nwi: current , previous and the next ith word from the current word.	

We have also observed that for different NE tags, different feature sets gives better results. The best F-score value is 80% found, in NEP tag because we have used it from the first name, middle name and the last name gazetteer lists. By using the prefix and suffix of the current name no result is changed. The f-score value will decrease when we will add person name list and location name will be add in a gazetteer list because some times person names and location names are same. In database, the designation of a person has been written with the person name.

5 Conclusion

CRF approach for NER for Hindi and all the 12 NE tags has been used. Experiments have been performed on training data from Sports domain because this data contains large number of named entities. Six features- Context Word Feature, Word Prefix, Word Suffix, POS Information, Named Entity Feature and various Gazetteer Lists have been used. The contextual window of the size seven, prefix and suffix length and NE information of the previous word, current word and different digit features have been used. After applying all these features an overall Precision, Recall and F-score of 72.78%, 65.82% and 70.45% respectively for the current NER Hindi system has been found. The performance of the system can further be improved by using the Part-Of-Speech (POS) information of the current and surrounding words also.

References

1. Saha, S.K., Chatterji, S., Dandapat, S., Sarkar, S., Mitra, P.: A Hybrid Approach for Named Entity Recognition in Indian Languages. In: Proceedings of the IJCNLP 2008 Workshop on NER for South and South East Asian Languages, Hyderabad, India, pp. 17–24 (January 2008)
2. Praveen Kumar, P., Ravi Kiran, V.: A Hybrid Named Entity Recognition System for South Asian Languages. In: Proceedings of the IJCNLP 2008 Workshop on NER for South and South East Asian Languages, Hyderabad, India, pp. 83–88 (January 2008)
3. Shishtla, P.M., Pingali, P., Varma, V.: Character n-gram Based Approach for Improved Recall in Indian Language NER. In: Proceedings of the IJCNLP 2008 Workshop on NER for South and South East Asian Languages, Hyderabad, India, pp. 67–74 (2008)
4. Gupta, P.K., Arora, S.: An Approach for Named Entity Recognition System for Hindi: An Experimental Study. In: Proceedings of ASCNT 2009, CDAC, Noida, India, pp. 103–108 (2009)
5. Ekbal, A., Haque, R., Das, A., Poka, V., Bandyopadhyay, S.: Language Independent Named Entity Recognition in Indian Languages. In: Proceedings of the IJCNLP 2008 Workshop on NER for South and South East Asian Languages, Hyderabad, India, pp. 33–40 (January 2008)

An N-Gram Based Method for Bengali Keyphrase Extraction

Kamal Sarkar

Computer Science & Engineering Department,
Jadavpur University,
Kolkata – 700 032, India
jukamal2001@yahoo.com

Abstract. Keyphrases provide the subject metadata that gives the clues about the content of a document. In this paper, we present a new method for Bengali keyphrase extraction. The proposed method has several steps such as extraction of n-grams, identification of candidate keyphrases and assigning scores to the candidate keyphrases. Since Bengali is a highly inflectional language, we have developed a lightweight stemmer for stemming the candidate keyphrases. The proposed method has been tested on a collection of Bengali documents selected from a Bengali corpus downloadable from TDIL website.

Keywords: Bengali keyphrase extraction, Information retrieval, Metadata.

1 Introduction

Keyphrases are sequence of words that capture the main topics covered in a document. Keyphrases are useful for many applications such as summarization, indexing, improving performance of retrieval engines etc.

In early works, the researchers have proposed a number of keyphrase extraction techniques discussed below.

A technique to choose noun phrases from a document as keyphrases has been proposed in [1] that uses the features such as phrase length, frequency and head noun.

Chien [2] introduced a PAT-tree-based keyphrases extraction system for Chinese and other oriental languages.

HaCohen-Kerner [3] and HaCohen-Kerner et al. [4] proposed a model for keyphrase extraction which uses a supervised machine learning method to combine the baseline methods. They applied the decision tree for effective feature combination.

Hulth et al. [5] developed a keyphrase extraction algorithm in which they integrated a hierarchically organized thesaurus and the frequency analysis.

A graph based model for keyphrase extraction has been proposed in [6].

A keyphrase extraction approach that uses a Neural Network for keyphrase extraction has been presented in [7].

Turney [8] has viewed the problem of keyphrase extraction as supervised learning task. Kea is a keyphrase extraction system, presented in [9][10], uses the Bayesian learning technique for keyphrase extraction task.

An n-gram based technique for filtering keyphrases has been presented in [11].

C. Singh et al. (Eds.): ICISIL 2011, CCIS 139, pp. 36–41, 2011.

All the keyphrase extraction approaches discussed above have been tested on the documents written in English language. In this paper, we have presented a new method for extracting keyphrases from Bengali documents using many features such as position of phrase's first occurrence, phrase's frequency, number of links of a phrase to other phrases, phrase length and phrase's inverse document frequency.

Since Bengali is a partially free order and highly inflectional language, candidate keyphrase identification is relatively tough and stemming of the candidate keyphrases is desirable. Stemming is required for computing the frequency of a phrase in a document. So, we have developed an n-gram based candidate keyphrase identification method and a lightweight stemmer for stemming the candidate keyphrases.

The paper is organized as follows. The description of the corpus has been discussed in section 2. In section 3, the proposed keyphrase extraction method has been presented. We present the evaluation and the experimental results in section 5.

2 Description of the Corpus

To build a test corpus, we have selected 41 documents from the corpus downloaded from TDIL website: "*www.tdil.mit.gov.in*". Keyphrases for each document have been created manually. Most of the keyphrases consist of one, two and three words. Keyphrases consisting of more than three words are few in number in our corpus. Average number of keyphrases per document= 9. Average number of sentences per document= 27.

3 Proposed Method

The proposed keyphrase extraction method has two major components: candidate phrase identification and ranking of candidate phrases for extracting the final set of keyphrases.

3.1 Candidate Keyphrase Identification

Our n-gram based candidate keyphrase identification technique extracts n-grams from the sentences of an input Bengali document. We treat the n-grams representing noun phrases as the initial set of candidate keyphrases. Each candidate phrase is stemmed using a lightweight stemming procedure.

Stemming. In our work, we use a lightweight stemmer for Bengali that strips the suffixes using a predefined suffix list, on a "longest match" basis, using the algorithm similar to that for Hindi [12]. We find that stemming procedure is useful to compute the frequency of a phrase.

N-gram Based Candidate Keyphrase Extraction. Our algorithm for N-gram based candidate keyphrase extraction, accepts a Bengali text document, a stop-word list, a Bengali verb suffix list and a Bengali noun suffix list as the input. The stop word list which consists of 275 stop words has been created manually from our corpus. The verb suffix list consists of 18 verb suffixes and the noun suffix list consists of 20 noun suffixes.

The algorithm initially computes unigrams, bigrams and trigrams from each sentence in a document. N-grams which start or end with a stop word are discarded. Similarly, the n-grams which start or end with a word having a verb suffix are also discarded to form the final list of candidate keyphrases. The candidate phrases are then stemmed using a lightweight suffix matching method [12] that uses the noun suffix list for this purpose.

3.2 Calculating Scores for Keyphrase Candidates

The score for a candidate keyphrase is defined by adding a factor S_F to a factor S_P. S_P is the score for a phrase for its first occurrence and S_F is the score for a phrase, which nonlinearly combines phrase's frequency (PF), number of links of a phrase to other phrases (LC), phrase length (PL) and phrase's inverse document frequency (IDF). Thus we have the following equation:

Score of a candidate phrase = $S_P + S_{F\_norm}$, where S_{F_norm} is the normalized value

of S_F. The score for a phrase for its first occurrence S_P is defined as: $S_P = \dfrac{1}{\sqrt{i}}$,

where i is the sentence number where the phrase has occurred first in a document.

The score for a phrase S_F, which nonlinearly combines phrase's frequency (PF), number of links of a phrase to other phrases (LC), phrase length (PL) and inverse document frequency (IDF), is defined as

$$S_F = \left(\sqrt{((1 + PF * PF)^E + LC)} \right) * IDF \qquad IDF = \log\left(\frac{N}{DF + 1} \right)$$

$$E = 1 + \frac{(PL * PL)}{(m * m)}$$

N: is the total number of documents in the collection
DF: the number of documents in the collection, which contain a phrase
m: maximum of the lengths of the candidate keyphrases belonging to a document

To normalize the value of S_F, we divide this value by the maximum of S_F values for the candidate keyphrases. The formula for computing S_F has four major components (1) *Phrase frequency*, whose value has been squared to emphasize more on the phrase's independent occurrence than its partial occurrence, (2) *Phrase's links to other phrases*, which indicates the number of partial occurrences of a phrase in a document. We consider a phrase P has a link to another phrase Q if P and Q share some common words, (3) *Phrase Length:* we hypothesize that phrase frequency has some non-linear relationships with the phrase length, because we observed in our corpus that the average frequency of one-word phrases is greater than that of two-word phrases and the average frequency of two-word phrases is greater than that of three-word phrases and so on, (4) *IDF (inverse document frequency):* It gives the global statistics about a term. IDF value of a phrase indicates that how rare a phrase in the natural language corpus.

3.3 Extracting Keyphrases

All the scores of keyphrase candidates are normalized to range from 0 to 1 after they are calculated. All candidate keyphrases for a document are then ranked in descending order by their scores. The keyphrases of a document can be extracted from the ranked list. Our system has a parameter for users to decide the number of keyphrases they want from a document. The number of extracted keyphrases for a document can be defined in a specific number of keyphrases to be extracted.

4 Comparisons to Existing Methods

We did not find any previous research work on Bengali keyphrase extraction for comparison to our work. So, we compare the proposed method with a hybrid method that linearly combines two baseline methods in English domain: TF*IDF based method (where TF is the number of times a phrase occurs in a document and IDF= log (N/(DF+1))) and position based method (discussed in subsection 3.2).

5 Evaluation and Experimental Results

To evaluate the Bengali keyphrase extraction methods, we use 41 Bengali articles downloaded from the TDIL website. Details on corpus development have been discussed in section 2. From each of 41 test documents, we extract the desired number of keyphrases using the proposed method (discussed in section 3) and the baseline method (discussed in section 4). The number of keyphrases to be extracted from a document is set to 5, 10 and 15.

We have evaluated the proposed keyphrase extraction system using the well known metrics: precision and recall. The precision and recall are defined as follows:

$$Precision = \frac{N}{K} \text{ and } Recall = \frac{N}{M}$$

Where, N = number of keyphrases matched, K = number of keyphrases generated by the system and M = number of keyphrases generated manually.

During the comparison between the manually created keyphrases and the system generated keyphrases, the keyphrases are also stemmed using the same procedure discussed in subsection 3.1 and a match between a manually created keyphrase and a system generated keyphrase is meant for a full match (no partial match is considered).

5.1 Results

Fig. 1 shows the manually created keyphrases for the test document number 3 in our corpus and fig. 2 shows the top 10 keyphrases extracted by the system that uses the proposed method for extracting keyphrases from the document number 3.

বিহারের মজঃফরপুরে, মূর্তি স্থাপন, শহিদ ক্ষুদিরাম, উপাচার্য্য নিমাইসাধন বসু, প্রফুল্ল চাকী, শিল্পী সেলিম মুন্সী, বোমা নিক্ষেপ, রাজীব গান্ধী. অর্থাভাব

Fig. 1. Manually created keyphrases for the document number 3 in our test corpus

মজঃফরপুরে, মূর্তি স্থাপন, শহিদ ক্ষুদিরাম, উপাচার্য্য নিমাইসাধন বসু, প্রফুল্ল চাকী, প্রফুল্ল চাকী মূর্তি, ক্ষুদিরাম মূর্তি পাশে, শহিদ, প্রফুল্ল, মূর্তি স্থাপন উদ্যোগ

Fig. 2. Top 10 keyphrases extracted by the proposed method from the document number 3 in our test corpus

We calculate the precision and recall for the baseline method and the proposed method when the number of extracted keyphrases is 5, 10 and 15 respectively and we show in table 1 the results after testing those methods on our dataset.

Table 1 shows the comparisons between the proposed method and the baseline method that uses TFIDF and positional features. We also conduct the statistical significance test on the difference between precisions of the two methods, as well as their recalls, using a paired t test. From table 1, we can find that, in respect to precision and recall, the proposed method performs better than a hybrid method combining two baseline methods TFIDF and position. The results are statistically significant at 95% confidence level in most of the cases.

Table 1. Comparisons between the proposed method and the baseline method that uses TFIDF and position features

Number of extracted keyphrases	Average Precision		Significance test on precision difference (ρ value)	Average Recall		Significance test on recall difference (ρ value)
	Proposed Method	Position +TFIDF method		Proposed Method	Position +TFIDF method	
5	0.25	0.19	<0.05	0.13	0.10	<0.05
10	0.19	0.15	<0.05	0.20	0.17	<0.05
15	0.15	0.13	<0.05	0.23	0.20	> 0.05

6 Conclusion and Future Work

In this paper we present a new method for Bengali keyphrase extraction that nonlinearly combines many features such as position of phrase's first occurrence, phrase's frequency, count of links of a phrase to other phrases, phrase length and phrase's inverse document frequency. The proposed method performs significantly better than the methods to which it is compared.

The proposed system can be enhanced by improving the two major components of the system mainly (1) stemming, (2) candidate keyphrase extraction.

References

1. Barker, K., Cornacchia, N.: Using Noun Phrase Heads to Extract Document Keyphrases. In: Hamilton, H., Yang, Q. (eds.) Canadian AI 2000. LNCS (LNAI), vol. 1822, pp. 40–52. Springer, Heidelberg (2000)
2. Chien, L.F.: PAT-tree-based Adaptive Keyphrase Extraction for Intelligent Chinese Information Retrieval. Information Processing and Management 35, 501–521 (1999)
3. HaCohen-Kerner, Y.: Automatic Extraction of Keywords from Abstracts. In: Palade, V., Howlett, R.J., Jain, L.C. (eds.) KES 2003. LNCS (LNAI), vol. 2773, pp. 843–849. Springer, Heidelberg (2003)
4. HaCohen-Kerner, Y., Gross, Z., Masa, A.: Automatic Extraction and Learning of Keyphrases from Scientific Articles. In: Gelbukh, A. (ed.) CICLing 2005. LNCS, vol. 3406, pp. 657–669. Springer, Heidelberg (2005)
5. Hulth, A., Karlgren, J., Jonsson, A., Boström, H.: Automatic Keyword Extraction Using Domain Knowledge. In: Gelbukh, A. (ed.) CICLing 2001. LNCS, vol. 2004, pp. 472–482. Springer, Heidelberg (2001)
6. Matsuo, Y., Ohsawa, Y., Ishizuka, M.: KeyWorld: Extracting Keywords from a Document as a Small World. In: Jantke, K.P., Shinohara, A. (eds.) DS 2001. LNCS (LNAI), vol. 2226, pp. 271–281. Springer, Heidelberg (2001)
7. Sarkar, K., Nasipuri, M., Ghose, S.: A New Approach to Keyphrase extraction using Neural Networks. International Journal of Computer Science Issues 7(2,3), 16–25 (2010)
8. Turney, P.D.: Learning algorithm for keyphrase extraction. Journal of Information Retrieval 2(4), 303–336 (2000)
9. Frank, E., Paynter, G., Witten, I.H., Gutwin, C., Nevill-Manning, C.: Domain-specific keyphrase extraction. In: Proceeding of the Sixteenth International Joint Conference on Artificial Intelligence, San Mateo, pp. 668–673 (1999)
10. Witten, I.H., Paynter, G.W., Frank, E., et al.: KEA: Practical Automatic Keyphrase Extraction. In: Fox, E.A., Rowe, N. (eds.) Proceedings of Digital Libraries 1999: The Fourth ACM Conference on Digital Libraries, pp. 254–255. ACM Press, Berkeley (1999)
11. Kumar, N., Srinathan, K.: Automatic keyphrase extraction from scientific documents using N-gram filtration technique. In: Proceeding of the Eighth ACM Symposium on Document Engineering, Sao Paulo, Brazil, pp. 199–208 (2008)
12. Ramanathan, A., Rao, D.D.: A lightweight stemmer for Hindi. In: Proceeding of Workshop of Computational Linguistics for South Asian Languages -Expanding Synergies with Europe, EACL 2003, Budapest, Hungary, pp. 42–48 (2003)

Feature Extraction and Recognition of Bengali Word Using Gabor Filter and Artificial Neural Network

Mahua Nandy (Pal) and Sumit Majumdar

Department of CSE, MCKV Institute of Engineering, 243, G.T.Road(N), Liluah, Howrah,
West Bengal, India
mahua.nandy@gmail.com, sm071985@gmail.com

Abstract. Character recognition is an emerging area of research in the field of image processing and pattern recognition. The objective here is to generate a Bengali dictionary and develop an artificial neural network based technique for matching an input word with the dictionary word. The technique uses the features of the words as a whole rather than the features of each character. For feature extraction purpose, we have used 2D Gabor filter. For the dictionary words, it shows 93.67% accuracy in matching and for non-dictionary words, it shows 83% accuracy in non-matching. The overall accuracy of the system becomes 91%.

Keywords: Gabor filter, iLeap, binarization, normalization, orientation matrix, back propagation algorithm.

1 Introduction

A language recognizing software may be thought as an application of artificial intelligence.

In case of English, the number of alphabets is limited and there is no hazard of combined character ("*yuktakshar*") or modifier character (↑,↿ etc.). In Bengali text, characters are complicated also as compared to English, which makes the Bengali language recognizing software a challenging task. There may be two effective approaches of feature extraction. First one is to find out structural features and second one is to find out frequency features.

Gabor filter is a frequency filter, which has already been applied for texture analysis, moving object tracking and face recognition. It has successfully been applied for character recognition [5] also.

2-D Gabor filters are local spatial filters that can extract orient-dependent frequency content.

In [6], response output of Gabor filter at four different orientations has been extracted and used for handwritten Bengali numeral recognition.

Two ways of finding out Bangla word from a Bangla dictionary have been discussed in [3] and [4].

Here, a novel approach of extracting features of a whole word using Gabor filter and text recognition using ANN will be discussed thoroughly.

C. Singh et al. (Eds.): ICISIL 2011, CCIS 139, pp. 42–47, 2011.
© Springer-Verlag Berlin Heidelberg 2011

The whole work is divided in five phases:

1. Image preprocessing, 2. Formation of Bengali dictionary, 3. Feature extraction of each word as a whole, 4. Train the neural network with dictionary words, 5. Recognition of Bengali text.

2   Gabor Filter

If λ is the wavelength of Gabor filter and θ is the orientation angle of Gabor filter. σ_x, σ_y is the standard deviation of Gaussian along the x – direction and y – direction respectively. If we set $\sigma_x = \sigma_y = \sigma$, Following is the impulse response [5] of 2D Gabor filter, we get,

$$h(x,y) = \frac{1}{\sqrt{2\pi}\sigma} e^{\frac{-(x^2|y^2)}{2\sigma^2}} e^{j\lambda(x\cos\theta + y\sin\theta)} \tag{1}$$

Applying this type of Gabor filter h(x, y) to an image u(x, y) we are getting the response output I(x, y) which is defined by the convolution sum:

$$I(x,y,\theta) = \sum_{x1=x-M/2}^{x+M/2} \sum_{y1=y-N/2}^{y+N/2} u(x1,y1)\, e^{\frac{-((x-x1)^2 + (y-y1)^2)}{2\sigma^2}} \qquad ......(2)$$

M and N are the dimensions of the filter.

The control parameters are λ, σ, M, N.

Here, the image of a word is convolved with a set of Gabor filters. This set consists of Gabor filters at different orientations where

$\{\theta_k \mid \theta_k = k\pi/n, k = 0, \dots, (n-1)\} \dots\dots (5)$

3   Present Work

3.1   Image Preprocessing

We have applied median filtering with 3×3 mask over the image to get rid of salt and pepper noise.

For further processing, the image needs to be binarized first. In the current work, after comparing different binarization methods [2], faster Sauvola's method [1] is applied for the purpose of binarization as it provides satisfactory result in case of noisy images also.

3.2   Generation of Bengali Dictionary Pages

Well circulated Bengali newspapers are used for collection of most frequently used words. These words are typed using the iLeap software. The iLeap file is sorted alphabetically using their ISCII (Indian Script Code for Information Interchange) values, printed on white papers and scanned to get the images of the dictionary pages (in .bmp format).

Fig. 1. Sorted Bengali words (.bmp image)

Fig. 1 shows a document page containing sorted Bengali words. These document images are preprocessed as discussed in section 3.1.

3.3 Thinning and Normalization of Each Boxed Word

The words usually have different aspect ratio (width divided by height of the box). They are normalized into a standard size of 20×100 pixel.

Thinning has been done following morphological thinning algorithm.

Figure 3, figure 4 and figure 5 show the binarized image of a particular Bengali word, corresponding normalized image and thinned image respectively.

Fig. 2. Binarized word   **Fig. 3.** Normalized word   **Fig. 4.** Thinned word

3.4 Feature Extraction and Calculation for Each Word Using Gabor Filter

For an X*Y image u, Gabor filter is applied to obtain the response output. The output responds maximally at those particular edges whose angle is θ. This property can be used to detect the edges of all orientations. The image of a word is convolved with four Gabor filters at angles 0, $\pi/4$, $\pi/2$, and $3\pi/4$.

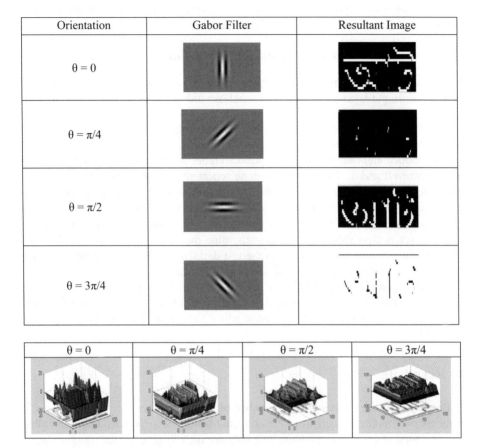

Fig. 5. Gabor Filters, Resultant Images and corresponding graphical representations

Now Gabor orientation matrix m is also an X*Y matrix, where m(x, y) is assigned the value k if u(x, y) obtain maximum response at an angle θ_k. In this case each element of m is assigned a value between 0-3.This X*Y vector is considered as feature vector.

The optimal values for λ, σ, M and N are taken as 1, 2, 11 and 11 respectively, which gives satisfactory results.

Gabor Orientation matrix is an X*Y matrix and each element m(x, y) of that matrix contains a value between 0 to n-1. Here,

$$\{ \theta_k \mid \theta_k = k\pi/n, k = 0, \dots , (n-1)\}$$

But besides orientation information, the position information is also important to the same extent. So, for each pixel, we are storing a value set comprises of three indices {r, c, k}. Thus for each word, 2000 three set feature values are obtained. Fig. 5 shows filtered images and corresponding graphical representations of that particular word obtained by using Gabor filter at four different orientations.

3.5 Training ANN Using Back Propagation Algorithm

In the current work multilayer perceptron is used with one input layer with 2000 nodes, one output layer with 1050 nodes and one hidden layer with nodes between 10 -50.

Insufficient no. of hidden nodes causes under fitting whereas excess no. of hidden nodes causes over fitting.

The hidden and output layer perceptron weights are initialized with random numbers between 0 and 1. The training is done repeatedly for all dictionary words for a predefined number of iterations and for a particular number of hidden layer neurons. Intermediately, the maximum accuracy thus obtained for a specific iteration is stored in the network file. In the same way, the network is trained for a specific range of hidden layer neuron values. For each such value the trained network is used to find the output efficiency when applied on the test dataset. The network giving maximum accuracy is stored in the network file and is used for matching purpose.

'Mean Squared Error' (MSE) has been used as performance parameter function, which is actually the average squared error between the actual output and the target output and the error is back propagated to adjust the weight factors. Learning rate is set to 0.1.

4 Results

A Bengali dictionary having 1050 number of words has been generated. 10 to 25 numbers of samples of each word are collected from newspapers leading to a collection of a total of 20000 words. Now 90% data are used to generate the training dataset and 10% are used to generate the test dataset.

We have collected fresh set Bengali words from newspapers, which includes 300 images of 150 words residing in the dictionary, termed hereafter as "class-1" words and 100 images of 50 words not residing in the dictionary, termed hereafter as "class-2" words. All these words are given as unknown word to the network to generate the output in terms of match or mismatch.

Out of the 300 samples of class-1 words, the network shows a matching for 281 samples, Out of the 100 samples of class-2 words, it shows a matching for 83 samples.

Table 1. Comparison of efficiency

Word Type	Percentage Efficiency
Class-I	93.67
Class-II	83
Overall Efficiency	91

References

1. Sauvola, J., Pietkainen Procedure, M.: Adaptive document image binarization. In: Pattern Recognition, pp. 225–236 (2000)
2. Nandy (Pal), M., Saha, S.: An Analytical Study of Different Document Image Binarization Methods. In: Proceedings of IEEE National Conference on Computing and Communication Systems (COCOSYS 2009), UIT, Burdwan, January 2-4, pp. 71–76 (2009)
3. Nandy (Pal), M., Saha, S.: A Novel Scheme for Searching a Bangla Word within a Bangla Dictionary. In: Proceedings of INDICON 2009, An IEEE India Council Conference, DAIICT, Gandhinagar, Gujarat, December 18-20, pp. 543–546 (2009)
4. Saha, S., Nandy (Pal), M.: ANN Based Approach for Searching a Bengali Word from within a Bengali Dictionary. International Journal on Computer Engineering and Information Technology (IJCEIT) 10, 7–11 (2010)
5. Hu, P., Zhao, Y., Yang, Z., Wang, J.: Recognition of Gray Characters using Gabor Filters. In: Proceedings of 5th International Conference FUSION 2002, Annapolis, USA, pp. 419–424 (2002)
6. Hamamoto, Y., Uchimura, S., Watanabe, M., Yasuda, T., Tomita, S.: Recognition of handwritten numerals using Gabor features. In: Proceedings of the 13th International Conference on Pattern Recognition, Part, vol. 3, pp. 250–253. IEEE Comput. Soc. Press, Los Alamitos (1996)

The Segmentation of Half Characters in Handwritten Hindi Text

Naresh Kumar Garg[1], Lakhwinder Kaur[2], and M.K. Jindal[3]

[1] GZS Collage of Engineering & Tech. Bathinda, Punjab, India
naresh2834@rediffmail.com
[2] Dept. of Computer Engineering, UCOE, Punjabi University, Patiala, Punjab, India
mahal2k8@yahoo.com
[3] Panjab University Regional Centre, Muktsar, Punjab, India
manishphd@rediffmail.com

Abstract. Character recognition is an important stage of any text recognition system. In Optical Character Recognition (OCR) system, the presence of half characters decreases the recognition rate. Due to touching of half character with full characters, the determination of presence of half character is very challenging task. In this paper, we have proposed new algorithm based on structural properties of text to segment the half characters in handwritten Hindi text. The results are shown for both handwritten Hindi text as well as for printed Hindi text. The proposed algorithm achieves the segmentation accuracy as 83.02% for half characters in handwritten text and 87.5% in printed text.

Keywords: Segmentation, half character, over segmentation.

1 Introduction

The simplest technique to segment the characters is to use inter-character gap between the characters. This technique cannot be applied on touching half characters. Also, the technique used to segment the printed characters cannot be applied to handwritten documents due to different writing styles, different sizes of characters and different shapes of characters in texts written by different people. The presence of half characters in handwritten text makes the problem of segmentation more complex.

2 Related Work

A good survey about OCR is given in [1]. Hindi is the official language of India. To the best of author's knowledge, no commercial OCR for handwritten Hindi text is available, yet. Many algorithms have been developed for segmenting touching characters in Indian scripts, but most of them are on printed text. Bansal and Sinha [2] had segmented the conjuncts (type of touching characters) based on structural properties

C. Singh et al. (Eds.): ICISIL 2011, CCIS 139, pp. 48–53, 2011.
© Springer-Verlag Berlin Heidelberg 2011

of text in printed Devanagari script. They segmented the conjuncts with an accuracy of 84%. Jindal *et al.* [3, 4] had segmented the touching characters in middle zone and upper zone of printed Gurmukhi script using structural properties of the script. Chaudhuri *et al.* [5] had used the principal of water overflow from a reservoir to segment touching characters in Oriya script. Garain and Chaudhuri [6, 7] had used a technique based on fuzzy multifactorial analysis to segment touching characters in printed Devnagari and Bangla scripts.

Tripathi and Pal [8] had worked on segmentation of touching characters in handwritten Oriya text using structural, topological and water reservoir features. But this technique cannot be directly applied to handwritten Hindi text due to presence of half characters touching the full characters(conjuncts). The work on line segmentation, consonant segmentation, upper modifier segmentation and lower modifier segmentation in Handwritten Hindi text were explained by us in [9, 10]. In this paper, we have explained a new method based on structural features for segmentation of half characters in handwritten Hindi text.

3 Database

All experiments were conducted on database constructed by taking handwritten data from 15 writers. The handwritten documents were reduced in size in paint to 35% to increase the speed of execution. The percentage of stretching of the document in horizontal and vertical direction was same. In some documents, up to 2 degree skew correction was done in paint. This was done on whole document and not on particular lines.

Figure 1 contain part of handwritten Hindi database.

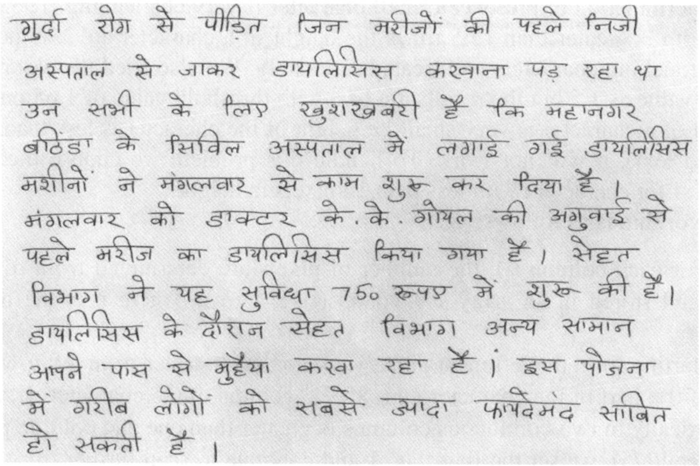
गुर्दा रोग से पीड़ित जिन मरीजों की पहले निजी
अस्पताल से जाकर डायलिसिस करवाना पड़ रहा था
उन सभी के लिए खुशखबरी है कि महानगर
बठिंडा के सिविल अस्पताल में लगाई गई डायलिसिस
मशीनों ने मंगलवार से काम शुरू कर दिया है।
मंगलवार को डाक्टर के. के. गोयल की अगुवाई से
पहले मरीज का डायलिसिस किया गया है। सेहत
विभाग ने यह सुविधा 750 रुपए में शुरू की है।
डायलिसिस के दौरान सेहत विभाग अन्य सामान
अपने पास से मुहैया करवा रहा है। इस योजना
में गरीब लोगों को सबसे ज्यादा फायदेमंद साबित
हो सकती है।

Fig. 1. Part of Database

4   Characteristics of Hindi Language

Devanagari is the script for writing Hindi language. Hindi is written from left to right and there is no concept of upper or lower case. The half characters may touch with full characters to make the characters called conjuncts.

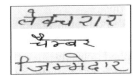

Fig. 2. Conjuncts

This paper deals with segmentation of these conjuncts. When two or more characters are combined to form a word, the horizontal lines touch each other and generate a header line called *shirorekha*. The vowels (modifiers) can be placed at the left, right (or both), top or bottom of the consonant.

5   Segmentation of Half Characters

For separation of Half characters (Conjuncts), the following algorithm has been developed.

After separating the lower and upper modifiers, the consonant separation is done. We determine vertical projection profile of the word and pass the separated characters through this algorithm. Let m is the matrix storing the conjunct character. Let r is number of rows and c is number of columns.

The determination of presence of half character is very challenging. The characters whose width is greater than 1.65 times the height of a character are assumed as conjuncts or touching characters and treated separately. We also tried the algorithm with threshold value as 1.5 but the results are best with threshold value of 1.65 only. But, if the height of a character is very small i.e height of the character is less than 12 pixels (in printed text), it will not work. To handle this problem we choose the threshold value as 1.4 for characters with less than 12 pixels in height.

The algorithm is as follows:

Step 1: For each column (i), the number of pixels are determined from row (r/7) to rth row, and stored in an array say vpixels(i), i=1 to n, where n is the number of columns.

Step 2: Starting from the left most pixel, we scan the character from left towards right upto first 70% part of the character i.e $(c \times 0.7)^{th}$ column of the character. If number of pixels vertically in two continuous columns is greater than one and column position is less than ceil(c/5), we set the flag flag_1 and continue to scan further till we get two continuous columns with single pixel. For printed text(r<12), the flag is set if number of pixels in any column is greater than one and column position is less than ceil(c/5).

Step 3: If we get two continuous columns with single pixel and column position is greater than $(c/5)^{th}$ column we set another flag flag_2 and continue to scan towards right till $(c \times 0.7)^{th}$ column.

Step 4: If flag_1 and flag_2 are set and we get the column with more than one pixel again, and column is between $(c/4)^{th}$ and $(c \times 0.7)^{th}$ column, we store the column-1 position in a variable say v1.

Step 5: In this step following conditions are checked to avoid over segmentation of other characters that satisfy the condition that their width is greater than 1.65 times the height of a character like अ, ढ़, अ, ऄ etc., which are generally written longer in width while writing the text.

    i)   The number of pixels in v1 column is less than half of the maximum no. of pixels in any column or there are more than three continuous columns with one pixel

    ii)   The maximum height of remaining columns(v1+1 to c) is greater than or equal to maximum height of any of the columns from 1 to v1.

    iii)   Presence of two continuous columns with pixels greater than two from v1 to v1+4 columns.

If all the above three conditions are true, we store the value of v1 for half character separation otherwise it is set to zero. If v1 is zero, than it is a problem of second type i.e. two consonants are touching each other (Segmentation is done separately).

Step 6: Starting from first left most pixel to the column v1, we copied the matrix to another matrix say m1 and copied the pixels from column v1 to end column, to matrix m2.

The step 2 is further modified to solve the problem of pen width in the starting of character. Instead of scanning the character from left most pixels we scan from third pixel position.

6   Results

The proposed algorithm is tested on both handwritten as well as on printed Hindi text to segment the half characters and it gives very good results (Table 1 and Table 2, figure 3).

Table 1. Accuracy of Segmentation of Handwritten Hindi Text

Total Words	Total Half Characters	% of Half Characters	Half Characters Correctly Segmented	% Accuracy of Half Character Segmentation
1294	106	8.18	88	83.02

Table 2. Accuracy of Segmentation of printed Hindi text

Total Words	Total Half Characters	% of Half Characters	Half Characters Correctly Segmented	% Accuracy of Half Character Segmentation
345	24	2.9	21	87.5

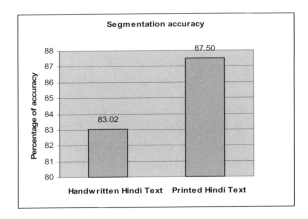

Fig. 3. Half Character Segmentation Results

Further, for visual inspection, some of the correctly segmented conjuncts are shown in figure 4 and incorrectly segmented conjuncts in figure 5.

Fig. 4. Correctly segmented conjuncts

Fig. 5. Incorrectly segmented figures

In figure 5(a), the half character is overlapped with consonant and not attached with during vertical separation of consonants. In figure 5(b), the half character and character are very much overlapped and no vertical single pixel columns present as required in step 3.

The main problem in half character separation is the overlapping of half character with the full character. The determination of presence of half character is even more difficult task to handle.

Multiple conditions specified in step five are put on data to avoid the over segmentation of other characters (like ध्र) with width greater than 1.65 times the height of a character. The same technique is also tried on printed Hindi text.

7 Discussion

From the above results, it is clear that the proposed technique used to segment the conjuncts (half characters) in Handwritten Hindi text is very useful. The study may be carried out in future in the following direction:

1. The above algorithm with some modification may be used to segment the touching characters in handwritten Hindi text.
2. The above technique may be used to segment touching characters in other Indian scripts.

References

[1] Mori, S., Suen, C.Y., Yamamoto, K.: Historical review of OCR Research and development. Proceedings of the IEEE 80(7), 1029–1058 (1992)
[2] Bansal, V.: Integrating knowledge sources in Devanagari text recognition. Ph.D. thesis, IIT Kanpur, INDIA (1999)
[3] Jindal, M.K., Lehal, G.S., Sharma, R.K.: On Segmentation of touching characters and overlapping lines in degraded printed Gurmukhi script. International Journal of Image and Graphics (IJIG) 9(3), 321–353 (2009)
[4] Jindal, M.K., Sharma, R.K., Lehal, G.S.: Segmentation of Touching Characters in Upper Zone in printed Gurmukhi Script. In: Proceedings of the 2nd Bangalore Annual Compute Conference, Bangalore, vol. (9). ACM, New York (2009)
[5] Chaudhuri, B.B., Pal, U., Mitra, M.: Automatic recognition of printed Oriya Script. In: Int. Conf. on Document Analysis and Recognition, pp. 795–799 (2001)
[6] Garain, U., Chaudhuri, B.B.: Segmentation of touching characters in printed Devnagari and Bangla scripts using fuzzy Multifactorial Analysis. IEEE Trans. on Systems, Man and Cybernetics. Part C 4(32), 449–459 (2002)
[7] Garain, U., Chaudhuri, B.B.: On recognition of touching characters in printed Bangla documents. In: Int. Conf. on Document Analysis and Recognition, Germany, pp. 1011–1016 (1997)
[8] Tripathi, N., Pal, U.: Handwriting segmentation of unconstrained Oriya Text. Sadhana 6(31), 755–769 (2006)
[9] Garg, N.K., Kaur, L., Jindal, M.K.: Segmentation of Handwritten Hindi Text. International Journal of Computer Applications (IJCA) 1(4), 22–26 (2010)
[10] Garg, N.K., Kaur, L., Jindal, M.K.: A new method for line segmentation of Handwritten Hindi Text. In: Proceedings of the 7th International IEEE Conference on Information Technology: New Generations (ITNG), pp. 392–397 (2010)

Finding Influence by Cross-Lingual Blog Mining through Multiple Language Lists

Aditya Mogadala and Vasudeva Varma

Search and Information Extraction Lab, IIIT Hyderabad, India
aditya.mogadala@research.iiit.ac.in, vv@iiit.ac.in

Abstract. Blogs has been one of the important resources of information on the internet. Now-a-days lot of Indian language content being generated in the form of blogs. People express their opinions on various situations and events. The content in the blogs may contain named entities–names of people, places, and organizations. Named entities also contain names of eminent personalities who are famous in or out of that language community. The goal of this paper is to find the influence of a personality among cross-language bloggers. The approach we follow is to collect information from blog pages and index the named entities along with their probabilities of occurrence by removing irrelevant information from the blog. When user searches to find the influence of a personality through a query in Indian language, we use a cross language lexicon in the form of multiple language parallel lists to transliterate the query into other Indian languages and mine blogs to return the influence of the personality across Indian language bloggers. An overview of the system and preliminary results are described.

Keywords: Cross-lingual, Blog analysis, multilingual.

1 Introduction

Blogs are considered to be one of the personal journals where people express their personal opinions on different aspects like movie reviews, travelling experiences, daily activities and current happenings in the society. Blog-tracking site *Technorati*[1] stated that blogosphere has doubled every six months for the last three years. This contains good percentage of blogs which are created in native languages. Work published previously explains designing frameworks [1] and new metrics for blog mining [2].Focus of them was to use new techniques, as it has been observed techniques that have been developed for standard information retrieval purposes are suboptimal when applied to blogs because of their high degree of quotation, brevity and rapidity of update. Another aspect of the blogs is that they are multi lingual; understanding them to building systems is important area of research. Some work was focused to get concern analysis from Multilingual Weblog Articles [3] and building collaboration system for intercultural collaboration [4].

---

[1] http://www.technorati.com

C. Singh et al. (Eds.): ICISIL 2011, CCIS 139, pp. 54–59, 2011.
© Springer-Verlag Berlin Heidelberg 2011

Analysis from Internet and Mobile Association of India (**IAMAI**)[2] and **Blogs.oneindia.in**[3] show increase in Indian language blogs. Proposed research paper focus to understand the commonality in thinking of these bloggers from different cultural backgrounds divided by language. Our approach is to find the influence of a personality across cross language bloggers. We propose a cross-lingual system that collects multilingual blog articles from Indian language blogs and extract keywords from articles to provide functions for (1) Eminent persons named entities (2) cross language reference and (3) Influence of eminent persons on cross language bloggers. Because we aim to facilitate users to find and compare influence of native language prominent people across other languages, the system transliterates keywords written in a language into other languages automatically, and shows influence across languages.

This paper is divided into following sections .Section2 describes overview of the system. Section3 is about preliminary results, Section4 explains related work. In Section5, we summarize the arguments and describe future work.Section6 is for references.

2 Overview

In this paper we try to find the influence of prominent persons across Indian community bloggers. Prominent means to be famous and important. It also means to be eminent. Therefore to be prominent indicates the act of standing out and towering. This also provides intuition that prominent people will show influence on others.

2.1 Understanding Influence

According to Dictionary.com[4] "Influence" is the capacity or power of persons or things to be a compelling force on or produce effects on the actions, behavior, opinions, etc. Blogs are usually maintained by individual's writing commentary on the news or events and there is chance to write something about eminent personalities from cross language community.

As Indian community bloggers are from different cultural backgrounds and their cross language knowledge may be limited, if anything written about a person who is outside of language community must have some influence .Our aim is to find this cross cultural influence through cross lingual blog mining using the influence scores for the named entities using the equation mentioned later in the section. Our system has two steps 1) An information retrieval step and 2) cross lingual search.

2.2 Information Retrieval

Initial preprocessing is done to get the indented content.

---

[2] http://www.iamai.in
[3] http://Blogs.oneindia.in
[4] http://dictionary.reference.com/browse/influence

2.2.1 HTML Parsing

Blogs produced by Blogger, Word Press and Live Journal are collected manually. They contain extra information besides the blog post in the form of readers' comments, previous posts, similar related pages, navigation bars, side bars, advertising, etc. Noise is reduced by taking content only from the body is into flat files. Each flat file consists of four blog articles that have been extracted.

2.2.2 Named Entity Recognition

Data is cleansed by removing stop words, whitespaces extra junk characters. Stop word dictionaries created for languages Telugu, Hindi, Bengali, Tamil, Marathi and Punjabi is used for removing stop words from the flat files. Then named entities are identified using NE dictionaries from the cleaned data.

2.2.3 Calculating Influence Scores

Influence scores are calculated for named entities. It is the probability of occurrence of named entity in total number of blog articles parsed. These scores decide the influence of that named entities among cross language bloggers. It is understood that if the named entity occurs in a particular article is valued less compared to the frequency of its occurrence in multiple blog articles. Equation below is used to calculate the influence score.

$$P\ (Per/Ar) = \quad P\ (FA/TA) * (Per\ /\ TotalNE) \tag{1}$$

Per = Personality whose influence to be calculated
Ar = In Total Articles.
FA= Found Articles
TA = Total Articles parsed up to then
TotalNE= Total Named entities.
P (Per/Ar) = Probability of influence of the personality
P (FA/TA) = Probability of finding the Personality in the Article

High value of P (Per/Ar) shows the more influence and there is no threshold for the same.

2.3 Cross Lingual Search

Cross lingual search involves finding influence of a personality across multiple Indian languages by querying the system.

2.3.1 Multiple Languages List

Multiple languages list stores named entities of all the eminent personalities in different Indian languages with English transliteration. User query in one Indian language is translated to other Indian languages to return the influence scores. This approach is new compared to the existing bilingual dictionaries [5, 6] which take a lot of permutations and combinations to maintain the parallel dictionaries. It also decreases the complexity of adding new languages.

Chandrababu: चंद्रबाबू, चंद्राबाबू, চন্দ্রবাবু, ಚಂದ್ರಬಾಬು, சந்திரபாபு, ਚੰਦ੍ਰਬਾਬੁ

चंद्रबाबू: Chandrababu चंद्राबाबू: Chandrababu চন্দ্রবাবু: Chandrababu

ಚಂದ್ರಬಾಬು: Chandrababu   சந்திரபாபு: Chandrababu ਚੰਦ੍ਰਬਾਬੁ:

Chandrababu

Fig. 2. Sample of Multi language list

2.3.2 Search Strategy

When a user enters a query about a personality it will be searched through the multiple language list to find the related translation or transliteration. First the English equivalent is retrieved and then the corresponding Indian language equivalents are extracted. नारायणा राजशेखर रेड्य,राजकुमार ராஜ்குமார் ಎಐಎ चंद्रबाबू are some of the sample query terms in Hindi, Marathi, Bengali, Tamil, Telugu and Punjabi respectively. These multi lingual keywords are then searched through Indian language parsed blog content to find the influence. Pseudo code written below depicts the sample parsing of a multiple language list used for the transliterations.

```
1. Read Input Value;   //Take Input from the user
2. From the Multi language list Map
3. Search for the input value;
4. If value found:
5. Take key from Map to find English Translation;
6. from the English Translation Key
7. Find the corresponding Indian language words;
8. Parse the Indian language words;
9. While the Indian language words not NULL:
     Search influence probabilities of those words;
10. If Found in any of the language data:
        Display the result;
```

3 Preliminary Results

In this section we describe about the preliminary results obtained by the system.

3.1 Data Collection

Total 350 in Hindi, 400 in Telugu, 300 in Punjabi, 400 in Marathi, 200 in Tamil and 300 in Bengali blog articles are collected randomly from different publishers like Blogger, Word press and live journal.

3.2 Sample NE Extracted and Their Influence Scores

Named entities extracted from the blogs are added with their influence scores.

Table 1. Sample keywords found in different language blogs

Languages	Words with Influence Scores
Telugu	ఇండియా:0.004566210, ఉత్తర:0.004566210, కొండ:0.002024291
Hindi	जम्मू:0.001290947, वर्मा:0.0003227368, तमिलनाडु:0.0001613684
Punjabi	ਕੰਡਿਆਂ:0.001736111, ਸੱਚ:0.003472222, ਸਪੀਕਰ:0.001736111
Tamil	ஆண்டுகள்:0.0004251701, தொழில்:0.0008503401, கடவுள்:0.0004251701
Marathi	लेक:0.001758087, मराठी:0.001758087, बाबा:0.0007032349
Bengali	দার্জিলিং:0.001319261,এল:0.001319261,পথ:0.002638522

3.3 Search Results

Search results obtained for the sample queries to find the popularity of eminent personalities among different community bloggers are listed below

Table 2. Results for the query entered in Marathi

Query	Transliteration to Other Languages	Search Results
सचिन	सचिन<======>सचिन Sachin सचिन,सचिन,সচিন,సచిన,சச்சின்,सचिन	After Parsing 31 Marathi Blog Articles probability of existence of [सचिन] in 32 is :: 0.003861004 After Parsing 43 Marathi Blog Articles probability of existence of [सचिन] in 44 is :: 0.01011236 After Parsing 131 Telugu Blog Articles probability of existence of [సచిన] in 132 is :: 0.0006963788

4 Related Work

Google provides linguistic tools such as *Google Trends*[5] that analyzes trends of keywords fed to Google search but it does not translate keyword strings into other

---

[5] http://www.google.com/trends

languages and display the results. With respect to cross-lingual analysis, there has be some work done prior to find the concerns reported in the blogs[7] and trend visualization through blog mining[8].But these works majorly concentrated on European, Chinese and Japanese languages. There has been severe dearth in analyzing the Indian language content generated in blogs.

5 Conclusion and Future Work

Blogs can be used to extract lot of information about society. This paper proposed an approach to analyze multilingual blogs by finding the influence of an eminent personality by cross lingual blog mining. In addition, it also proposed multiple language lists for storing all the translations into a single dictionary instead of storing them in parallel dictionaries for different languages.

Our future works for cross-lingual blog analysis continue in the direction of finding the influence by incorporating multilingual sentiment analysis techniques. This helps us in knowing whether the personality has a positive or negative influence on the bloggers.

References

1. Joshi, M., Belsare, N.: Blog Harvest: Blog mining and search framework. In: Lakshmanan, L.V., Roy, P., Tung, A.K. (eds.) Proceedings of the 13th International Conference on Management of Data (COMAD), pp. 226–229. Computer Society of India, Delhi (2006)
2. Brian, U., Ken, B., Amy, M.: New Metrics for Blog Mining. In: Proceedings of SPIE Defense & Security Symposium 2007, Orlando, FL, vol. #6570(657001) (2007)
3. Tomohiro, F., Takehito, U., Hiroshi, N.: Cross-Lingual Concern Analysis from Multilingual Weblog Articles. In: Proc. 6th Inter. Workshop on Social Intelligence Design, pp. 55–64 (2009)
4. Ishida, T.: Language Grid: An Infrastructure for Intercultural Collaboration. In: IEEE/IPSJ Symposium on Applications and the Internet (SAINT 2006), pp. 96–100 (2006); keynote address
5. Pardeep, K.: Development of Hindi-Punjabi Parallel Corpus Using Existing Hindi-Punjabi Machine Translation System and Using Sentence Alignments. International Journal of Computer Applications (0975 – 8887) 5(9) (August 2010)
6. Lisa, B., Bruce, C.W.: Phrasal Translation and Query Expansion Techniques for Cross-Language Information Retrieval. In: ACM SIGIR 1997, pp. 84–91 (1997)
7. Hiroyuki, N., Mariko, K., Sayuri, Y.: Visualizing Cross-Lingual/Cross-Cultural Differences in Concerns in Multilingual Blogs. In: Proceedings of the Third International ICWSM Conference, pp. 270–273 (2009)
8. Andreas, J., Elisabeth, L.: Cross language Blog Mining and Trend Visualisation. In: Proceedings of the 18th International Conference on World Wide Web, WWW 2009, Madrid,Spain, pp. 1149–1150 (2009)

Renaissance of Opinion Mining

Ankur Rana[1], Vishal Goyal[2], and Vimal K. Soni[3]

[1] Department of Computer Science, Punjabi University Patiala
[2] Assistant Professor, Department of Computer Science, Punjabi University Patiala
[3] Sr. Lecturer Department of CS & IT, Krishna Engineering College, Ghaziabad
{ankurrana628,vishal.pup,ruvimals}@gmail.com

Abstract. Everyone has short span of time and the information to be analyzed is too large. Opinion Finder or sentiment analysis provides the quick response to user that whether the sentence follow positive or negative opinion. As WWW is growing more rapidly more and more information is available on web. Various sites provide daily routine facilities like shopping, blogs and consultancy etc. On shopping site various users provide the reviews for the particular product with rating. But to read each review (where 1000's of review has been posted by users) is difficult and time consuming. Sentiment analysis or Opinion finder provides a summarization and overall opinion for all the reviews. Sentiment can be positive or negative and favorable or unfavorable. In this paper, we will discuss research work done by various researchers related to sentiment analysis.

Keywords: Sentiment analysis, Opinion finder, Polarity of sentence, Summarization.

1 Introduction

Sentiment analysis and opinion finders are important for majority of the organizations and users. Government used this tool to analyze the previous work done, e-commerce site also used to give user overview of the product when a large amount of reviews are given. Also sentiment of the person depends on the pitch of voice while speaking. But we cannot get to know in written text. In this paper we will discuss works done various researchers on sentiment analysis for the written text. Many researcher train their system using reviews from various site, like e-commerce site, movie reviews etc. Little work has been done in sentiment analysis of Indian languages or any other languages except English. Many factors affect the polarity of the sentence. So it is not easy to predict the sentiment of the sentence. There are many modifiers which affects the sentiment of the sentence. Also it is necessary to know in what sense or combination adjective or adverb are used in the sentence. Sentiments are related to some features. E.g. 'mobile has long battery life' has a positive polarity whereas 'Weighing machine takes long time to measure the weight' has negative polarity. As the same word 'long' has been used in both the sentences but polarity depends upon the context in which it has been used. There are many techniques have been developed for this.

C. Singh et al. (Eds.): ICISIL 2011, CCIS 139, pp. 60–67, 2011.
© Springer-Verlag Berlin Heidelberg 2011

2 Related Work

Verma and Bhattacharayya[21] used SentiWordNet [20] in their sentiment analysis implementation. The input document is preprocessed and the sentiment scores are calculated using the words appearing in the input document using the SentiWordNet [20]. They make one threshold value of the sentiment score. Those words which have higher score than threshold values are selected and others are rejected. Sometime some attribute features are more important to tell the sentiment. So, the importance of attribute/features for particular class of attribute also calculated. Following formula is used to calculate the Information gain (measure the importance of an attribute(X) w.r.t. class attribute(Y))

$$\text{InforGain}(X, Y) = \text{Entropy}(Y)\text{-Entropy}(Y|X) \tag{1}$$

Where X is the attribute and Y is the class to which attribute belongs.

$$Entropy(Y) = -\sum_{i=1}^{n} P(Y = yi)Entropy(Y = yi) \tag{2}$$

$$Entropy(Y \mid X) = -\sum_{j=1}^{m} P(X - xj)Entropy(Y \mid X = xj) \tag{3}$$

Then document vector is created in four steps: a) pre processing, b) sentiment score based pruning (c)TF-IDF (Term Frequency - Inverse Document Frequency) vector creation and d) information gain based pruning. The authors have used websited www.rottenmaotes.com and www.imdb.com for gathering review corpus used in training set.

Goldensohn et al. [19] developed hybrid model based summarizer that uses both lexicon based and machine learning algorithm. Text extractor has been used for breaking the reviews text into a set of text pieces. Using WordNet, all the synonym and antonym for all the features words on which sentiment depends were extracted. They assigned +1 for the positive sentiment word, -1 for the negative sentiment word and 0 for the neutral word. If string x= (w_1, w_2... w_N), the formula for classifying the sentiment is

$$\text{raw-score}(x) = \sum_{i=1}^{n} si \tag{4}$$

They used the simple lexical negation detector to reverse the sign of S_i if the word proceeded with negative term like no, not or never etc. If the raw-score(x) is below some threshold value then it is neutral otherwise positive or negative depending upon the sign of the raw score.

Wiebe's [8] developed algorithm for finding whether the sentence is subjective or objective which further helps in the sentiment analysis of that sentence. WordNet is very useful to identify whether the word is subjective or objective in nature. The author simply divided the corpus into 10 different random sets, and then they identify all the adjectives of strength 3 from top 20 entries. He used Hatzivassiloglou's and

Mckeown's [5] method for recognizing the semantic orientation or polarity of adjectives. The words having desirable or feasible state have been assigned positive polarity while other has been assigned negative polarity. For this purpose, they had extracted adjectives and conjunctive words from large corpus upon which the polarity depends. They also generated a list of 73 adverbs and noun phrase which is used as grading modifier (words which change the orientation of the words). The result of this work produced the intersection of seed set and lexicon feature set were at least 9% point.

Hu and Lu [11] proposed a technique for summarizing customer reviews. They identified the features of the product about which customer has given his/her reviews and then made the summary using this information about the product. In their technique, first they used crawler to get all the reviews posted on the website by different customers. They have used POS (part of speech tagging) to know which element is noun, noun phrase, verb, adjectives, adverb etc. Then the frequent features of the product are identified and the feature pruning is used to remove the features that repeat itself in the sentence. Then the opinion words were found. For finding the orientation of the word they used the WordNet [23] for collecting all the synonyms and antonym of the word. If the word is not found then the user manually assigns the orientation to the word. The authors also found the infrequent features. But the searching for the infrequent feature pose some problem i.e. it may find the noun or noun phrase which is irrelevant to product feature. Also if two opinion words are found in one sentence then nearest features or opinion word is given more weightage because it occurs most of the time. At last, summary is generated to show which are positive and negative words along with the count which show how many reviews or sentences are positive or negative. They performed opinion sentence extraction on many products and found average 0.693 and 0.642 recall and precision respectively.

Hu and Lu [12], developed a system for feature-based opinion summarization. The input to the system is the product name and an entry page for all the reviews of the product. They have used POS tagging to know which element is noun, noun phrase, verb, adjectives, adverb etc. After processing the sentence, it is saved in the review database along with POS tagging output. Then the item set is extracted from the sentence and phrase. Item set is the set of word or a phrase that occur together. For extraction of the features or item set form, the sentences association rule mining is used. Association Mining rule is implication of the form $X \rightarrow Y$, where $X \subset I, Y \subset I$ and $X \cap Y = \phi$. The rule $X \rightarrow Y$ holds in D with confidence c if c% of transactions in D that support X also support Y, Where D is a set of dataset. Apriori [1] algorithm is used to find all the frequent items or features. First it finds all the frequent item sets from a set of transactions that satisfy a user-specified minimum support and then Generate rules from the discovered frequent item sets. Then feature pruning is used to remove the incorrect features. It works on the principle - "For each sentence in the review database, if it contains any frequent feature, extract the nearby adjective". If such an adjective is found, it is considered an opinion word. A nearby adjective refers to the adjacent adjective that modifies the noun/noun phrase that is a frequent feature. If no feature is found in the sentence then it is known as infrequent feature. Sentence which contains the opinion word, near to the noun or noun phrase then added to the dataset as infrequent feature of the product. At the end orientation of the review is identified with bootstrapping technique and using WordNet. Explicit features are manually tagged. They achieved average recall of 80% and average precision of 72%.

Srivastaval et al [17] used grammatical dependencies and dependency structure for sentiment analysis. A syntactic structure consists of lexical items, linked by binary asymmetric relations called Binary Grammatical Dependencies (BGD). They represented BGD by terms governor and dependent, where governor is a superior term and dependent is an inferior term. They collect the reviews from sites amazon.com and epinio.com. They developed the DIOWL (Domain Independent Opinion Words Lexicon). DIOWL considered only adjectives as opinion words. They also found and counted the explicit product features seeds, which are mostly the noun or noun phrase in the sentence. For the implicit features, they made the corpus which contained the entire implicit feature which may describe the product features. In their proposed approach, they used Stanford Typed Dependency [10] and identified the binary grammatical relationship among the words of a sentence. System incorporates the SDs (Stanford Typed Dependency), EPFS (Explicit Product Feature Seed) and IFTC (Implicit Feature Tag Corpus) for finding the infrequent, compact, and other features in the review. The opinion words in DIOWL have polarity ambiguity. If the sentence or phrase contain two opinion words and one word is of unknown polarity and connected by cooperative conjunction then the opinion word with unknown polarity adopts the same polarity of the prior polarity opinion word conjoined with it. On the same way if two opinion words are conjoined with contrary conjunction then opinion word with unknown polarity adopts the reverse polarity of the prior polarity. To determine the polarity of the sentence, they have used the Adverb Adjective Combination (ACC) scoring method which is based on linguistic classification of adverb of degree. Every opinion can be found only of identifying the presence of adjective and adverbs.

Ding, Liu and Yu [24], used the general term to identify the opinion in the sentence. The objects have other parts like features or attribute of the features. They identified the implicit and explicit features and represented these features with features set F= $\{f_1, f_2 ... f_n\}$. Each feature f_i in F expressed a finite set of words. W= $\{w_1, w_2 ... W_n\}$ is the set of synonyms of n features. Opinion holder j commented on a subset of features $S_j \subseteq F$. They choose the word from set W to described the feature and gave a positive, negative or neutral opinion on it. In their proposed technique, they made a list of words having particular state which may be positive and negative. Such a list is known as opinion lexicon. They also make a database of idioms which may be positive or negative. A positive word is assigned +1 otherwise -1. All the scores are then summed as equation (5):

$$score(f) = \sum_{wi:wi \in S \wedge wi \in V} \frac{Wi.SO}{dis(wi, f)} \qquad (5)$$

Where V is set of all opinion words, S is sentence containing the feature f, and dis(W_i,f) is the distance between feature f. If the final score is positive then the opinion of the sentence is positive otherwise negative. They also give the negation rule to reverse the sentiment opinion. They observed that without context dependence, F score dropped to 87% because many features are assigned the neutral orientation.

Das, Bandyopadhyay [2], developed the algorithm for identifying opinionated and non-opinionated sentences Bengali and English language. They used the rule based approach for opinion subjectivity. They extracted the private state in the phrase by making direct subjective frames and expressive subjective elements.. Bengali

SentiWordNet has been used which is created using English to Bengali dictionary. They found the noun, adjective and adverb and verb using POS tagging in the sentence. Then identified the theme to which sentence or phrase belongs to. They proposed the opinion units or the words as quadruple i.e. Subject, Aspect, Opinion hold and Evaluation. They used the POS tagger developed by IIIT-H. Then they identified Subject phrase (noun combinations), Aspect phrase (attribute of subject) and evaluation phrases (positive or negative). Accuracy of these three with Bengali text has been reported to be 79.11%. They also created a list of 205 suffixes for Bengali language because in some sentences opinion words may be present in inflected form. Stemming clustering is used for feature analyzes prefixes and suffices word that is identified in a document. Also verb inflection list with 50 entries was generated. For English language Standard Porter Stemmer algorithm is used for this purpose. After performing all the above steps, frequency of the words is counted. Title of the document is also found because it always carries some meaningful subjective information. They measured the distribution value which is done by counting the distance between the occurrences of a thematic word measure. They achieved the Precision and Recall of 51% and 61% for English and 49.86% and 58.66% for Bengali respectively.

Wilson et al [22] worked on finding the subjectivity of the sentence. Subjectivity of the sentence identifies the various aspects of the phrase to predict the opinion. Their Opinion Finder operates in one large pipeline. It has two parts. First part performs mostly general purpose document processing. Second part performs the subjectivity analysis. They used Sundance partial parser to get semantic class tag, identify Named entities, and match extraction patterns that corresponds to subjectivity language. Then OpenNLP 1.1.0 [6] is used to tokenize, sentence splitting and POS tagging the data. SCOL [7] version 1G is used to stem. Clue finder is used to identify the words and phrases from a large subjective language lexicon. For the subject sentence classification they used the Naive Bayes classifier [14] to know which are subjective and objective. They used large corpus of un-annotated data to train their system. Then system identifies speech events and direct subjective expression. The third component is a source identifier that combines a conditional random field sequence tagging model and extraction pattern learning to identify the source of speech events and direct subjective expression. They used two classifiers. First classifier identifies sentiment expression and second classifier identifies those that are positive and negative. They used MPQA opinion corpus [13].

Prabowo and Thelwall [18] used three approaches viz. rule based, Support Vector machine and hybrid. In *rule based* approach, a rule is an antecedent and its associated consequent is in 'if then' relationship. There are various rule based classifiers like GIBC, RBC, and SBC. GIBC is General Inquirer Based Classifier which has 3672 pre classified rules. Out of which 1598 are positive and others are negative. It is applied to classify document. IRBC is the rule based classifier in which a second rule set is built by replacing each proper noun found within each sentence with '?' or '#' to form a set of antecedents, and assigning each antecedent a sentiment. For statistics based classifier (SBC) following formula was used:

$$S^+ = \sum Closeness(antecedent, word\ i+) \tag{6}$$

$$S^- = \sum Closeness(antecedent, word\ i+) \tag{7}$$

If $S^+ > S^-$ then it is positive otherwise it is negative. If $S^+ = S^-$ then its neutral. In *SVM (Support Vector Machine)*, given two training sets a positive sample set $T_r^+ = \sum_{i=1}^{n}(di,+1)$, and a negative training set $T_r^- = \sum_{i=1}^{n}(di,0-1)$. SVM gives the hyper plane to separate the two set the maximum margin i.e. the maximum distance. Whereas in *Hybrid Classification* apply various classifiers in sequence. A list of possible hybrid classification given by them is as follows:

1. RBC → GIBC
2. RBC → SBC
3. RBC → SVM
4. RBC → GIBC → SVM
5. RBC → SBC → GIBC
6. RBC → SBC → SVM
7. RBC → SBC → SVM
8. RBC$_{induced}$ → SBC → GIBC → SVM etc.

They collected large corpus from various sites like Pang (2007) and MySpace (2007).

Ahmed et al. [4], all the adjective, adverb and noun phrase of the sentence mostly describes the sentiment of the word. First they pre-processed the document tagged with POS. Then feature selection phase find the features such as noun or noun phrase. Uni-gram [15] found noun, and bi-gram [15] found adjective followed by Noun. Also polarity list made. Polarity depends on the feature and the context in which it is used. They referred inclusive features as Local Polarity term. General Enquirer used to build global polarity list manually. Modifier have important role in sentiment classification. It reverses the polarity of the term. As negation terms (not, couldn't, never) change the associated polarity to its opposite meaning using WorldNet's antonyms. They set a window size based on which Part-of-Speech allowed in between the negation and the polarity term. In last the polarity values of the terms identified both in global and local list as the ration of a term's occurrence in the positive or negative reviews to the total number of occurrence. Polarity values calculated using ratio in either positive or negative reviews using the equation (8) and (9)

$$P_{+ve} = \frac{Count_{+ve}}{Count_{+ve} + Count_{-ve}} \tag{8}$$

$$P_{-ve} = \frac{Count_{-ve}}{Count_{+ve} + Count_{-ve}} \tag{9}$$

Check each review features for the training set. For each feature a score is calculated by summing up all the polarity values concern to that feature. e.g $FS_1 = P_1 + P_2 + ...P_{n,}$ where FS_1 is score of the first feature, P_i is the polarity expressed in a particular sentence. The Overall polarity calculation formula is:

$$RS = Positive \ if \ FS_1 + FS_2 + ... + FS_n > \Omega \tag{10}$$

$$RS = Positive \ if \ FS_1 + FS_2 + ... + FS_n < \Omega \tag{11}$$

Where RS is the sentiment of the online review and Ω is the threshold value. For the experiment they got corpus from www.eopinions.com.

Table 1. Experimental Result of Ahmed et al. System

Method	Class	Accuracy
Term Count (TC)	Positive	76.00
	Negative	72.00
TC+Neg	Positive	78.50
	Negative	83.00
Tc+Mod	Positive	77.50
	Negative	76.00
TC+Neg+Mod	Positive	79.00
	Negative	86.00

Hiroshi and et. al [9], categorized sentiment features as: favorable, unfavorable, question and request. A sentiment unit is as « <sentiment unit> := <sentiment> <predicate> <argument>+ <surface>».To know the sentiment of Japanese sentence they converted sentence into English language. Top down parser is used to parse the sentence and generate the parse tree. Then top node examined to know the combination relation with other words. They examined three types of patterns: principal patterns (subject is excluded from the arguments), auxiliary patterns (expands the scope of matching), and nominal patterns (used to avoid a formal noun being an argument). Word sense disambiguation was used to get the polarity of the word when adjective and verb occur together in the sentence. Aggregation of synonymous expressions had done to organize extracted sentiment units. Following table show the precision and recall result of sentiment unit extraction from 200 sentences:

Table 2. Experimental Result of Deeper sentiment analysis

	(A) MT	(B) Lexicon only
Weak prec.	100% (31/31)	80% (41/51)
Strong prec.	89% (31/35)	44% (41/93)
Recall	43% (31/72)	57% (41/72)

3   Conclusion

We have studied the work carried by researchers in the field of sentiment analysis. SentiWordNet was used by various researchers to know the positive and negative score for the words in sentence. Some used statistical method or rule based method to know the polarity of the sentence. Sentiment analysis for English, Bengali, Chinese and Japanese language has been done.

References

1. Aggrawal, R., Srikant, R.: Fast algorithm for mining association rules in Large Databases. In: VLDB, pp. 487–499 (1994)

2. Das, A., Bandyopadhyay, S.: Theme Detection an Exploration of Opinion Subjectivity. In: Proceeding of Affective Computing & Intelligent Interaction, pp. 1–6. IEEE, Los Alamitos (2009)
3. Riloff, E., Phillips, W.: An Introduction to the Sundance and AutoSlog Systems. Technical report, School of Computing, University of Utah (2004)
4. Ahmed, F., Ashok, B., Mukherjee, S., Murugeshan, M.S., Sampath, A.: Effect of Modifiers for Sentiment Classification of Reviews. In: 6th International Conference on Natural Language Processing (2008)
5. Hatzivassiloglou, V., Mckeown, K.: Predicting the semantic orientation of adjectives. In: Proceedings of the Joint ACL/EACL Conference, pp. 174–181 (1997)
6. OpenNLP, http://opennlp.sourceforge.net/
7. SCOL, http://www.vinartus.net/spa/
8. Wiebe, J.M.: Learning Subjective Adjectives from Corpora. In: American Association for Artificial Intelligence, pp. 735–740. ACM, New York (2000)
9. Hiroshi, K., Tetsuya, N., Hideo, W.: Deeper Sentiment Analysis using Machine Translation Technology. In: Proceedings of the 20th international conference on Computational Linguistics, pp. 494–500. AAAI Press, Menlo Park (2004)
10. de Marneffe, M.-C., Manning, C.D.: Stanford type dependencies manual. Technical report (2008)
11. Hu, M., Lu, B.: Mining and Summarizing Customer Reviews. In: Proceeding of International Conference on Knowledge Discovery and Data Mining Seattle, pp. 168–177. ACM, New York (2004)
12. Hu, M., Lu, B.: Mining opinion Features in Customer Reviews. In: Proceedings of the 19th National Conference on Artificial Intelligence, pp. 755–760 (2004)
13. MPQA Opinion Corpus,
http://nrrc.mitre.orgcc/NRRC/publication.html
14. Naïve Bayes Classifier,
http://en.wikipedia.org/wiki/Naive_Bayes_classifier
15. N-gram, http://en.wikipedia.org/wiki/N-gram
16. Precision and Recall,
http://en.wikipedia.org/wiki/Precision_and_recall
17. Srivastaval, R., Bhatia, M.P.S., Srivastava, H.K., Sahu, C.P.: Exploiting grammatical Dependencies for Fine-grained Opinion Mining. In: International Conference on Computer & Communication Technology, pp. 768–775. IEEE, Los Alamitos (2010)
18. Prabowo, R., Thelwall, M.: Sentiment Analysis: A Combined Approach. Journal of Informetrics 3(2), 143–157 (2009)
19. Blair-Goldenshon, S., Hannan, K., McDonald, R., Neylon, T., Reis, G.A., Reynar, J.: Building a Sentiment Summarizer for Local Services Reviews. In: NLP Challenges in the Information Explosion Era (NLPIX 2008), Beijing China, April 22 (2008)
20. SentiWordNet, http://SentiWordNet.isti.cnr.it/
21. Verma, S., Bhattacharyya, P.: Incorporating Semantic Knowledge for Sentiment Analysis. In: 6th International Conference on Natural Language Processing, ICON, India (2008)
22. Wilson, T., Hoffmaan, P., Somasundaran, S., Kessler, J., Wiebe, J., Choi, Y., Cardie, C., Rilfoff, E., Patwardhan, S.: Opinion finder: A system for subjectivity analysis. In: Proceeding HLT-Demo 2005 Proceedings of HLT/EMNLP on Interactive Demonstrations, pp. 34–35. ACM, New York (2005)
23. WordNet, http://wordnet.princeton.edu/
24. Ding, X., Liu, B., Yu, P.S.: A holistic Lexicon Based Approach to Opinion Mining. In: The Proceeding of WSDM, pp. 231–240. ACM, New York (2008)

OpenLogos Machine Translation: Exploring and Using It in Anusaaraka Platform

Sriram Chaudhury, Sukhada, and Akshar Bharati

Language Technologies Research Center,
IIIT-Hyderabad
{sriram_c,sukhada}@research.iiit.ac.in

Abstract. OpenLogos is the open source version of the Logos Machine Translation System. The current system translates from English and German into the European languages (French, Italian, Spanish and Portuguese). This papers deals with extracting parse and useful linguistic information from English-German OpenLogos MT system. Understanding and extracting useful information from linguistic rich diagnosis file is explained in detail. Various parse relations such as POS, clause boundary, dependency, constituent information is extracted and mapped to Paninian format for use in English to Hindi MT system Anusaaraka.

Keywords: OpenLogos, Anusaaraka, Panini, Scon, Tran, Pada.

1 Introduction

OpenLogos [2] is the Open Source version of the Logos Machine Translation System. The current system translates from English and German into the European languages (French, Italian, Spanish and Portuguese). Currently we are using the English to German translation system. The system generates diagnosis file at 3-levels (i.e Short, Long and Deep). For our purpose, we are using *Long* diagnosis file for extracting the various parse information of the source language (i.e English).

2 Extracting Informations from OpenLogos System

Various methods of extracting informations from the Open-Logos systems are described below.

2.1 Part of Speech (POS)

The part of speech values in Open Logos are mainly present in *RESOLVED SWORK RECORDS* table in its wc (word-class) column where each number represents POS value of the word. The numbers and their corresponding POS values are given in Table 1.

C. Singh et al. (Eds.): ICISIL 2011, CCIS 139, pp. 68–73, 2011.

Table 1. POS information from Word Class mapping

WC	Description	WC	Description	WC	Description
1	Noun	8	Clausal construction	16	Arithmates
2	Verb	11	Preposition (locative/conceptual)	17	Negatives
3	Adverb (locative/time/place)	12	Auxiliary/ modal verbs	18	Conjunctions
4	Adjectives	13	Preposition (locative)	19	Relatives/ Interrogatives
5	Pronouns (Personal/Indefinite)	14	Definite articles/ Demonstratives	20	Punctuations
6	Adverbs (manner/agency/degree)	15	Indefinite articles		

Overloading of the word class values gives rise to ambiguity. For example, wc 1 is given to proper nouns, noun modifiers (adjectives) and verbal nouns. In case of an adjective, this happens only when the adjectives are parts of some NP. Such cases are resolved by consistency checking of form values for the word given in the SAL code. For example, if a word has wc **1** and the form **23** then the POS of that particular word is adjective because the form **23** represents adjectives. Similarity, form **54** represents adjectives which are also past participial adjectives (i.e. left and wet), form **60** represents only past participial adjectives and form **70** represents past participial adjectives that are also nouns (e.g. cut, input, set). Unknown first capital words have subset value **859**, using this information we make their POS as proper-noun. Table 2 shows the POS info for English sentence *I met Mohan in a beautiful garden*.

Table 2. POS information for the English sentence *I met Mohan in a beautiful garden*

WC	Form	Subset	Word	Description
5	1	795	I	(pronoun)
2	7	142	met	(verb)
1	33	**859**	Mohan	(proper-noun)
13	3	481	in	(preposition)
15	1	315	a	(determiner)
1	23	185	beautiful	(adjective)
1	1	26	garden	(noun)

2.2 Extracting Parse Information

OpenLogos system provides the detail parse information of the source sentence in a diagnosis file, from which the equivalent dependency and constituent information is extracted.

2.2.1 Extracting Dependency Relations

The diagnosis file gives the relations of subject, object, preposition and prepositional object with the verb in SEMWRK VALUES[1] . For example, a verb in SEMWRK VALUES is represented by word class (wc) 2 and its subject is represented by the form 91. Though the object is not represented explicitly but we assume the noun as object if its form is not 91 and it is not a prepositional object. A prepositional object is easily identifiable because it is always kept next to the preposition in SEMWRK VALUES.

Fig.1 represents the SEMWRK VALUES for the sentence *Rama gave a book to Mohan..* It contains 7 groups of numbers and each group consists of a set of 4 numbers. These set of 4 numbers represents word-class (wc), subset, form and word-id respectively. Openlogos adds an additional BOS to the beginning of the sentence. Fig.2 shows the words and their corresponding ids in our example sentence.

```
***SEMWRK VALUES
    2   494   37    0          2   494    7    3          1     1   91    2
    1    76   30    5         13   945    4    6          1   859   33    7
   20    10    1    8
```

Fig. 1. Semwork values

Word Ids:	1	2	3	4	5	6	7	8
Sentence:	bos	Rama	gave	a	book	to	Mohan	.

Fig. 2. English sentence with its corresponding word ids

In SEMWRK VALUES, the first set {**2 494 37 0**} is not of our interest because its fourth number represents word id less than 2. The second set {**2 494 7 3**}, with first number **2** represents word class of a verb and the fourth number 3 represents id of the verb *gave*. The third set {**1 1 91 2**} with its third number (form) 91 says the word-id **2,** *Rama* is the subject of the verb *gave*. The set next to the subject is assumed to be object if its wc is not 13 i.e preposition. And if the wc is 13 then it becomes prepositional object. In our case the fifth set represents preposition having word-id 6 and the sixth set represents prepositional object having word-id 7.

The within pada [1] (phrase/group) relations such as modifier-modified relations are extracted from the swork tables of tran1, tran3 and *RESOLVED SWORK RE-CORDS* table. For example, in the pada *a book*, *a* is a determiner modifier of *book*. To get a modified-determiner_modifier relation among *book* and *a* we have to compute it from the *OUTPUT TARGET ARRAY IN tran1*. Fig.3 gives the tran1 table for the head word *book*. In the row starting with *SCONPO*, ids greater than 0 (zero) and less than the head id are assumed to be the premodifiers of the head and the ids greater than the

---

[1] From SEMWRK VALUES, predicate argument structure of the words in a sentence can be extracted.

head id and less than the length of the sentence are considered to be postmodifiers of the head. The information whether the id present in the *SCONPO* is a determiner modifier or a adjectival modifier is decided by looking at the wc of each word id in *RESOLVED SWORK RECORDS* table.

SWORK0 =	1	76	30	5	book			25	36			
OPADRO	-102	-107	-107	-103	-105	-108	-106	-101	5	-104	-112	-110
SCONPO	50	4	51	52	53	54	55	56	5	57	58	59
HFDOPO	0	4	0	0	0	0	0	0	0	0	0	0

Fig. 3. Analysis of the head and components in tran1

The inter clausal relations are extracted form tran3. For example, in the sentence *The dog which Chris bought is really ugly.* there are two different clauses: The main clause, *The **dog** is really ugly* and the dependent clause, ***which** Chris bought*. In these clauses *dog* and *which* are related with each other but this information is not explicitly marked in SEMWRK VALUES. The SEMWRK VALUES for these clauses are given in Fig. [4] and Fig. [5].

```
***SEMWRK VALUES
    2 886 60 0    2 886  3 7    1 126 91 3    1 185  90 9    20 10 1 10
```

Fig. 4. Semwrk values for the main clause: *The dog is really ugly*

```
***SEMWRK VALUES
    2 709 41  0    2 709 7 6    1 126 94  4    1 859 91 5
```

Fig. 5. Semwrk values for the dependent clause: which Chris bought

Such relations are extracted from tran3 using the subset and form ids, where subset id of modified is same as the subset id of its modifier. As in Fig. [6] the subset id for both *dog* and *which* is **126** and the object of the relative clause (i.e *which*) is marked by its form **94**.

```
***** THE SWORK TABLE IN tran4 *****

20 900  2 1    1 126  17 3    2 886  3 7    1 185  90 9    20 10  1 10
   BOS            dog            is            ugly           EOS
20 103 18 12    1 126  94 4    1 859 33 5    2 709  7 6    20  10 18 13
 * CLS-BOS *      which         Chris         bought       * CLS-EOS *
```

Fig. 6. Tran3 output

2.2.1.1 Insertion of Relations for Missing Words in the Sentence. Sometimes new words are inserted in the sentence for proper translation. Along with the insertion of new words we need to give relations of that word with the other words present in the sentence. OpenLogos uses SWITCH68 to solve such cases. For example, in the sentence *The dog who I chased was black.* can also be written as *The dog I chased was black*, where the word *who* is dropped and still the sentence remains grammatically correct but in Hindi presence of *who* is a must for the sentence to be grammatically correct. For e.g *The dog I chased was black* will be correctly translated into Hindi as *vaha kuttaa jisakaa maine piichaa kiyaa kaalaa thaa*. We have to insert who(jisakaa)[2] for the correct translation.

```
          ***** THE SWORK TABLE IN tran4 *****
  20 900 2 1     1  126  17 3    2 886 10 7     1 609 90 8    20  10  1  9
    BOS              dog             was            black         EOS
  20 103 18 11    1  126  94 4    1 795 1 5      2 889  7 6    20  10  18 12
  * CLS-BOS *     * SWITCH68 *       I             chased       * CLS-EOS *
```

Fig. 7. Diagram showing insertion of SWITCH68 in OpenLogos

For inserting the extra pada, we use the *SWITCH68* information given in the *SWORK TABLE IN tran4*. The wc, subset, and form of SWITCH68 tell us about the type and place of the word to be inserted.

Fig.7 shows the **SWORK TABLE IN tran4** for our example sentence. It shows the presence of * *CLS-BOS* * and * *SWITCH68* * in the analysis. Whenever there is an insertion of SWITCH68 a word is assumed to be inserted exactly at this place in the sentence. The ids representing wc, subset and form help us to identify where and what type of word is to be inserted. In our example, * *CLS-BOS* * and the number present above show that the relative clause begins here and presence of * *SWITCH68* * after it suggests the place of the word at the beginning of the relative clause. The subset number *126* which is similar to subset of *dog*} says that the word to be inserted belongs to the same subset that of dog and the form *94* says that it is an object. Using this information we can insert the word *who* and give a modifier relation between *dog* and *who* and object relation between *chased* and *who*.

2.2.2 Extracting Constituent Information

Constituent information is extracted and represented in *Pada* [1] notation for the further use in Anusaaraka[6] system.

Hindi *Pada* is the minimal group of words, from which moving any word out of the group makes the sentence ungrammatical. According to Paninian grammar two types of padas (i.e *Subanta* and *Tinganta*) padas exist in a sentence. Noun phrases (NP) and preposition phrases (PP) come under *Subanta pada* and verb phrases (VP) come under *Tinganta pada*. Due to the language divergence between English and Hindi the grouping rules have to take account of the differences while forming the padas.

OpenLogos provides grouping information of NPs and PPs in the *OUTPUT TARGET ARRAY* of tran1, tran2 and tran3 level [2].The pada module constructs the subanta padas (NP and PP) using the collective information from these 3 level of analysis. Tinganta padas (VP) are constructed from the analysis of *SWORK TABLE IN tran4, RESOLVED SWORK RECORDS, SCON FOR tran1, tran2, tran3* and applying some additional heuristics.

---

[2] The word **who** is translated as *jisakaa* in Hindi after word sense disambiguation.

```
                   ***** THE SWORK TABLE IN tran4  *****
  20 900 2 1      1 70 30 4        12 710 7 5       2 835 46 7      1 43 2 8
    BOS             boy               had             eating          fruit
  20 10 1 9
    EOS
```

Fig. 8. Analysis of SWRK TABLE at the tran4 level in OpenLogos

Fig.8 shows the SWORK TABLE IN tran4 for the sentence *A fat ugly boy had been eating fruits.* We compute the *subanta pada*(VP) by looking at the verbs (wc=2) present in the analysis. Here the word *eating* has the set of numbers {2 835 46 7} representing word class, subset, form and word-id respectively. The form 46 determines the verb group to be\textquotedblleft had been eating\textquotedblright and the *ting*TAM [3] part to be *had been ing*.

In this way the *ting* part and the root part of the verb are computed separately and given to the generation module for final generation of Hindi pada(VP).

The above extracted information is represented in CLIPS fact format, in accordance with Paninian framework. These sets of initial facts are given input to the central expert system. Initial facts are processed by different modules and new output facts are generated until the target language generation is reached.

3 Conclusion

In this paper we have studied the Open-Logos system and successfully extracted various linguistic and parse information of source language (I.e English) and implemented it in English-Hindi Anusaaraka MT system.

Acknowledgment. Authors want to acknowledge Shree Bud Scott and Prof. Vineet Chaitanya for their precious help and guidance in exploring the Open-Logos system and using the extracted information in Anusaaraka.

References

1. Sukhada, Pada: The concept of phrase in Panini and its applicability in machine translation, http://anusaaraka.iiit.ac.in/sites/default/files/pada.pdf
2. Scott, B(B.): The Logos Model: An Historical Perspective, Machine Translation. Springer, Netherlands (2003)
3. Bharati, A., Chaitanya, V., Sangal, R.: Natural Language Processing A Paninian Perspective, ch. 4, pp. 49–58 (1994)
4. .Anusaaraka, http://anusaaraka.iiit.ac.in
5. OpenLogos, http://logos-os.dfki.de/
6. Chaudhury, S., Rao, A., Sharma, D.M.: Anusaaraka:An Expert system based MT System. In: The Proceedings of IEEE Conference on Natural Language Processing and Knowledge Management, IEEE-NLPKE 2010, Beijing China, pp. 448–453 (2010)

Role of e-Learning Models for Indian Languages to Implement e-Governance

Avinash Sharma[1] and Vijay Singh Rathore[2]

[1] Research Scholar, Shree Karni Collge affiliated to SGV University, Jaipur
[2] Director & Research Guide, Shree Karni College, Jaipur, Rajasthan, India

Abstract. E-learning is becoming dominant delivery method in workplaces around the globe in various sectors and of varying sizes. E-learning had become now a three decade old technology in comparison to computer based training and education. There is essential need of design of new and efficient e-learning models which can incorporate all Indian languages spoken across for successful implementation of e-governance. Basically, e-learning models are attempts to develop a generalized framework to address the concerns of the learner and challenges presented by the technology so that online learning can take place effectively. The growth of e-learning changes the very nature of education, how it is designed, administered, delivered, supported, and evaluated. In this paper, of e-learning models has been proposed for Indian languages to implement e-governance to develop school networks, to upgrade non-formal systems to improve literacy and life skills, for teacher education, for development of policy in information and communications technology systems, and for modernizing curricula and learning methods.

Keywords: Enterprise learning, Targeted learning, e-learning.

1 Introduction

As far as the relationship of e-learning to a country's development is concerned, the creation of the national wealth of the country comes from the way it uses its investment and labor capital. Many developing countries are not able to provide students and citizens with the knowledge and skills training needed to compete in the increasingly sophisticated global workforce. Many students are not able to take advantage of learning and training due to vast distances from learning centers. Through e-learning, people in rural areas especially in India can now remain in their communities and access a world-class education. Thus, e-learning can provide a cost-effective solution to geographic gaps in education. E-learning can be the fuel to propel India's economies to new and greater heights. Many countries in Asia and the Pacific have shown their interest in the field of e-learning. Some have even embraced this technology and it has become a way of life. The private sector will always be a key element in developing e-learning technology and implementation of e-governance. However, the public sector also needs to do its part by establishing policies that embrace this type of education, and setting standards and regulations.

C. Singh et al. (Eds.): ICISIL 2011, CCIS 139, pp. 74–80, 2011.

E-learning covers a wide set of applications, but people mostly focus on the "e," technical aspect rather than on the second part, which is learning. Some important aspects of e-learning include the specific needs of the target audience, the course content, the delivery mechanisms, and the tutorial and technical support. There are many debates on e-learning as it is the trade-off between traditional, instructor-led training and e-learning. Some of the major advantages of e-learning are individualization, flexibility, active participation, continuous availability, and cost effectiveness. India is now at the forefront as active users of e-learning. Many states in India are not making significant progress in e-learning and are falling behind their neighboring states due to language barriers as there are more than 26 Indian languages: Angika, Awadhi, Bagelkhandi, Bengali, Bhojpuri, Bishnupriya Manipuri, Bundelkhandi, Chhattisgarhi, Gujarati, Hindi, Kannada, Kashmiri, Konkani, Maithili, Malayalam, Manipuri, Marathi, Marwadi, Oriya, Punjabi, Rajasthani, Tamil, Telugu.

2 Role of e-Learning in e-Governance Implementation

An e-government uses technology to deliver services based on customer, rather than administrative, convenience and by transformation rather than automation. There are three applications where e-learning can help this transition. One is to facilitate cultural and organizational challenges faced by governments in transforming their structures, processes, and internal employee culture to drive e-government development. Second is to deploy e-learning in the community to raise the level of technology and application-user skills, thereby lowering the cost of access to and raising demand for e-government applications and services. Third is to effectively leverage investments in an e-government platform to complement e-learning frameworks in the formal and non-formal educational system. Governments integrating e-learning have to manage the learning transformation—changing the way an organization addresses its learning strategy, processes, and supporting infrastructure. This includes evaluating the impact of current learning programs on human and organizational performance, and redesigning instructional processes, content, and delivery mechanisms. There are a number of benefits e-learning can contribute for e-government. It eliminates the barriers that have prevented people from different departments acquiring high-quality education and support services. It also makes learning pervasive, continuous, and relevant. Finally, e-learning propagates knowledge sharing through access to expertise and collaboration between employees and partners as well as improving the performance and productivity of employees. E-learning is a key enabler of e-government success. E-learning can change people's acquisition of skills through access to knowledge technology, eliminate barriers that hinder people from accessing high–quality technology[5], and enable organizations to be more adaptive to the changing environment. Implementing e-learning will result in more pervasive, continuous, relevant, and collaborative learning that will deliver faster, measurable results. Governments in India are realizing the change; however, significant challenges still remain. Learning challenges fall into three categories, namely:

• Enterprise learning, Targeted learning and Infrastructure for learning

Enterprise learning is the establishment of an enterprise approach to learning with the goal of integrating and aligning learning with organizational priorities. Targeted learning is the development of high-impact targeted learning initiatives that focus on performance improvement. Finally, infrastructure for learning is the implementation of an open, reliable, and scalable infrastructure to support learning initiatives that can be easily integrated with other enterprise systems. Governments are still struggling with the e-government concept and for some the transformation is very daunting. However, it cannot be denied that e-learning would be a very valuable tool for governments wanting to shift to a paradigm of e-government. Government projects should start small; patience would be required in bringing everyone to the same mindset of e-government. Finally, one should not underestimate the amount of money, time, effort, and support needed because these are critical for e-government implementation and sustainability. Through e-learning, learning will be more pervasive, continuous, relevant, and collaborative. Are we all ready for the future? As citizens of any country, need to realize how to move forward.

There are many factors that shape the future of learning.

- **Workforce:** In the current generation, many are in the habit of conducting numerous tasks at the same time. They are used to multitasking and a fast lifestyle. Multitasking can facilitate the improvement of skills.
- **Technology:** Technology also is a factor because it creates pervasive and intuitive innovations [7]. Learners are empowered to shape their learning experience through the use of technology. The next-generation workforce would be knowledge hungry, interactive, and would value time, all of which imply that learning would be relevant and available, accessible beyond institution boundaries, integrated, and dominated by collaboration.

The Combination of learner empowerment and organizational learning results in embedded learning. This is the ultimate goal of organizations involved in e-learning. Collaborative learning will enable innovation. Learning will start from the individuals linking into teams and then to the organization level where it can foster creation of ideas and growth. It is important to determine the "e-status" of the organization or institution and its priorities because this will determine where to start transforming innovations. IBM transformed itself from a hardware company to a services-driven organization. This change required rebuilding and re-skilling their employee base. IBM integrated a four-tiered approach to e-learning. Tier one is learning from information, which requires persons to read, hear, or see information that comes their way. Tier two is learning from interaction, which requires trying and experiencing games and simulations for interactive learning. Tier three is collaborative learning or learning from peers. This involves virtual classroom, live conferences, teaming, real-time awareness, and collaborative sessions. The last tier is experience-based learning. This requires learning from co-location or face-to-face learning that includes role-playing, mentoring, coaching, case studies, etc. The important question is how to integrate these creations. Leaders and designers should plan such integration carefully. They should not duplicate what other countries are doing because there are factors that are not applicable to all countries. Leaders, policymakers, planners, teachers, and other stakeholders have to be consistent and vigilant about enforcing standards. Countries must continually evolve by creating and recreating strategies that will bring them to the global revolution.

3 e-Learning and e-Governance in India

E-learning and E-governance in India facing lot of changes due to technology and barriers of vast Indian languages spoken especially in rural part of India. e-governance and e-learning is now been practiced in urban part of India due to awareness and availability of technology at affordable cost. Technology is not a key item in India; how to use technology is the biggest challenge. In response to this challenge, the Government has partnered with private organizations, such as IBM. Their e-learning program focuses on providing computer literacy in the rural areas. As per 2001 Population Census of India, the Literacy rate of India has shown as improvement at 65.38%. It consists of male literacy rate 75.96% and female literacy rate is 54.28%. Kerala with 90.86% literacy rate is the top state in India. Mizoram and Lakshadweep are at second and third position with 88.80% and 86.66% literacy rate respectively. Bihar with 46% literacy rate is the last in terms of literacy rate in India. Government of India has taken several measures to improve the literacy rate in villages and towns of India. State Governments has been directed to ensure and improve literacy rate in districts and villages where people are very poor. There has been a good improvement in literacy rate of India in last 10 years but there is still a long way to go. There is a great digital divide between urban and rural India; thus, India needs to take information technology to the masses. There are approx. 5000 universities and engineering colleges providing computer education at the degree/diploma level. The output of trained manpower in IT at this level has consistently been increasing. Mastery over quantitative concepts coupled with English proficiency has resulted in a skill set that has enabled the country to take advantage of the current international demand for IT. Still, there are many places in India, where both e-learning and e-governance are just a dream.

4 Action Plans to Implement e-Learning and e-Governance

Workforce skills development is imperative for 21st century workforce requirements. [5] Officials and citizens need to have basic skills training in order to survive in the global knowledge economy. Action plans should incorporate establishing a local training center. This entails new infrastructure but will improve education and enable a focus on training and improvement of skills. Many dramatic changes are happening in higher education: rising costs, competition between universities internationally, jobs requiring post-secondary education, etc. These changes are facts and institutions need to address them. One proposal is to enhance and encourage collaborative learning and integrate a management system. Information Technology learning project planning needs careful attention. [2] E-learning planning requires broad capabilities that help to advance learning effectiveness and efficiency to produce real education/training value. Aspects to consider include learning strategy, content development and management, learning delivery, learning technology, learning integration, and learning outsourcing. These are the key aspects for implementing a vast initiative that delivers measurable outcomes and enables real time learning. Good learning project planning consists of policy development, scope, project plan, and implementation. Once a plan is completed, it is important that the proposed policies be shared with other people. There are key elements identified in making successful e-learning models for implementing e-governance: Strategic planning and vision, Curriculum and Content, Use of the Internet, Acceptable Use of Policies, ICT and Education Reform, Quality Assurance and Accreditation, Connectivity Infrastructure, Networks & Professional Development, Intellectual Property and Rights, Intra-governmental Issues, Cost, Finance and Partnership

Issues for Strategic planning and vision

- Lack of strategic planning and vision
- Quality gap between private and public school education
- Lack of awareness of problems and solutions
- No appraisal of available technologies and needs of the country
- Need to identify acceptable changes, cost-effective planning
- Need for strong leadership and target pilot projects

Strategies for Strategic planning and vision

- Reach consensus on the vision &Share the vision nationwide
- Set up basic ICT infrastructure & Develop e-learning societies
- Keep fees for educational use of e-learning, especially in schools low
- Clearly define agency responsibilities
- Prepare national plans by government and private sector experts jointly
- Provide high-level support to determine priorities and make clear Statements

Issues for Curriculum and Content

- Materials are outdated and Materials may be biased politically
- Little capacity is available to develop content Strategies
- Create an enabling environment to foster development of local content
- Set up mechanisms for ongoing reviews of curriculum and content

Issues for Use of the Internet and Acceptable Use of Policies

- Lack of computer and language skills to use the Internet effectively
- Incorporate "firewalls" for control purposes & Promote private sector

Issues ICT and Education Reform

- Lack of teachers' materials and Provide e-learning materials for teachers
- Teachers lack time to be (re)trained and Lack of funds Strategy

Issues in Quality Assurance and Accreditation

- Mechanisms of compliance implementation & Standards of quality
- Institutional capacity needs to be matched with work/project requirements
- Encourage self-regulation to avoid excessive government regulation
- Develop an accreditation scheme for e-learning by a professional body
- Encourage recognition of e-learning by requiring e-learning courses
- Develop several levels of accreditation, such as diploma and degree
- Adopt international standards & Develop an effective QA and monitoring
- Enforce standards, also in projects and procurement
- Allow accreditation by the private sector for non formal education

Issues in Connectivity Infrastructure and Networks

- Costs of connectivity and infrastructure are high
- Lack of basic infrastructure (electricity and telecommunications)
- Access should be via a national "backbone" and networks should be secure

Strategies in Connectivity Infrastructure and Networks

- Create a national budget to subsidize ICT costs in education
- Public-private partnerships should cooperate with ICT providers
- Allow competition in telecommunication sector to reduce costs
- Introduce IP networks with multimedia capacity and Introduce open learning

Issues in Professional Development

- Need for qualified teachers and Need for training of trainers
- Need for high-quality ICT instructional material, also in local languages

Strategies in Professional Development

- Develop a system of continuing education and accept international standards

Issues in Intellectual Property and Rights

- Absence of "cyber" laws and Lack of enforcement where cyber laws exist
- High costs of some brands of software and Lack of coordination
- Emphasis on copyright, sharing of content and Need for strong enforcement

Strategies in Intellectual Property and Rights

- Copyright to be recognized and registered
- Make government-sponsored development (software, content, etc.)
- Negotiate an international pricing model based on gross domestic product
- Find solutions based on innovative partnerships/financing models

Issues in Intra-governmental Issues

- Need for political will to enforce compliance regarding usage and licensing
- Need for inter-government collaboration on enforcement of cyber laws

Strategies in Intra-governmental Issues

- Develop a strong interdepartmental coordination mechanism in ICT
- Use ICT for information sharing/communication between government Develop an international (government) portal for sharing information

7 Conclusion

This paper concludes as for successful implementation of e-governance, the various issues and strategies must be considered seriously while drafting the action plan. The role of e-learning models for Indian Languages to implement e-governance will be to incorporate all the issues and strategies.

References

1. Sharma, A.: Social Benefits of e–learning in India. In: National Conference on e-Learning: An Innovative Knowledge Oriented Framework, pp. 105–111 (2009)
2. Singh, V., Sharma, A.: Information Technology and e-learning: A Comparative Analysis of Impacts. International Journal and Communication, 147–153
3. Banduni, M.: The future of e-learning in India. Express Computer (November 2005)
4. Kumar, K.: Central Institute of Education, University of Delhi, Quality Education at the Beginning of the 21st Century- Lessons from India
5. National Policy on Information and Communication Technology(ICT) in School Education Department of School Education and Literacy Minister of Human Resource Development, Government of India (2009)

A Compiler for Morphological Analyzer Based on Finite-State Transducers

Bhuvaneshwari C. Melinamath, A.G. Math, and Sunanda D. Biradar

B.L.D.E.A.'s Engg College and Tech. Bijapur 586103,
Karnataka
bmelinamath@yahoo.co.in,
agmath@yahoo.com,
sunanda_biradar@rediffmail.com

Abstract. Morphological analyzers are an essential parts of many natural language processing (NLP) systems such as machine translation systems. They may be efficiently implemented as finite state transducers. This paper describes a morphological system that can be used as stemmer, lemmatizer, spell checker, POS tagger, and as E-learning tool for Kannada learning people giving detailed explanation of various morphophonemics changes that occur in saMdhi. The language specific components, the lexicon and the rules, can be combined with a runtime engine applicable to all languages. Building Morphological analyzer/generator for morphologically complex and agglutinative language like Kannada is highly challenging. The major types of morphological process like inflection, derivation, and compounding are handled in this system.

Keywords: FST, POS, NLP, DFA, NFA.

1 Introduction

Morphological analyzers are essential parts of many natural language processing systems such as machine translation systems. Morphological analysis reads the inflected surface form of each word in a text and writes its lexical form consisting of of a canonical form of the word and a set of tags showing its syntactic category and morphological characteristics. The analyzer relies on two sources of information: a dictionary of valid lemmas of the language and a set of rules for inflection handling.

Finite state transducers (FST) is a most efficient approach to morphological analysis (M. Mohri 1997: Oncina et al. 1993) a class of finite state automata, is a complete example using an intuitive pattern matching approach which tries first to decompose the word in number of stem inflection pairs which are subsequently validated. There are a number of tools for the construction of FST based morphological analyzers the best known being developed at Xerox (Karttmen 1994: Karttmen 1993: Chanod 1994) for a review in Spanish on finite state morphology.

In this work a FST based morphological analyzer is developed. It is a compiler that reads a morphological dictionary containing static description of lemmas and inflections and writes a PERL program that implements a compact FST based analyzer performing the task. This allows the linguist to focus on describing the lexicon and morphology of the language in question in a simple format.

C. Singh et al. (Eds.): ICISIL 2011, CCIS 139, pp. 81–85, 2011.

2 Finite State Transducers

The morphological analyzers are based on finite state transducers; in particular, we use string or pattern transducers instead of letter transducers (Roche & Schabes 1997). Any finite-state transducer may always be turned into an equivalent letter transducer. Instead of transition on letters we have transitions on sequences of letters i.e, strings, and generally valid suffixes in the language. The machine starts in the specified initial state and reads in a string of symbols from its alphabet. The automaton uses the state transition function to determine the next state using the current state, and the symbol just read or the empty string. However, "the next state of an NFA depends not only on the current input event, but also on an arbitrary number of subsequent input events. Until these subsequent events occur it is not possible to determine which state the machine is in. If, when the automaton has finished reading, it is in an accepting state, the NFA is said to accept the string, otherwise it is said to reject the string. When the last input symbol is consumed, the NFA accepts if and only if there is *some* set of transitions that will take it to an accepting state. Equivalently, it rejects, if, no matter what transitions are applied, it would not end in an accepting state. Unlike a DFA, it is non-deterministic in that, for any input symbol, its next state may be any one of several possible states. Thus, in the formal definition, the next state is an element of the power set of states.

The transducer is defined as $T = (Q, L, \delta, qI, F,)$ where Q is a finite set of states, L a set of transition labels, $qI \in Q$ the initial state, $F \subseteq Q$ the set of final states, and $\delta : Q \times L \to 2^Q$ the transition function (where 2^Q represents the set of all finite sets of states). The set of transition labels is $L = (\Sigma \cup \{\varepsilon\}) \times (\Gamma \cup \{\varepsilon\})$ where Σ is the alphabet of input symbols, Γ the alphabet of output symbols, and ε represents the empty symbol. According to this definition, state transition labels may therefore be of four kinds: $(\sigma : \gamma)$, meaning that symbol $\sigma \in \Sigma$ is read and symbol $\gamma \in \Gamma$ is written $(\sigma : \varepsilon)$, meaning that a symbol is read but nothing is written; $(\varepsilon : \gamma)$, meaning that nothing is read but a symbol is written; and $(\varepsilon : \varepsilon)$ means that a state transition occurs without reading or writing. The last kind of transitions are not necessary neither convenient in final FSTs, but may be useful during construction. It is customary to represent the empty symbol ε with a zero ("0"). A letter transducer is said to be deterministic when $\delta : Q \times L \to Q$. Note that a letter transducer which is deterministic with respect to the alphabet $L = (\Sigma \cup \{\varepsilon\}) \times (\Gamma \cup \{\varepsilon\})$ may still be non-deterministic with respect to the input Σ.

A string $w' \in \Gamma^*$ is considered to be a transduction of an input string $w \in \Gamma^*$ if there is at least one path from the initial state qI to a final state in F whose transition labels form the pair $w : w'$ when concatenated. There may in principle be more than one of such paths for a given transduction; this should be avoided, and is partially eliminated by determinization. On the other hand, there may be more than one valid transduction for a string w (in analysis, this would correspond to lexical ambiguity; in generation, this should be avoided). In analysis, the symbols in Σ are those found in texts, and the symbols in Γ are those necessary to form the lemmas and special symbols representing morphological information, such as <noun>, <feminine>, <first person p1. second person p2. Third person p3> for pronouns, etc. In generation, Σ and Γ are exchanged. The general definition of letter transducers is completely parallel to that of non-deterministic finite automata (NFA) and that of deterministic letter transducers, parallel to that of DFA; accordingly, letter transducers may be determinized and minimized (with respect to the alphabet L) using the existing algorithms for NFA and DFA (Hopcroft & Ullman 1979; Salomaa 1973; van de Snepscheut1993). Transitions labeled (ε : ε) may be eliminated during determinization using a technique parallel to ε closure. For all one writes if and only if q can be reached from p by going along zero or more ε arrows. For any , the set of states that can be reached from p is called the **epsilon-closure** or ε **-closure** of p, and written as

$$E(\{p\}) = \{q \in Q : p \xrightarrow{\varepsilon} q\}.$$

For any subset, define the ε -closure of P as for any P$\subset$Q subset, define the ε-closure of P as

$$E(P) = \bigcup_{p \in P} E(\{p\}).$$

Which allows a transformation to a new state without consuming any input symbols. For example, if it is in state 1, with the next input symbol an a, it can move to state 2 without consuming any input symbols, and thus there is an ambiguity: is the system in state 1, or state 2, before consuming the letter a. Because of this ambiguity, it is more convenient to talk of the set of possible states the system may be in. Thus, before consuming letter a, the NFA-epsilon may be in any one of the states out of the set {1,2}. Equivalently, one may imagine that the NFA is in state 1 and 2 'at the same time': and this gives an informal hint of the power set construction 2^Q.

Unlike other compilers like Karttunen's (1993), the compiler described in this paper builds transducers having no cycles (transitions form a directed acyclic graph) which, in addition, have a unique final state. The absence of cycles is due to the fact that only concatenations and alternations are allowed in the morphological dictionary (see section 3)1. To minimize the resulting transducer, we use an algorithm described by van de Snepscheut (1993), which has two identical steps which may be summarized as follows: in each step, the transition arrows in the letter transducer are reversed, so that the final state is initial and the initial state is final, and the resulting transducer is determinized with respect to L (that is, new states are formed with sets of old states so that the new δ is δ : Q×L$\rightarrow$ Q). The transducer resulting from the double reversal determinization process is minimal. This algorithm is particularly efficient in the case of acyclic letter transducers. Moreover, the two steps have a simple interpretation: the first step joins common endings (finds regularities in suffixes) and the second one joins common beginnings of transduction (finds regularities in prefixes). FST-based analyzers output all possible analyses.

3 Morphological Dictionary

The morphological dictionary is a text file. Any text starting with "#" is ignored and may be used as a comment. The dictionary has the following three sections: 1. The symbol declaration section representing the actual root words in the language. 2. Second section represent categories of the word followed by morphological features such as (such as <feminine> or <singular>) are explicitly declared. First second fields are separated by two vertical bars "||". 3. The information section, where any morph relevant information like real u, past participle form of irregular verbs information etc. are declared: when lemmas in the dictionary share a common infection pattern, this pattern may be given a name in another file handling the inflection of the words in the language. While generating the inflections for a word the information field is looked upon. Rules may be indefinitely nested, that is, the names of rules previously defined may be used to define derived forms of the words. (Category wise set of rules are compiled into sub transducers that are then integrated to build the complete transducer). The transition on suffix 'a' in Kannada stand for a genitive case of nouns also for negation for verbs and another sense of imprecate meaning of verb in some case. Hence we get more than one analyzes. All are valid transitions. Since context is not considered here.

taavu || PRO-REF-P23.MFN.PL-NOM
tamma || PRO-REF-P23.MFN.PL-GEN||N-COM-COU-M.SL-NOM::TYPE-kinship
biMdu || N-COM-COU-N.SL-NOM::LV-real-u
cakshu || N-COM-COU-N.SL-NOM::LV-real-u
caru || N-COM-COU-N.SL-NOM::LV-real-u
daaru || N-COM-UNC-N.SL-NOM::LV-real-u
dattu || N-COM-COU-N.SL-NOM::LV-real-u

Fig. 1. Sample of the morphological dictionary

The null transitions are allowed since it is a NFA automata. We are not preserving any fixed length pattern strings since we perform transitions on suffixes not on single letter; Separate rule file is used to holds set of orthographic rules, it is not a part of dictionary. The idea behind holding separate file is to make the system language independent, the code is not hard wired i.e, SaMdhi rules governing insertion or deletion of vowels are put in separate file and not written as part of code. And another advantage is the same code can be used for other languages too just by replacing orthographic rule file with their language morphophonemic. The valid transitions may be an entry in the another file called FST transitions file.

4 The Compiler

The compiler has been developed under Linux using Perl. Which reads in the morphological dictionary file and combines the partial transducers corresponding to the declared paradigms and the dictionary entries into a single transducer containing one initial and one final state using (ε: ε) transition.

Error messages are designed to help the linguist correct possible errors in the format of the morphological dictionary. The back end minimizes the resulting transducer and combines the resulting code with a standard skeleton to produce a Perl program which is ready to be used on its own or included in a larger application such as a machine translation system.

5 Experiments and Comparisons

We perform the experiments to evaluate the heuristic used by the compiler and used the morphological dictionary of 10000 words compiled by us following hierarchical tag set for Kannada which covers more detailed analysis of the word, design of Hierarchical tag set for Kannada was an another stage towards developing the morphological analyzer for Kannada, and for testing, the first 5000 most frequent Kannada words of Department of Electronics (DoE), Central Institute of Indian Languages (CIIL) corpus words are selected which included all the nominal, pronominal, adjectival and verbal inflections of the Kannada language. Around 80% words were analyzed correctly, manually checked and verified, remaining 20% words were a mixture of spelling variations, dialect variations, compound words, hence such words in the raw corpus selected for testing are not analyzed, regarding ambiguity error is very less and missing entries in dictionary. Currently we are not handling compound words and dialectic variations. This system is first of its kind for Kannada using FST.

6 Concluding Remarks

A compiler to automatically build finite state transducer based morphological analyzers using set of rules consisting of features for categories and another file with FST transition rules and morphological dictionaries has been described. This tool may be of great interest when building natural language processing systems such as machine translation programs. When the linguist does not supply an explicit alignment between surface forms and lexical forms, the compiler uses a simple heuristic to produce an alignment that has been experimentally shown to be equally efficient. We are currently testing an extended version the program which to handle spelling variations, few dialectic variations and to improve dictionary as per the need of hierarchical design of tag set.

References

1. Karttunen, L.: Finite-state lexicon compiler. Technical Report ISTL-NLTT, Xerox Palo Alto Research Center, Palo Alto, California (1993-04-02)
2. Oncina, J., García, P., Vidal, E.: Learning subsequential transducers for pattern recognition interpretation tasks. IEEE Transaction on Pattern Analysis and machine Intelligence 15, 448–458 (1993)
3. Hopcroft, J.E., Ullman, J.D.: Introduction to automata theory, languages and computation. Addition -Wesley, Reading (1979)
4. Mohri, M.: Finite-state transducers in language and speech processing. Computational Linguistics 23(2), 269–311 (1977)
5. Roche, E., Schabes, Y.: On the use of sequential transducers in natural language processing. In: Finite State Language Processing, pp. 353–382. MIT Press, Cambridge (1997b)
6. Salomaa, A.: Formal Languages. Academic Press, New York (1973)
7. Van de Snepscheut, J.L.A.: What computing is all about. Springer, New York (1993)
8. Chanod, J.-P.: Finite state composition of French verb morphology. Technical Report Technical Report MLTT-005, Xerox Research Centre Europe, Meylan, France (1994)

On Multifont Character Classification in Telugu

Venkat Rasagna, K.J. Jinesh, and C.V. Jawahar

International Institute of Information Technology,
Hyderabad 500032, India

Abstract. A major requirement in the design of robust OCRs is the invariance of feature extraction scheme with the popular fonts used in the print. Many statistical and structural features have been tried for character classification in the past. In this paper, we get motivated by the recent successes in object category recognition literature and use a spatial extension of the histogram of oriented gradients (HOG) for character classification. Our experiments are conducted on 1453950 Telugu character samples in 359 classes and 15 fonts. On this data set, we obtain an accuracy of 96-98% with an SVM classifier.

1 Character Classification

Large repositories of digitized books and manuscripts are emerging worldwide [1]. Providing content-level access to these collections require the conversion of these images to textual form with the help of Optical Character Recognizers (OCRs). Design of robust OCRs is still a challenging task for Indian scripts. The central module of an OCR is a recognizer which can generate a class label for an image component. Classification of isolated characters and thereby recognizing a complete document is still the fundamental problem in most of the Indian languages. The problem becomes further challenging in presence of diversity in input data (for example, variations in appearance with fonts and styles.)

Characters are first segmented out from page or word images. A set of appropriate features are then extracted for representing the character image. Features could be structural or statistical. Structural features are often considered to be sensitive to degradations in the print. A feature-vector representation of the image is then classified with the help of a classifier. Multilayer neural network, K nearest neighbour, support vector machines (SVM) etc. are popular for this classification task. Classification of Indian scripts is challenging due to (i) large number of classes (compared to Latin scripts) (ii) many pairs of very similar characters. (See Figure 1.)

In a recent work, Neeba and Jawahar [2] had looked into the success rates of character classification problem in an Indian context. Though their results are primarily on Malayalam, they are directly extendible to other scripts. They successfully solved the character classification problem (even in the presence of large number of classes) for limited number of fonts popularly seen in print. They had argued that (i) multiclass classification solution can be made scalable by designing many pair-wise classifiers. (ii) use of large number of features (of the order of few hundred) makes the problems better separable and solvable with simple classifiers. (iii) when the dimensionality of

C. Singh et al. (Eds.): ICISIL 2011, CCIS 139, pp. 86–91, 2011.
© Springer-Verlag Berlin Heidelberg 2011

Fig. 1. Challenges in character classification of Telugu. First row shows similar character pairs from Telugu. Second row shows how the same character gets rendered in different fonts.

the feature is made reasonably high, even the simple features like raw-pixels or PCA-projections provide satisfactory results.

A strong requirement of any robust character recognition system is the high classification accuracy, in the presence of multiple and diverse font sets. In this paper, we explore the problem of character classification in a multifont setting. Though our studies are for Telugu script, we believe that these results are also extendible to other languages. Our objective is to demonstrate the utility of the histogram of gradients (HoG) [3] sort of features for character classification. We also show that the linear SVM with DDAG sort of classifier fusion strategy provides equivalent results to an Intersection kernel SVMs. We validate our experimental results on 1453950 Telugu character samples in 359 classes and 15 fonts.

Telugu Script: Telugu is a south Indian language with its own script. Like most other Indian scripts, there are consonants, vowels and vowel-modifiers. In addition, there are also half consonants which get used in consonant clusters. Though the script is getting written from left to right in a sequential manner, many of these modifiers often gets distributed in a 1.5D (not purely left to right; they are also written top-to-bottom at places) manner. Compared to most other Indic scripts, Telugu has large number of basic characters/symbols. Many of them are also similar in appearance. This makes the character classification problem in Telugu very challenging. (See Figure 1)

Telugu character recognition has been attempted in the past with various features. Negi et al. [4] used fringe maps as the feature. The method was tested on 2524 characters. Jawahar et al. [5] did a more thorough testing (of close to one Million samples) of the character classifiers but with limited font variations as well as degradations. They used PCA, LDA etc. as the possible feature extraction scheme for Telugu character classification.

2 Features and Classifiers

Recent years have witnessed significant attention in development of category level object recognition schemes with many interesting features. Histogram of oriented gradients (HOG), which was successfully used for detecting pedestrians [3], is one of the prominent and popular features for capturing the visual data, when there are strong edges. Naive histogram representation looses the spatial information in the image. To address this, spatial pyramid matching was proposed [6]. Similar to [7], we also employ a feature vector which captures spatial information and histograms of oriented gradients.

We are motivated by the recent classification experiments in multifont data sets [8] and handwritten MNIST and USPS digit data sets [7]. Many of these studies are limited to handwritten digits. There have been many studies in this area (i) focusing on generalization of classification results to unknown fonts, and thereby solving the character 'category' recognition problem [8]. (ii) accurately solving the handwritten digit recognition with many machine learning concepts [7]. (iii) development of recognition algorithms with fewer training data or lesser resource usage.

Character/Symbol images are first normalized to a fixed size of 28×28 and histograms are constructed by aggregating the pixel responses within the cells of various sizes. Our cell sizes include 14×14, 7×7 and 4×4, with overlap of half the cell size. The histograms at different levels are multiplied by weights 1, 2 and 4. The entire sets of histograms are finally concatenated to form a single histogram. We refer this feature as SPHOG in the paper.

Based on the conclusions obtained in our earlier work on character classification [2], we use SVM classifiers. SVM classifiers are the state of the art in machine learning, to produce highly accurate and generalizable classifier. The classification rule for a sample x is

$$sign(\sum_{i=1}^{nSV} \alpha_i \kappa(x, s_i) + b)$$

where s_i s are the support vectors and $\kappa()$ is the kernel used for the classification. The Lagrangians α are used to weigh the individual kernel evaluations. The complexity of classification linearly increases with the number of support vectors. To make the classification fast, we can do the following [9]: (i) Use linear kernels instead of nonlinear ones. (ii) Store the weight vector instead of the support vectors (iii) Use binary representation as well as appropriate efficient data structures and (iv) Simplification of repeating support vectors in a decision making path consisting of multiple pair wise classifiers.

It was shown that the intersection kernel can be evaluated fast in many practical situations [10]. However, the comparisons are with that of complex kernels like RBF Kernel. Such classifiers are appropriate when the classes are not well separable. In the case of large class character recognition data set, most of the pair wise classifiers could be linearly separable. The overall classification accuracy reduces due to (i) cascading effects in the multiple classifier systems (ii) some of the pairs are difficult to separate with simple features. In this work, we compare the IKSVM with linear SVM and prefer to go for linear SVMs due to the computational and storage advantages of the linear SVM over IKSVM.

Based on the experimental results presented in the next section, we argue that (i) object category recognition features are useful for the character recognition especially in presence of multiple fonts. (ii) linear SVMs perform very similar to IKSVM for most of the character classification tasks. (iii) Use of SPHOG sort of features can successfully solve the multifont character classification problem in Indic scripts.

3   Results and Discussions

We start by investigating the deterioration of performance with the number of fonts. For this purpose, we collected a character level groundtruthed Telugu data set in fifteen fonts. Number of classes which is common to all these fonts is 359. We first investigate the utility of raw pixels as a feature with a linear SVM classifier. For this experiment, we consider only the first 100 classes.

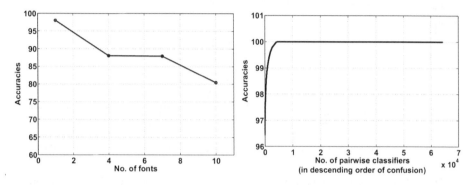

Fig. 2. (a) Variations of accuracies with increase in fonts (b) Accuracy and confused pair wise classifiers

Results of the variation of accuracy are plotted in Figure 2(a). It may be seen that with only one or limited fonts, the accuracies are acceptable, however, with the number of fonts increasing, the accuracy comes down significantly.

We now quantitatively show the results on a 100 class subset of the Telugu characters in 15 different and popular fonts. We show that the naive features, like raw pixels or PCA, are unable to address the significant font variation present in the dataset.

Table 1. Comparative results on a smaller set of Telugu Multifont Data Set

Classifier	RawPixels	PCA	SPHOG	PCA-SPHOG
	d=784	d=500	d=2172	d=500
LSVM(OneVs All)	91.91	90.19	98.10	96.70
LSVM(DDAG)	94.19	93.84	97.25	97.51
IKSVM(OneVs All)	92.26	96.369	98.71	98.39

Table 1 compares the performance of the four features in presence of two different SVM classifiers – Linear SVM(LSVM) and Intersection Kernel SVM (IKSVM). Linear SVMs are also implemented as One Vs All as well as DDAG [2]. It may be noted that the raw image features are not able to perform well when the number of fonts increases. This is expected because of the variation in the styles and shapes of the associated glyphs. It is surprising that the PCA, which was performing reasonably well for limited number of fonts [2] is also not able to scale well for the multifont situation. A graph which shows the variation of the number of eigen vectors (principal components) selected Vs the accuracy obtained is shown in Figure 3 (a). The plot of magnitudes of eigen values of the covariance matrix (used in PCA) is shown in Figure 3(b).

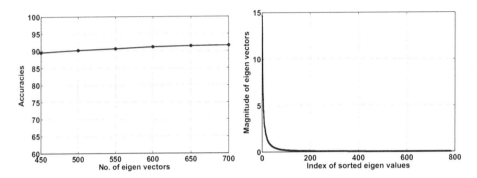

Fig. 3. (a) Accuracy and number of eigen vectors (b) Eigen vectors and their magnitude

These graphs explain that with an increase in the number of PCs the accuracy monotonically improves. However, the accuracy saturates at a level 91%, which is not an acceptable level of accuracy, we are looking for an OCR. On the contrary, the SPHOG features are performing consistently well for the large font data set, as can be seen in Table 1. PCA has been applied on the SPHOG feature as the dimensionality of the feature is large. Even with 23% of the SPHOG feature vector, accuracy close to the SPHOG result has been obtained.

In short, it is clear from the experiments conducted on a 100 class data set, that SVM classifier with SPHOG and PCA-SPHOG features provide the most accurate classifiers. We have extended the results obtained for a full Telugu character set consisting of 359 classes. They summarize as follows:

Obtaining an accuracy of 96.4 on a truly challenging multifont data is significant. However, we would like to see the possibility of enhancing the accuracy further. For this, we analyze the confusions associated with all the pair wise classifications. As can be seen from Figure 2(b), the errors are associated with only certain pairs. In Figure 2(b), we plot the cumulative accuracy over all pair wise confusions. If we can address the errors in these pairs with the help of an additional classifier (we call them as post-processing classifier), we can enhance the accuracy. We propose to use an RBF based SVM classifier for this purpose. The detailed design and analysis of the post-processing classifier are beyond the scope of this paper. It is observed that with the help of a few robust post-processing classifiers, one can enhance the accuracy to 98%.

Table 2. Classification accuracy: No of classes = 359, No of samples = 1453950

Raw pixels with Linear SVM classifier results in an accuracy	81.05
SPHOG with Linear SVM classifier results in an accuracy	96.41
PCA-SPHOG with Linear SVM classifier results in an accuracy	92.95

4 Conclusions

We show that high classification accuracies can be obtained for character classification problem with the help of SPHOG-SVM combination. Left out confusions is associated only to a small percentage of the classifier and a post-processing classifier with an uncorrelated feature set can successfully boost the overall classification performance.

References

1. Sankar, K.P., Ambati, V., Pratha, L., Jawahar, C.V.: Digitizing a million books: Challenges for document analysis. In: Bunke, H., Spitz, A.L. (eds.) DAS 2006. LNCS, vol. 3872, pp. 425–436. Springer, Heidelberg (2006)
2. Neeba, N.V., Jawahar, C.V.: Empirical evaluation of character classification schemes. In: Seventh International Conference on Advances in Pattern Recognition (ICAPR), pp. 310–313. IEEE, Los Alamitos (2009)
3. Dalal, N., Triggs, B.: Histograms of oriented gradients for human detection. In: Conference on Computer Vision and Pattern Recognition (CVPR), vol. 1, pp. 886–893. IEEE, Los Alamitos (2005)
4. Negi, A., Bhagvati, C., Krishna, B.: An OCR system for telugu. In: Proceedings of Sixth International Conference on Document Analysis and Recognition (ICDAR), pp. 1110–1114. IEEE, Los Alamitos (2002)
5. Jawahar, C.V., Kumar, P., Kiran, R., et al.: A bilingual OCR for hindi-telugu documents and its applications. In: Proceedings of Seventh International Conference on Document Analysis and Recognition (ICDAR), pp. 408–412. IEEE, Los Alamitos (2003)
6. Lazebnik, S., Schmid, C., Ponce, J.: Beyond bags of features: Spatial pyramid matching for recognizing natural scene categories. In: Conference on Computer Vision and Pattern Recognition (CVPR), vol. 2, pp. 2169–2178. IEEE, Los Alamitos (2006)
7. Maji, S., Malik, J.: Fast and accurate digit classification. Technical Report UCB/EECS-2009-159, EECS Department, University of California, Berkeley (2009)
8. De Campos, T.E., Babu, B.R., Varma, M.: Character recognition in natural images. In: Proceedings of International Conference on Computer Vision Theory and Applications (VISAPP), INSTICC, pp. 273–280 (2009)
9. Ilayaraja, P., Neeba, N.V., Jawahar, C.V.: Efficient implementation of SVM for large class problems. In: 19th International Conference on Pattern Recognition (ICPR), pp. 1–4. IEEE, Los Alamitos (2009)
10. Maji, S., Berg, A.C., Malik, J.: Classification using intersection kernel support vector machines is efficient. In: IEEE Conference on Computer Vision and Pattern Recognition (CVPR), pp. 1–8. IEEE, Los Alamitos (2008)

Parallel Implementation of Devanagari Document Image Segmentation Approach on GPU

Brijmohan Singh, Nitin Gupta, Rashi Tyagi, Ankush Mittal, and Debashish Ghosh

Astt. Professor, Computer Science Deptt., College of Engineering Roorkee, Roorkee-247667,
Uttarakhand, India
bmsingh1981@gmail.com

Abstract. Fast and accurate algorithms are necessary for Optical Character Recognition (OCR) systems to perform operations on document images such as pre-processing, segmentation, extracting features, training-testing of classifiers and post processing. The main goal of this research work is to make segmentation accurate and faster for processing of large numbers of Devnagari document images using parallel implementation of algorithm on Graphics Processing Unit (GPU). Proposed method employs extensive usage of highly multithreaded architecture and shared memory of multi-cored GPU. An efficient use of shared memory is required to optimize parallel reduction in Compute Unified Device Architecture (CUDA). Proposed method achieved a speedup of 20x-30x over the serial implementation when running on a GPU named GeForce 9500 GT.

Keywords: OCR, Segmentation, Parallelization, GPU, CUDA.

1 Introduction

Research on Devnagari character and word recognition is very difficult due to its challenging properties. This area of research is still open for further research due to the extent of variation among writing styles, speed, thickness of character and direction of different writers, Real-world handwriting is a mixture of cursive and noncursive parts, which makes the problem of recognition and synthesis more difficult, Similar looking characters may give ambiguity, Characters segment may touch where they should not or vice versa, variations and noises introduced during scanning and continuously increasing demand for accuracy, fast recognition, cheap and more practical to implement recognition system.

The applications [1-3] of OCR such as form processing, automatic mail sorting, bank checks processing, and office automation for text entry. In handwritten document processing, half of the errors are due to segmentation. Segmentation process is challenging due to touching components, wide variety of handwriting styles, text line segmentation, location of word boundaries, overlapping of characters, identification of physical gaps between words and characters. Some elaborate studies on line, word and character segmentation are in [4-9]. In this work, we modified the traditional profiling based segmentation method and parallelized to make it faster using CUDA.

C. Singh et al. (Eds.): ICISIL 2011, CCIS 139, pp. 92–97, 2011.

2 Introduction to nVidia CUDA

NVIDIA® CUDA™ [10] is a general purpose parallel computing architecture introduced by NVIDIA. It includes the CUDA Instruction Set Architecture (ISA) and the parallel compute engine in the GPU.

3 Proposed Segmentation Method

The proposed segmentation method consists of two distinct stages. In the first stage, a preliminary segmentation was performed that executes line segmentation using modified histogram profiling method which uses the horizontal density of black pixels along with an axis. This process leads to the isolation of sub images corresponding to each line of the complete text. These sub images contain more than one word. In the second step, word segmentation was done using vertical density of black pixels along with an axis. These separate images of words can be further used for recognition purpose.

3.1 Sequential and Parallel Implementation of Line Segmentation Method

First, we implemented the sequential code and then parallelized code of proposed method. The following pseudo code outlines the structure of proposed parallelized modified horizontal profiling method implementations for line segmentation:

```
Input    -   2 D image, Threshold Value
Output   -   Segmented Lines
M        :   Image intensity matrix
Density:  An  array  having  number  of  pixels  less  than
             threshold value.
Minima :  An array having the position from where image is
             to be segmented
     Subroutine main()
         Define a block and grid
         For each row of M
             Call Density_Kernel(Image)
         End For
   Set: i=0
         For each horizontal row (j) of M
             If (Density[j] ==0 && (Density[j - 1] > 0  ||
  Density[j + 1] > 0))
                 Then minima[i++]=j;
         Modify_Kernel()
      End For
  While(minimaLength !=null)
      Calculate height between two rows
  Call Output_ Kernels(image)
         End While
         End Subroutine
  Density_kernel(image)
```

```
      If  Intensity is less than threshold
      Then increment the density for each row
Output_kenel(image)
  Use to create an image
Modify_kernel()
```

Calculate average of all the vertical intensities and then check the leftover parts between neighbour segmented lines. Divide it in to two parts and compare the density of upper and lower half's, using neighbour pixels and according to this results either neglect these part or embed with segmented image.
      End Function

3.2 Sequential and Parallel Implementation of Word Segmentation Method

Proposed method for word segmentation for sequential processing was implemented first and then in CUDA to get more efficiency. Parallelized algorithm is more efficient than sequential in terms of time. The parallel implementation followed the structure shown in the pseudo code below:

```
Input    - 2 D image, Threshold Vale
Output   -  Segmentated Words
M        : Image intensity matrix
Density: An array having number of pixels less than
           threshold value.
Minima : An array having the position from where image is
         to be segmented
  Subroutine main()
        Define a block and grid
        For each column in M
             Call Density_Kernel(Image)
      End For
       Set i:=0
        For each column (j) in M
            If (Density[j] ==0 && (Density[j - 1] > 0  ||
  Density[j + 1] > 0))
  Then minima[i++]=j;
  End For
  While(minimaLength !=null)
      Calculate height between two rows
  Call Output_ Kernels(image)
        End While
  End Subroutine
Density_kernel(image)
       If Intensity is less than threshold
         Then increment the density for each row
  Output_kenel(image)
```

```
Use to create an image
Modify_kernel()
```

Calculate average of all the horizontal intensities and then check the leftover parts between neighbour segmented lines. Divide it in to two parts and compare the density of upper and lower half's, using neighbour pixels and according to this results either neglect these part or embed with segmented image.
End Function

4   Results and Discussions

To test the proposed approach, we have collected a dataset of 10 document images of old newspapers and some are written by different writers. The results shown that modified profiling method works better than traditional profiling method. The comparison results are shown in figure 1and 2. The traditional profiling method works poor in overlapping text; it fails in calculating minima and maxima. Thus to avoid such problems we modified the method of finding minima and maxima for a text line. To make faster the profiling method, we parallelized proposed algorithm on CUDA and achieved a speedup of 20x-30x (on 10 images) over the serial implementation when running on a GPU named GeForce 9500 GT having 30 cores. Table 1 shows the comparison of execution time of proposed algorithm on CPU over GPU. Hence proposed method proved that it works better in overlapping cases and run faster on GPU. Our method does not give good result when there is a more noise in background, but it gives good result as compared to traditional profiling method as shown in fourth image in figure 1.

S.N.	Input Image	Traditional Profiling Method	Proposed Method
1.	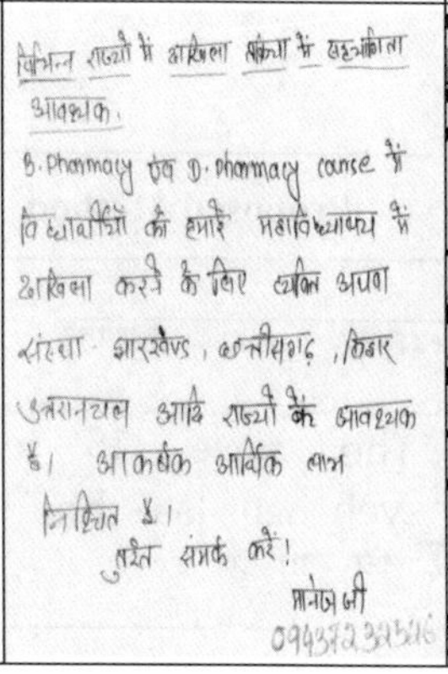	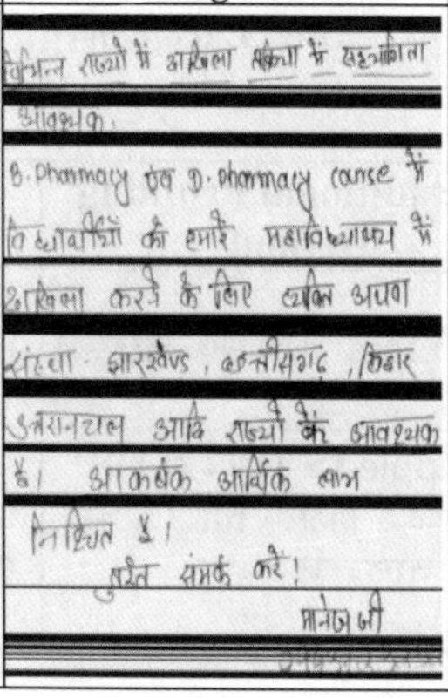	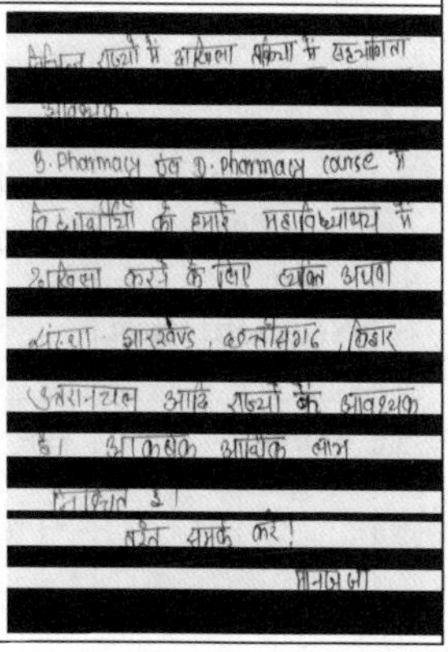

Fig. 1. Shown line detection and segmentation in document images

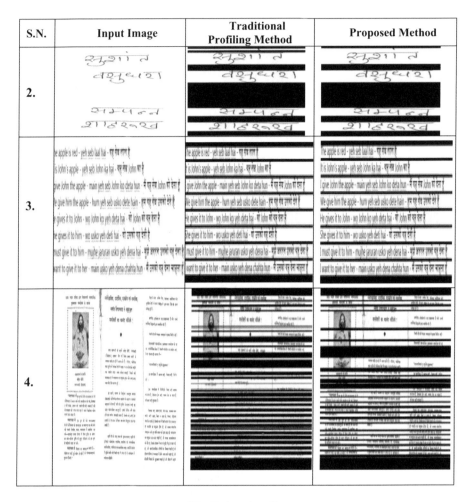

Fig. 1. (*continued*)

Method Task	Traditional Profiling Method	Proposed Method
Line Segmentation	The apple is red - yeh seb laal hai - यह सेब लाल है	The apple is red - yeh seb laal hai - यह सेब लाल है
Word Segmentation	The apple is red - yen seb laal ha l - यह सेब लाल ह	The apple is red - yeh seb laal hai - यह सेब लाल है

Fig. 2. Shown a sample of line and word segmentation

Table 1. Shown comparison of execution time CPU over GPU

| Image | Execution time on CPU (Sec) | Execution time on GPU(Sec) | Average (Sec) | | Speedup |
			CPU	GPU	
1.	1.24	.0413	1.24	.0413	30x
	1.24	.0423			
	1.23	.0413			
2.	.61	.0277	.61	.0277	22x
	.62	.0279			
	.61	.0278			
3.	1.12	.0412	1.12	.0412	28x
	1.13	.0412			
	1.12	.0413			
4.	1.91	.0706	1.91	.0707	27x
	1.93	.0709			
	1.91	.0705			

5   Conclusion

In this work a fast and modified profiling based segmentation algorithm has been presented and analyzed. The results of proposed method on the graphics device are very fast, with large two dimensional images than sequential algorithm. This algorithm serves as an excellent framework to solve a diverse array of problems. Parallelization plays important role in OCR research to speed up any algorithm to make faster processing.

References

1. Marinai, S.: Introduction to Document Analysis and Recognition. SCI, vol. 90, pp. 1–20 (2008)
2. Tang, Y.Y., Suen, C.Y., Yan, C.D., Cheriet, M.: Document Analysis and Understanding: a Brief Survey. In: First ICDAR 1991, France, pp. 17–31 (September-October 1991)
3. Plamondon, R., Srihari, S.N.: On-line and Off-line Handwritten Recognition: A Comprehensive Survey. IEEE Trans. on PAMI 22, 62–84 (2000)
4. Lecolinet, E., Crettez, J.: A Grapheme Based Segmentation Technique for Cursive Script Recognition. In: Proceeding of First ICDAR 1991, France, pp. 740–748 (September-October 1991)
5. Yanikoglu, B., Sandon, P.A.: Segmentation of Off-Line Cursive Handwriting using Linear Programming. Pattern Recognition 31(12), 1825–1833 (1998)
6. Plessis, B., Siscu, A., Menu, E., Moreau, J.W.V.: Isolated Handwritten Word Recognition for Contextual Address Reading. In: Proc. USPS 51h Advanced Technology Conference, vol. 27(1), pp. 158–164 (November 1992)
7. Paquet, T., Lecourtier, Y.: Handwritten Recognition: Application on Bank Cheques. In: Proceeding of First ICDAR 1991, France, pp. 749–757 (September/October 1991)
8. Leroux, M., Salome, J.C., Badard, J.: Recognition of Cursive Script Words in a Small Lexicon. In: Proc. of first ICDAR 1991, France, pp. 774–775 (September/October 1991)
9. Chaudhuri, B.B., Pal, U.: Skew Angle Detection of Digitized Indian Script Documents. IEEE Trans. on Pattern Analysis and Machine Intelligence 19(2), 182–186 (1997)
10. Nvidia Corporation, CUDA Zone, http://www.nvidia.com

A Rule Based Schwa Deletion Algorithm for Punjabi TTS System

Parminder Singh[1] and Gurpreet Singh Lehal[2]

[1] Dept. of Computer Sc. & Engg, Guru Nanak Dev Engg. College, Ludhiana (Pb.), India
parminder2u@rediffmail.com
[2] Dept. of Computer Science, Punjabi University, Patiala (Pb.), India
gslehal@gmail.com

Abstract. Phonetically, schwa is a very short neutral vowel sound, and like all vowels, its precise quality varies depending on the adjacent consonants. During utterance of words not every schwa following a consonant is pronounced. In order to determine the proper pronunciation of words, it is necessary to identify which schwas are to be deleted and which are to be retained. Schwa deletion is an important step for the development of a high quality Text-To-Speech synthesis system. This paper specifically describes the schwa deletion rules for Punjabi written in Gurmukhi script. Performance analysis of the implemented rule based schwa deletion algorithm, evaluates its accuracy to be 98.27%.

Keywords: Punjabi schwa deletion, Text-to-speech synthesis, Speech synthesis, Punjabi vowels and consonants.

1 Introduction

Schwa is a mid-central vowel that occurs in unstressed syllables. Phonetically, it is a very short neutral vowel sound, and like all vowels, its precise quality varies depending on its adjacent consonants. Each consonant in Punjabi (written in Gurmukhi script) is associated with one of the vowels. Other vowels, except schwa ('ਅ' the third character of Punjabi alphabet and written as [ə] in International Phonetic Alphabet (IPA) transcription), are overtly written diacritically or non-diacritically around the consonant; however schwa vowel is not explicitly represented in orthography. The orthographical representation of any language does not provide any implicit information about its pronunciation and is mostly ambiguous and indeterminate with respect to its exact pronunciation. The problem in many of the languages is mainly due to the existence of schwa vowel that is sometimes pronounced and sometimes not, depending upon certain morphological factors. In order to determine the proper pronunciation of words, it is necessary to identify which schwas are to be deleted and which are to be retained. *Schwa deletion* is a phonological phenomenon where schwa is absent in the pronunciation of a particular word, although ideally it should have been pronounced [1]. The process of schwa deletion is one of the complex and important issue for grapheme-to-phoneme conversion, which in turn is required for the development of a high quality text-to-speech (TTS) synthesizer. In order to produce natural

C. Singh et al. (Eds.): ICISIL 2011, CCIS 139, pp. 98–103, 2011.

and intelligible speech, the orthographic representation of input has to be augmented with additional morphological and phonological information in order to correctly specify the contexts in which schwa vowel is to be deleted or retained [2].

Mostly phonological schwa deletion rules have been proposed in literature for Indian languages. These rules take into account morpheme-internal as well as across morpheme-boundary information to explain this phenomenon [3]. The morphological analysis can improve the accuracy of the schwa deletion algorithm which is a diachronic and sociolinguistic phenomenon [1, 4]. The syllable structure and stress assignment in conjunction with morphological analysis can also be used to predict the presence and absence of schwa [5].

1.1 Punjabi Language and Schwa

Punjabi is an Indo-Aryan language spoken by more than hundred million people. Like other Indian languages, Punjabi includes segmental phonemes (vowels and consonants), but not supra-segmental phonemes (stress, intonation, juncture, nasality and tone) in its alphabet. In Gurmukhi script, which follows the *one sound-one symbol* principle, Punjabi language has thirty eight consonants, ten non-nasal vowels (ਇ, ਈ, ਏ, ਐ, ਅ, ਆ, ਔ, ਉ, ਊ, ਓ) and same numbers of nasal vowels (ਇੰ, ਈਂ, ਏਂ, ਐਂ, ਅੰ, ਆਂ, ਔਂ, ਉੰ, ਊਂ, ਓਂ).

Vowels can appear alone in orthography (known as full vowels) however consonants can appear along with vowels only. Vowels, except schwa ([ਅ]), are represented diacritically when these come along with consonants (known as half vowels), otherwise as such. The consonant sound varies according to the vowel attached to consonant. For example, consonant [ਸ] conjoined with vowel [ਈ] (having diacritic ੀ) results a single orthographic unit "ਸੀ", having pronunciation of a consonant-vowel sequence /ਸ+ਈ/ (/si/) however when this consonant comes with vowel [ਆ] the resulting single unit [ਸਾ] will be pronounced as /ਸ+ਆ/ (/sā/).

Consonants represented in orthography without any attached diacritic, basically have the associated inherent schwa vowel that is not represented diacritically. While pronouncing the written word, the speaker retains the intervening schwa vowel associated with a consonant where required and eliminate it from pronunciation where it is not required. In Punjabi the inherent schwa following the last consonant of the word is elided. For example, Punjabi word "ਸੜਕ" ([sədəkə] means road) pronounced as \ ਸ ਅ ੜ ਕ \ (\s ə d k\) is represented orthographically with only the consonant characters [ਸ], [ੜ] and [ਕ]. Schwa following the last consonant [ਕ] is deleted as per rule said above and the deletion the schwa following the second consonant [ੜ] makes the word monosyllabic of type CVCC (Consonant-Schwa-Consonant-Consonant).

2 Schwa Deletion Algorithm

Schwa vowel is not explicitly represented in orthography, so the schwa deletion algorithm basically turns to be *schwa insertion* at the morphological processing grounds in the text analysis component of the TTS system. The developed algorithm for schwa deletion basically consists of the two tasks: *vowel-consonant pattern generation* and *schwa deletion (/insertion) in vowel-consonant pattern*. These modules of the developed system have been discussed in the following subsections.

2.1 Module I: Vowel-Consonant Pattern Generation

This module generates the Vowel-Consonant pattern of the input word. It locates the full/half vowels (nasal or non-nasal) and the consonants positions; and hence identifies the schwa locations in the word. The output sequence corresponding to the input word consists of a string of symbols: C, V, v and n for the consonants, full-vowels, half-vowels and nasal-morphemes (bindī [ੰ] / tippī [ੰ]) respectively. This sequence of symbols will be helpful for marking the schwa positions. For example, the vowel-consonant pattern for the word "ਕਿਤਾਬ" ([kitāb] means book) is CvCvC and that of "ਅਧਿਆਪਕ" ([adiāpək] means teacher) is VCvVCC.

The further processing of the input word emphasises on finding the presence or absence of the schwa vowel sound during word's pronunciation. For example, the vowel-consonant pattern for the Punjabi word "ਮਰਦ" ([mərəd] means man) is CCC. Grammatically, there must be schwa vowel following each consonant in Punjabi but the word's pronunciation specifies the existence of schwa [ə] /ਅ/ sound after the first consonant only. So, schwa following the first consonant [ਮ] will be retained, however schwa vowels following the second [ਰ] and third [ਦ] consonants will be deleted.

2.2 Module II: Schwa Deletion (/insertion) in Vowel-Consonant Patterns

As already discussed the schwa is not represented orthographically. For the processing of the input text, schwa needs to be represented symbolically like other vowels. So, the schwa deletion process is taken up as schwa insertion and rules have been developed for the same. These rules are based on mainly three parameters: grammatical constraints, inflectional rules and morphotactics of Punjabi.

For these rules, let $\Sigma = \{V, v, C, S, b\}$ be the set of all the symbols, where V = set of full vowels, v = set of half vowels, C = set of consonants, S = schwa vowel and b = blank space. A set of eleven rules has been designed and are discussed below. The underlined symbol in the following rules represents the current consonant that is under consideration.

Rule I: VCCC → VCCSC. If a consonant is preceded by VC syllable and followed by a single consonant (C), then schwa is inserted after that consonant. For example, in word "ਇਕਦਮ" ([ikdəm] means immediate) schwa is inserted after consonant 'ਦ' ([d]) as per this rule, however not after 'ਕ' ([k]) and 'ਮ' ([m]) being at the syllable and word boundary respectively. In terms of schwa deletion we can say that schwa vowels after the consonants 'ਕ' and 'ਮ' are being deleted and that after 'ਦ' is being retained.

Rule II: VCCCC → VCCCSC. If a consonant is preceded by VCC syllable and is followed by single consonant (C), then schwa is inserted after that consonant. For example, in word "ਆਕਰਸ਼ਨ" ([ākərəshən] means attraction) schwa is inserted only after the consonant 'ਸ਼' ([sh]). No schwa after consonants 'ਕ' [k] and 'ਰ' [r] is inserted (being at syllable and word boundary respectively) and thus making this word disyllabic.

Rule III: $\underline{C}CCv(\Sigma - b) \rightarrow \underline{CSC}Cv(\Sigma - b)$. If a consonant at word starting position is followed by a consonant cluster (CC), a half vowel (v) and one more character except the word boundary (Σ - b), then schwa is inserted after that consonant. For example, in word "ਹੜਤਾਲ" ([hədtāl] means strike), the schwa is inserted after consonant 'ਹ' ([h]) as per this rule, however not after 'ੜ' being at syllable boundary.

Rule IV: $\underline{C}CCC(\Sigma - b) \rightarrow \underline{CS}CCC(\Sigma - b)$. If a consonant at the word starting position is followed by consonant triplet (CCC) and one more character except the word boundary (Σ - b), then schwa is inserted after that consonant. For example, the consonant 'ਗ' ([g]) in word ਗਰਦਨਾਂ ([gərdnā]) satisfies this rule and so schwa is inserted after this consonant.

Rule V: $\underline{C}C\Sigma(V/b) \rightarrow \underline{CS}C\Sigma(V/b)$. If a consonant cluster at word starting position is followed by any character (V, v or C) and a full vowel or word boundary, then schwa is inserted after the consonant at first position. For example, in the words "ਕਹਿਆ" ([kəhiā] means said), "ਰਹੇ" ([rəhe] mean doing), "ਤਰਨ" ([tərn] means swimming), "ਕਰ" ([kər] means do), schwa is inserted after the first consonant of each word.

Rule VI: $\underline{C}CvC \rightarrow \underline{CS}CvC$. If a consonant at word starting position is followed by a CvC syllable, then schwa is inserted after that consonant. For example in word "ਸਵਾਲ" ([səwāl] means question) schwa is inserted after the first consonant 'ਸ' ([s]).

Rule VII: $(V/v)C\underline{C}CC \rightarrow (V/v)C\underline{CS}CC$. If a consonant is preceded by full or half vowel and one consonant and is followed by a consonant cluster, then schwa is inserted after that consonant. For example, in words "ਅਸਚਰਜ" ([aschərj] means strange), schwa is inserted after the consonant 'ਚ' ([ch]).

Rule VIII: $CvC\underline{C}CvC \rightarrow CvC\underline{CS}CvC$. If a consonant is preceded by a CvC syllable and is followed by CCvC syllable, then a schwa is inserted after that consonant to make the word tri-syllabic. For example in word "ਰਾਸ਼ਟਰਵਾਦ" ([rāshtərwād] means nationalism), the consonant 'ਟ' ([t]) is at the position that satisfies the above said rule, so schwa is inserted after that consonant making the word tri-syllabic having syllables ਰਾਸ਼ (CvC), ਟਰ (CvC, where 'v' is schwa 'ə') and ਵਾਦ (CvC).

Rule IX: $(V/v/S)C\underline{C}CCv(V/b) \rightarrow (V/v/S)C\underline{CS}CCv(V/b)$. If a consonant is preceded by the VC or CvC syllables and is followed by consonant cluster, half vowel and (/or) full vowel, then a schwa is inserted after that consonant. For example, the schwa insertion in words "ਤਿਲਕਣਗੀਆਂ" ([tilkəngiyā] means to be slipped) and "ਉਸਰਨਗੀਆਂ" ([usrəngiyā] means to be constructed) after the consonants 'ਕ' ([k]) and 'ਰ' ([r]) respectively.

Rule X: <u>C</u>CCvC → <u>C</u>SCCvC. If a consonant at the word starting position is followed by the CCvC syllable, schwa is inserted after that consonant. For example, in the word "ਤਕਰਾਰ" ([təkrār] means argument) the consonant 'ਤ' ([t]) at the word starting position, satisfies the above said rule and so schwa is inserted after this consonant.

Rule XI: (V/Cv)C<u>C</u>CC → (V/Cv)C<u>C</u>SCC. If a consonant is preceded by VC or CvC syllable and is followed by consonant cluster (CC), then schwa is inserted after that consonant. For example, in words "ਉਪਕਰਣ" ([upkərn] means instrument) and "ਨਾਮਕਰਣ" ([nāmkərn] means naming), schwa is inserted after the consonant 'ਕ' ([k]) in both words.

2.3 Algorithm

The schwa deletion algorithm that has been basically implemented as schwa insertion can be described briefly as below.

Input: vowel-consonant pattern of the input word.
Output: vowel-consonant pattern with inserted (/deleted) Schwa.
Algorithm:
  i.   Set variable *CVpattern* to consonant-vowel pattern of the input word.
  ii. Set *currSymbol* to first symbol of the *CVpattern*.
  iii. Repeat steps (iv) and (v) while *currSymbol* < > ""
  iv. If *currSymbol* is a consonant then
       (a) Seach rule base for the *currSymbol*.
       (b) If the *currSymbol* satisfies any of the rules then insert schwa after that consonant in the word.
  v.   Set variable *currSymbol* to the next symbol of *CVpattern*.
  vi. Return current word.

3   Performance Analysis

The developed algorithm has been tested on ten thousand most frequently used words of Punjabi. These words have been selected from a Punjabi corpus having 104425741 total and 232565 unique words. The set of most frequently used words have been generated on the basis of their frequency of occurrence in the above said corpus. Output of the algorithm for these words has been checked manually. Out of ten thousand, 173 words have been found with wrong schwa insertion. This results the accuracy of the algorithm to be 98.27%. It has been observed that most of the words for which the algorithm is giving wrong results are those containing *addak* (ੱ) specifying gemination in Punjabi and the words having *consonant conjuncts* those appear at bottom of the barer consonant.

4   Conclusions

Schwa plays an important role in the correct pronunciation of a language and hence for the development of a high quality TTS system. The decision for retention or deletion of schwa is very much obvious for a native speaker, but for machine processing purpose this decision will be based on language specific rules. A set of eleven rules have been developed and discussed in this paper for Punjabi. These rules are based on grammar rules, inflectional rules and morphotactics for Punjabi. The algorithm developed on the basis of these rules has good accuracy which will definitely improve the quality and hence naturalness of the output speech.

References

1. Choudhury, M., Basu, A., Sarkar, S.: A Diachronic Approach for Schwa Deletion in Indo Aryan Languages. In: Workshop of the ACL Special Interest Group on Computational Phonology (SIGPHON), pp. 20–27. ACL, Barcelona (2004)
2. Narasimhan, B., Sproat, R., Kiraz, G.: Schwa-Deletion in Hindi Text-to-Speech Synthesis. Intl. J. Speech Tech. 7(4), 319–333 (2004)
3. Bali, K., Talukdar, P.P., Krishna, N.S., Ramakrishnan, A.G.: Tools for the development of a Hindi speech synthesis system. In: 5th ISCA Speech Synthesis Workshop, Pittsburgh, pp. 109–114 (2004)
4. Choudhury, M., Basu, A.: A rule based algorithm for schwa deletion in Hindi. In: International Conf. on Knowledge-Based Computer Systems, Navi Mumbai, pp. 343–353 (2002)
5. Tyson, N.R., Nagar, I.: Prosodic Rules for Schwa-Deletion in Hindi Text-to-Speech Synthesis. Intl. J. Speech Tech. 12(1), 15–25 (2009)
6. Singh, P.: Sidhantik Bhasha Vigyan, pp. 376–378. Madaan Publications, Patiala India (2002)
7. Bhatia, T.K.: Concise encyclopaedia of languages of the world. In: Brown, K., Ogilvie, S. (eds.), pp. 886–889. Elsevier Ltd., Oxford (2009)

Clause Based Approach for Ordering in MT Using OpenLogos

Sriram Chaudhury, Arpana Sharma, Neha Narang, Sonal Dixit, and Akshar Bharati

IIIT-Hyderabad,
AIM & ACT Banasthali Vidyapith
sriramchaudhury@gmail.com, arpanasharma04@gmail.com,
neha.rohtak@gmail.com, sonaldixit07@gmail.com

Abstract. We are proposing an approach to improve the final output coming English-Hindi Anusaaraka System. To improve final output, we need to refine the way of ordering of the Hindi sentence in existing system. The basic idea is to improve the target language reordering by marking the clausal level information using OpenLogos diagnosis file. This paper presents, how to use clausal information along with pada, relations information (already used by present Anusaaraka system) to get Hindi sentence reordering correctly.

Keywords: Anusaaraka, OpenLogos, Swork Table, Word Class, CLS-MARKER.

1 Introduction

Anusaaraka is one of the well known machine translation system which gives result in two ways, one in the layered form and second the final ordered output. Layered output is significant in order to preserve the information coming from source text. The final ordered output lacks in proper ordering which causes incorrect Hindi translation. During the various analyses of the source sentences and the target sentences in the Anusaaraka, it is noticed that the Hindi ordering for the target sentence lacks due to improper ordering of *pada* and their *vibhakti*. Here our main focus is to solve this problem by extracting and implementing the knowledge of the clause boundary from the diagnosis file produced by open-logos and using it for target language reordering in the Anusaaraka system.

1.1 Anusaaraka

As described above, Anusaaraka is an English-Hindi language accessor cum machine translation software. The approach of the Anusaaraka System lies in the fact that it uses various Open Source Softwares such as Link, Stanford parsers, Appertium and Open-Logos system for the parsing of the English Sentences. The information coming from these parsers is used along with predefined rules to get Hindi translation.

C. Singh et al. (Eds.): ICISIL 2011, CCIS 139, pp. 104–109, 2011.
© Springer-Verlag Berlin Heidelberg 2011

1.2 OpenLogos Diagnosis File

OpenLogos is a free machine translation system, in which currently German and English are the source languages available. The target languages for English include the major European languages (such as French, Italian, Spanish and Portuguese). We are using diagnosis file of the system in which English and German are configured as source and target language respectively. It contains information of source sentence analysis in various tabular forms.

SWORK. A Swork is created for every word in the text and is sent to Tran. It consists of Word class, Form field, Type, Tran-id.

TRAN. TRANs are modules that progressively redefine analyses of input sentences and store information pertinent to the synthesis of output text, keep collecting information about the source language, putting phrases together and reducing them to their heads, extracting subordinate clauses and gathering the information necessary for the generation of the target in Tran4 where the final decision is made.

CLS-MRKR. It is introduced in table whenever there is modifier clause exist in the input sentence. It indicates the location where that modifier clause is to be placed.

WORD CLASS. The Word class corresponds to the part of speech of the word which is indicated by 2 digits. There are 20 word classes i.e. 01 Noun, 02 Verb, 03 Adverb, 04 Adjective etc.

2 Existing Approach

If we see layered output of Anusaaraka it gives the gradual analysis of the English sentence in the translation process. But if we see final output it contains almost all word meanings with proper *vibhakti* (case marker) but the target language ordering is not correct in most of the cases. Anusaaraka is performing Hindi ordering through set of rules. As Hindi is free word order language and we cannot bind it in rules. Our main focus is to improve the final translation output using clausal information from diagnosis file.

3 Proposed Approach

The final ordered output in Anusaaraka lacks in proper ordering which causes incorrect Hindi translation. During the various analyses of the source sentences and the target sentences in the Anusaaraka, it is noticed that the Hindi ordering for the target sentence lacks due to improper ordering of *pada* and their *vibhakti*. Here our main focus is to solve this problem by extracting and implementing the knowledge of the clause boundary from the diagnosis file produced by open-logos and using it for target language reordering in the Anusaaraka system.

3.1 Approach

Sentences can be grouped on three levels. i.e. 1. Pada level[1] 2. Clause level 3. Sentence level. In pada level we only identify the pada present in the sentence and then place those pada as one unit. Pada give meaning with their vibhkati and preposition. On the other hand a clause represents the part of sentence which has some meaning and it may have subject and predicate or either one of them. After recognizing the clauses, within clause level ordering is done. Each clause will act as one unit. For final translation we need join different clauses and perform inter clause level ordering using tran4 information.

3.2 Algorithm

Input: English sentence
1.   Find pada in the given sentence including there category (VP or PP).
2.   Extract clauses information from the SWORK TABLE tran4.
    a.   To extract clauses identify word class of each word. If it is 20 then it acts as a connector.
    b.   All the words which are coming between two words whose word class is 20 are considered in   one clause.
    c.   Repeat this check of word class and identify all clauses of sentences.
3.   Now take every clause one by one and fill the missing word ids in increasing order using RESOLVED SWORK RECORD table.
4.   Group pada in each clause.
5.   Apply subject object verb (SOV) rule for each clause. Move the pada having category VP to the last position of the clause.
6.   Now place all the clauses in the order they were in tran4/tran3.
Output: Final order of the sentence

Special Case. In case if *CLS-MRKR* comes in any clause switch to tran3.During inter clause ordering modifier clause will place at *CLS-MRKR* place. Modifier clauses are the one whose beginning word is preceded by the word CLS-BOS and id of CLS-BOS is equal to *CLS-MRKR*'s id plus one.

3.3 Description of Algorithm Using Example

Example 1. In below translation of example sentence 1, the location of connector "because: क्योंकि"   is not in the proper order due to lack of clause information.

English Sentence 1: The group of tourists decided to have lunch in the village because the van needed repairs.

---

[1] Pada grouping is done in LWG module (local word grouping) which groups Subanta padas (Noun phrases and Prepositional phrases) and Tinganta pada (Verb phrases) differently using various Tran level information from the Openlogos diagnosis file.

Anusaaraka Output (Romanized): Paryaṭakō kē samūha nē gām̆va mēṁ dōpahara kā khānā kyōm̆ki khānē kā niścaya kiyā vaina kī durustī āvaśyakatā

Implementation of Algorithm. The words having swork containing WC 20 acts as connector (Con) and rest of words coming between two words of word class type 20, act as one clause, as shown in Fig 1.

20 900	2 1 1 91 17 3	2 304 7 6	2 710 38 8	1 43 1 9
BOS	group	decided	have	lunch

1 481 86 12	20 927 19 13	1 611 17 15	2 974 7 16	1 749 2 17
village	because	van	needed	repair

20 10 1 18
EOS

Fig. 1. Information from the SWORK TABLE in tran4

Table 1. Connectors and Clauses with their corresponding chunks

Clause & Connector level	Chunks
Con1	BOS
Clause1	group decided have lunch
Con2	because
Clause2	van needed repairs
Con3	EOS

The connectors and clauses with their corresponding chunks are formed as shown in Table 1. Now this information can be used to reorder the clauses as given in Table 2.

Table 2. Reordering of Clauses

Steps	Description
Step 1	Head Ids from tran4 of clause1-3 6 8 9 12
Step 2	All ids including missing-2 3 4 5 6 7 8 9 10 11 12
Step 3	Group pada units-(2,3) (4,5) (6) (7,8) (9) (10,11, 12)
Step 4	As id 6 pada form VP in clause, it must come at last.
Step 5	Now Clause1Order (C1) - 2 3 4 5 7 8 9 10 11 12
Step 6	Reorder Clause 2 using above steps
	Clause2 Order (C2) - 14 15 17 16
Step 7	Final Order becomes - Con1 C1 Con2 C2 Con3
Step 8	Output: Paryaṭakō kē samūha nē gām̆va mēṁ dōpahara kā khānā khānē kā niścaya kiyā kyōm̆ki vaina kō maram'mata kī āvaśyakatā thī

Example 2. Following example illustrates the case when CLS-MRKR appears in SWORK Table and also shows that conjunctions are not properly placed in existing system.

English sentence 2: We learn what we have said from those who listen to our speaking.

Anusaaraka   output (In Romanized): Hama hama kyā sīkhatē haiṁ unasē kaha cukā hai kauna hamārē bōlanē kē liyē sunatā hai

Implementation of Algorithm. As in Fig. 2, we can see the tran4 contains the *CLS-MRKR* then, we switch to the tran3 for clause and connector information as shown in Table 3

20 900 2 1	1 796 2 2	2 579 1 3	20 393 19 4	1 796 2 5
BOS	we	learn	what	we
12 710 1 6	2 312 54 7	1 679 43 9	1 175 18 16	20 10 1 15
Have	said	those	* CLS-MRKR *	EOS
20 103 18 17	1 103 2 10	2 573 1 11	1 945 53 14	20 10 18 18
* CLS-BOS *	who	listen	speaking	* CLS-EOS *

Fig. 2. Information from the SWORK TABLE in tran4

The connectors and clauses with their corresponding chunks are formed as shown in Table 3.

Table 3. Connectors and Clauses with their corresponding chunks

Clause & Connector level	Chunks
Con1	BOS
Clause1	we learn
Con2	what
Clause2	We have said those *CLS- MRKR*
Con3	EOS
Con 4	*CLS-BOS*
Clause 3	who listen speaking
Con 5	*CLS-EOS*

Now above information can be used to reorder the clauses as given in Table 4.

Table 4. Reordering of Clauses

Steps	Description
Step 1	Head Ids from tran4 of clause1-2 3
Step 2	All ids including missing-2 3
Step 3	Group pada units-(2,3)
Step 4	If any VP is there apply SOV on it else the order remains same. Now Clause1Order (C1) - 2 3
Step 5	Head Ids from tran4 of Clause 2-5 6 7 9
Step 6	All ids including missing - 5 6 7 8 9
Step 7	Group pada units - (6 7) (8 9)
Step 8	Check for VP, as (6, 7) is VP, place it at last in clause. Now Clause1 Order (C1) - 5 8 9 6 7 Now Clause2 Order (C2) - 14 15 17 16.
Step 9	As clause2 contains *CLS-MRKR. Mark clause marker's id as i, and jump to the word whose id is i+1 which is to be placed in the position of the *CLS-MRKR*.
Step 10	Reorder Clause 3 using the same method above
Step 11	Now Clause3 Order (C3) - 10 12 13 14 11
Step 12	Final Order Con1 C1 Con2 C2 Con4 C3 Con5 Con3
Step 13	Final Output-Hama sīkhatē haiṁ jō hama unasē kaha cukē haiṁ jō hamārī bātō kō sunatē haiṁ

4 Conclusion and Future Work

The proposed method gives correct result for only 70% of cases. Future work will involve handling of inter-clause ordering heuristics rules. We must give our special thanks to Dr.Vineet Chaitanya for his guidance and knowledge during this work.

References

1. Bharati, A., Chaitanya, V., Sangal, R.: Natural Language Processing: A Paninian Perspective. The Prentice-Hall of India, Englewood Cliffs (1995)
2. Scott, B., Barreiro, A.: OpenLogos MT and the SAL Representation Language.logossystemarchives.homestead.com
3. Narayana, V.N.: Anusaaraka: A Device to Overcome the Language Barrier. The Ph.D. thesis, Dept. of CSE, I.I.T. Kanpur, India (1994)
4. http://ltrc.iiit.ac.in/~anusaaraka/

Comparison of Feature Extraction Methods for Recognition of Isolated Handwritten Characters in Gurmukhi Script

Dharam Veer Sharma[1] and Puneet Jhajj[2]

[1] Asstt. Prof., Dept. of Computer Science, Punjabi University, Patiala
dveer72@hotmail.com
[2] MTech (CS), Dept. of Computer Science, Punjabi University, Patiala
puneet158@gmail.com

Abstract. The present paper is a comparative study of different feature extraction techniques for recognition of isolated handwritten characters in Gurmukhi script. The whole process consists of three stages. The first, feature extraction stage, analyzes the set of isolated characters and select the set of features that can be used to uniquely identify characters. For the selection of stable and representative set of features of character under consideration in this problem Zoning, Directional Distance Distribution (DDD) and Gabor methods have been used. The second stage is classification stage which uses features extracted in the first stage to identify the character. For classification Support Vector Machine (SVM) has been used to identify the character. In the third stage, feature extraction methods have been compared with respect to recognition rate. An annotated sample image database of isolated handwritten characters in Gurmukhi script has been prepared which has been used for training and testing of the system. Gabor based feature extraction proved to be better as compared to others.

Keywords: Feature extraction methods, Zoning, Gabor filters, DDD, handwritten character recognition, Gurmukhi script.

1 Introduction

Optical character recognition (OCR), is the mechanical or electronic translation of scanned images of handwritten, typewritten or printed text into machine-encoded such as ASCII code. The potential of OCR systems is enormous because computer systems armed with OCR system improve the speed of input operations, reduce data entry errors, reduce storage space required by paper documents and thus enable compact storage, fast retrieval, scanning corrections and other file manipulations. OCR have applications in postal code recognition, automatic data entry into large administrative systems, banking, automatic cartography, 3D object recognition, digital libraries, invoice and receipt processing, reading devices for blind and personal digital assistants. Accuracy, flexibility and speed are the three main features that characterize a good OCR system.

C. Singh et al. (Eds.): ICISIL 2011, CCIS 139, pp. 110–116, 2011.

Recognition of isolated handwritten characters is the process of identifying individual characters. It is useful in wide range of real world problems like documentation analysis, mailing address interpretation, bank check processing, signature verification, documentation verification etc. Due to applications of recognition, it is one of the most challenging areas of pattern recognition. It has been topic of research for a long period of time. Work has been done in recognizing handwritten Chinese, Arabic, Devnagari, Urdu and English characters, recognizing handwritten numerals and handwritten digits. In the problem of recognition of isolated handwritten characters the input is isolated characters. Word segmentation provides isolated characters. Characters can be in upper zone, middle zone or lower zone. The rest of the paper is divided into three sections. Section 2 contains the existing work done in this area; section 3 covers feature extraction methods, section 4 covers classification and section 5 covers results and discussions.

2 Literature Survey

Pattern recognition is a field of research since a long period of time. Lazzerini and Marcelloni [1], presents EYE, a fuzzy logic based classifier for recognition of isolated handwritten characters. EYE is based on a new linguistic classification method. The method describes characters in terms of linguistic expressions and adopts a purposely defined operator to compare these expressions. Hanmandlu et al. [2], presents an innovative approach called box method for feature extraction for the recognition of handwritten characters. In this approach, the character image is partitioned into a fixed number of sub images called boxes. The features consist of normalized vector distance and angle from each box to a fixed point. The recognition schemes used are back propagation neural network (BPNN) and fuzzy logic. The recognition rate is found to be around 100% with the fuzzy based approach on the standard database. Zhang et al. [3], proposed a handwritten character recognition feature based on the combination of gradient feature and coefficients of wavelet transform. The gradient feature represents local characteristic of a character image properly, but it is sensitive to the deformation of handwritten character. The wavelet transform represents the character image in multiresolution analysis and keeps adequate global characteristic of a character image in different scales. Gary and Joe [4], proposed a new method to measure the similarity between two fuzzy attributed graphs of a known character class and an unknown character. In the recognition stage, when this similarity measure is applied, an input character can be correctly classified. Liolios et al. [5], presents a system capable of recognizing isolated handwritten characters using shape transform method. The shape transform approach is based on the calculations the cost of transforming the image of a given character into that of another, thus taking into account local geometrical similarities and differences.

3 Feature Extraction Methods

Different types of features can be extracted depending on the representation forms of characters.

3.1 Zoning

The frame containing the character is divided into several overlapping or non-overlapping zones and the densities of object pixels in each zone are calculated. Density is calculated by finding the number of object pixels in each zone and dividing it by total number of pixels. Densities are used to form a representation. For binary images, value of each pixel is either 1 or 0. We have considered pixels having value BLACK (0) as object pixels. This feature is extracted from the scaled (normalized) character matrix of the character. The original character image (matrix) is first scaled to Normalized window of size 48*48.

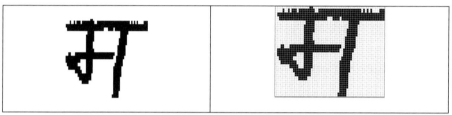

Fig. 1. Original Image of Gurmukhi character sassa (ਸ)

Fig. 2. Scaled Image of Gurmukhi Character sassa (ਸ)

The Zoning feature set consists of 64 values. The values in feature vector are normalized in the range 0 to 1. Normalization is done by dividing all the values by the largest value in the feature set.

3.2 Directional Distance Distribution

DDD, proposed by Oh and Suen [9], is based on the distance information computed for both black pixels and white pixels in 8 directions. It regards input pattern array as circular. By regarding the array as being circular when computing distance, we could get a better discriminating power of the feature. As both distances from black pixel to white pixel and from white pixel to black pixel are computed, the feature contains both the black/white distribution and the directional distance distribution. To each of the pixels in the input binary pattern array, two sets of 8 bytes which W (White) set and B (Black) set are allocated. For a white pixel, the set W is used to encode the distances to the nearest black pixels in 8 directions. The set B is simply filled with value zero. For a black pixel, the set B is used to encode the distances to the nearest white pixels in 8 directions. The set W is filled with value zero. The 8-direction codes are 0(E), 1(NE), 2(N), 3(NW), 4(W), 5(SW), 6(S), 7(SE). The distances of nearest black/white pixels in each direction for pixels (0,0) and (4,37) have been given in Table2 and Table3. After computing WB encoding for each of the pixel, we have divided the input array into four equal zones both horizontally and vertically, hence producing 16 zones. In each of the sixteen grids an average for each of 16 bytes in WB encodings is computed. So, we finally get a 16(16 bytes in WB)*16 (4*4 grids) feature vector.

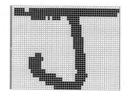

Fig. 3. Scaled image of size(48*48) of character hahha(ਹ)

Table 1. WB Encoding for white pixel (0,0) in Figure 3

W0	W1	W2	W3	W4	W5	W6	W7	B0	B1	B2	B3	B4	B5	B6	B7
34	2	4	2	6	4	34	40	0	0	0	0	0	0	0	0

Table 2. WB Encoding for Black pixel (4,37) in Figure 3

W0	W1	W2	W3	W4	W5	W6	W7	B0	B1	B2	B3	B4	B5	B6	B7
0	0	0	0	0	0	0	0	6	4	5	5	3	3	22	7

The DDD (Directional Distance Distribution) feature set consists of 256 values. The values in feature vector are normalized in the range 0 to 1. Normalization is done by dividing all the values by the largest value in the feature set.

3.3 Gabor

A Gabor filter is a kind of local narrow band pass filter and selective to both orientation and spatial frequency. It is suitable for extracting the joint information in two-dimensional spatial and frequency domains and widely applied in the fields such as character recognition, face and texture recognition, etc. Gabor filters are applicable to both binary images and   gray-scale images, and are immune to image noise. Gabor filters were utilized to extract the basic structures of the character, which combined the advantages of image analysis based on spatial domain and spatial-frequency domain. The Gabor feature set consists of 252 values. The values in feature vector are normalized in the range 0 to 1. Normalization is done by dividing all the values by the largest value in the feature set.

4 Classification Methods

Classification stage uses the features extracted in the feature extraction stage to identify the text segment. It is concerned with making decisions concerning the class membership of pattern in question. The task here is to design model using training data which can classify the unknown pattern based on that model. For training purposes we have used isolated Gurmukhi characters written in different forms. Feature vector for all training data is produced and stored in files. SVM (Support Vector Machine) is a useful technique for data classification [6,7,8]. The Support Vector Machine (SVM) is learning machine with very good generalization ability, which has been applied widely in pattern recognition, regression estimation, isolated handwritten character recognition, object recognition speaker identification, face detection in images and text categorization.

5   Results and Discussions

An annotated sample image database of isolated handwritten characters in Gurmukhi script has been prepared. The database contains name of source image, size of image and character value of the image. We have experimented the system on 2050 images of Gurmukhi characters contained in the database. The system is analyzed using different combinations of feature extraction methods and classification methods. We have used 3075 images to train the system and 2050 for testing. The recognition accuracy obtained by using different combinations of feature extraction methods and classifiers is given in the Table 3. The recognition accuracy is obtained by dividing the correctly recognized characters to total number of character images which are actually present in the database.

Table 3. Performance of different combinations of feature extraction method and classification techniques

Feature extraction method	Classifier	Total Images	Correctly recognized	Recognition Accuracy
Zoning	SVM(Linear Kernel)	2050	1490	72.68%
Zoning	SVM(Polynomial Kernel)	2050	1497	73.02%
Zoning	SVM(RBF Kernel)	2050	1493	72.83%
DDD	SVM(Linear Kernel)	2050	1490	73.21%
DDD	SVM(Polynomial Kernel)	2050	1497	73.65%
DDD	SVM(RBF Kernel)	2050	1493	73.36%
Gabor	SVM(Linear Kernel)	2050	1490	73.90%
Gabor	SVM(Polynomial Kernel)	2050	1497	74.29%
Gabor	SVM(RBF Kernel)	2050	1493	74.00%

Gabor with SVM (Polynomial kernel) gives the best results of all the combinations of feature extraction methods and classification methods as is evident from the Table 3.

5.1   Reasons of Failure

Sometimes the characters are also wrongly classified. It happens due to many reasons like

- The variability of writing styles, both between different writers and between separate examples from the same writer overtime. For example:

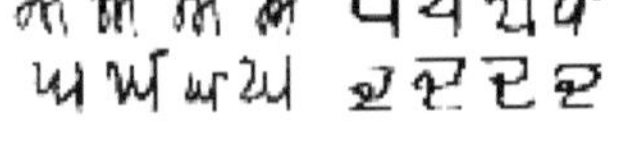

Fig. 4. Variability in writing style

- Some characters have similar topological structures.

Table 4. Similar Characters

ਮ and ਮ	ੲ and ੲ	ਸ and ਮ	ੲ and ੮	ੲ and ੲ	ਅ and ਘ	ਫ and ਫ
ਗ and ਘ	੩ and ੜ	ਜ and ਜ	ਪ and ਧ	ਲ and ਲ਼	ਹ and ਰ	੮ and ੲ
ਨ and ਠ	੩ and ੜ	ਬ and ਬ	ਪ and ਖ	ਚ and ੲ	ਗ and ਗ੍ਰ	ਬ and ਖ

- The possible low quality of the text image. For example:

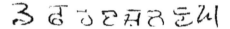

Fig. 5. Low quality image

- The unavoidable presence of background noise and various kinds of distortions (such as poorly written, degraded, or overlapping characters) can make the recognition process even more difficult. For example

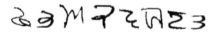

Fig. 6. Distorted images

System sometimes confuses the character with some other character and does not recognize it correctly. In the Table5 the confusion matrix caused when Gabor and SVM (Polynomial Kernel) have been used is given. We have taken 2050 images, 50 images of each character. The characters ੲ , ਗ, ਘ , ੲ, ਨ , ਲ are recognized with higher accuracy and characters ਅ , ੲ , ਧ , ੲ are recognized with least accuracy.

References

[1] Lazzerini, B., Marcelloni, F.: Fuzzy classification of handwritten characters. In: 18th Int. Conf. of the North American, pp. 566–570 (1999)
[2] Hanmandlu, M., Murali Mohan, K.R., Chakraborty, S.: Fuzzy logic based handwritten character recognition. In: Int. Conf. on Image Processing, vol. 3, pp. 42–45 (2001)
[3] Zhang, W., Tang, Y.Y., Xue, Y.: Handwritten Character Recognition Using Combined Gradient and Wavelet Feature. In: Int. Conf. on Computational Intelligence and Security, vol. 1, pp. 662–667 (2006)
[4] Gary, M.T.M., Poon, J.C.H.: A fuzzy-attributed graph approach to handwritten character recognition. In: 2nd IEEE Int. Conf. on Fuzzy Systems, vol. 1, pp. 570–575 (1993)
[5] Liolios, N., Kavallieratou, E., Fakotakis, N., Kokkinakis, G.: A new shape transformation approach to handwritten character recognition. In: 16th Int. Conf. on Pattern Recognition, vol. 1, pp. 584–587 (2002)

[6] Ming, T., Yi, Z., Songcan, C.: Improving support vector machine classifier by combining it with k nearest neighbor principle based on the best distance measurement. In: Proc. of IEEE in Intelligent Transportation System, vol. 1, pp. 373–378 (2003)

[7] Cao, L.J., Chong, W.K.: Feature extraction in support vector machine: a comparison of PCA, XPCA and ICA. In: 9th Int. Conf. on Neural Information Processing, vol. 2, pp. 1001–1005 (2002)

[8] Burges, C.J.C.: A Tutorial on Support Vector Machines for Pattern Recognition, pp. 1–43 (1998)

[9] Oh, I.S., Suen, C.Y.: A Feature for character recognition based on Directional Distance Distributions. In: 4th Int. Conf. on Document Analysis and Recognition, vol. 1, pp. 288–292 (1997)

Dewarping Machine Printed Documents of Gurmukhi Script

Dharam Veer Sharma[1] and Shilpi Wadhwa[2]

[1] Dept. of Computer Science, Punjabi University, Patiala
dveer72@hotmail.com
[2] MTech (CS), Dept. of Computer Science, Punjabi University, Patiala

Abstract. During the scanning of bound documents, some part of the document image is curled near the corners or near the binding resulting in bending of text lines. This hard to tackle distortion makes recognition very difficult. A method has been proposed for estimation and removal of line bending deformations introduced in document images during the process of scanning. The estimation of bend involves determining the side of the document on which curl is present and direction of the bend. The method has been tested on varieties of printed document images of Gurmukhi containing the bent text-lines at page borders. The method consists of three stages. In the first stage, a decision methodology is proposed to locate the site of deformation and the direction of deformation. An elliptical approximation model is derived to estimate the amount of deformation in the second stage. Finally, a transformation process brings out the correction. Experiments show that the method developed works well under conditions where pixel distribution is uniform.

Keywords: Dewarping, Machine printed, Gurmukhi script, OCR.

1 Introduction

One of the major issues while preparing a document image for processing is to produce a quality document image for further image analysis. Noise is a prevalent artifact introduced in document images by image acquisition device or due to poor quality of document media. Skew is the orientation introduced while placing the document into scanning device. Unless these two problems are handled properly in the document images, it is very difficult to proceed with the other sequence of activities in DIA (digital image acquisition). Generally, 'noise' components in a document image are referred as salt-and-pepper noise or impulse and speckle noise or just dirt. One more typical type of noise introduced while scanning a document image is due to:

(1) Copying a page of a thick bound book because of non-planar surface created by the book on the flat copying surface

(2) non-linearity in copying the contents at the start and finish ends of scanning. These result in 'bending of text-lines' at the page borders. Such bent text–lines are elliptical in shape. Samples of such bending of text-lines document images are shown in fig 1.

C. Singh et al. (Eds.): ICISIL 2011, CCIS 139, pp. 117–123, 2011.
© Springer-Verlag Berlin Heidelberg 2011

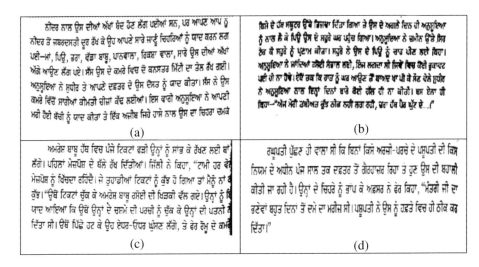

Fig. 1. Sample of line bending in document images (a) left upward (b) left downward (c) right upward (d) right downward

Bending of text-lines in document images may occur on left or right borders of the document image and deformation may be in either upward or downward direction. Hence it is required to know the side and direction of this deformation in document image viz., left upward or left downward or right upward or right downward to take further correction work. A decision process is adopted to locate the border and direction of deformation and our method also assumes that the document images are free from noise and skew.

Rest of the paper has been organized as follows: previous work has been discusses in section 2, section 3 presents the proposed solution, experiments and results are given in section 4, the work has been concluded in section 5.

2   Previous Work

T. Vasudev et. al[1] focused on the problem of bending of text-lines observed in document images. The authors developed a method for estimation and removal of line bending deformations introduced in document images during the process of scanning. This method consists of three stages. In the first stage, a decision methodology is proposed to locate the site of deformation and the direction of deformation. An elliptical approximation model is derived to estimate the amount of deformation in the second stage. Finally, a transformation process brings out the correction. The process of estimation of deformation area is made using ellipse and line drawing algorithms. A digital differential analyzer (DDA) line drawing algorithm draws a line between any two specified points. A midpoint ellipse drawing algorithm draws an elliptical arc for given center, x-radius and y-radius. An imaginary elliptical arc and an imaginary line based on the side and direction of deformation from the position of deformation to edge of document helps in further processing. A series of elliptical arcs and lines are drawn until a suitable arc and a line are encountered such that the arc and line enclose the deformed region.

The approach proposed by B. Gatos et. al[2] is based on the construction of outer skeletons of text images. The main idea of this algorithm is based on the fact that it is easy to mark up long continuous branches that define inter-linear spaces of the document in outer skeletons. Such branches can be approximated by cubic Bezier curves to find a specific deformation model of each inter-linear space of the document. On the basis of a set of such inter-linear space approximations, the whole approximation of the document is built in the form of a two-dimensional cubic Bezier patch. Then, the image can be dewarped using the obtained approximation of the image deformation. The main idea behind the algorithm is that in an outer skeleton of a text document image, one can easily find branches that lie between adjacent text-lines. Then, one can use this separation branches to approximate deformation of inter-linear spaces on the image.

In their technique, Xu-Cheng Yin et. al[3], enhance the quality of documents captured by a digital camera relying upon

1. Automatically detecting and cutting out noisy black borders as well as noisy text regions appearing from neighboring pages
2. text-lines and words detection using a novel segmentation technique appropriate for warped documents
3. a first draft binary image de-warping based on word rotation and translation according to upper and lower word baselines
4. a recovery of the original warped image guided by the draft binary image de-warping result.

In their approach, black border as well as neighboring page detection and removal is done followed by an efficient document image de-warping based on text-line and word segmentation. The methodology for black border removal is mainly based on horizontal and vertical profiles. First, the image is smoothed, and then the starting and ending offsets of borders and text regions are calculated. Black borders are removed by also using the connected components of the image. Noisy text regions appearing from neighboring page is detected with the help of the signal cross correlation function. At a next step, all words are detected using proper image smoothing. Then, horizontally neighboring words are consecutively linked in order to define text-lines. This is accomplished by consecutively extracting right and left neighboring words to the first word detected after top-down scanning. For every detected word, the lower and upper baselines are calculated, which delimit the main body of the word, based on a linear regression which is applied on the set of points that are the upper or lower black pixels for each word image column. The slope of each word is derived from the corresponding baselines slopes. All detected words are then rotated and shifted in order to obtain a first draft estimation of the binary de-warped image. Finally, a complete restoration of the original warped image is done guided by the draft binary de-warping result of the previous stage. Since the transformation factors for every pixel in the draft binary de-warped image have been already stored, the reverse procedure is applied on the original image pixels in order to retrieve the final de-warped image. For all pixels for which transformation factors have not been allocated, the transformation factors of the nearest pixel are used.

B. Fu et al.[4] in their method have used a coordinate transform model and document rectification process for book dewarping. This model assumes that the book surface is a cylinder. It can handle both perspective distortion and book surface warping problems. The goal is to generate a transformation to flatten the document image to its original shape. The transformation is a mapping from the curved coordinate system to a Cartesian coordinate system. Once a curved coordinate net is set up on the distorted image as shown in Figure 8, the transformation can be done in two steps: First, the curved net is stretched to a straight one, and then adjusted to a well-proportioned square net. According to the transform model, two line segments and two curves are needed to dewarp a cylinder image. Therefore, the left and right boundaries and top and bottom curves in book images are found for the rectification as shown in Figure 4. The rectification process involves three steps: 1) the text-line detection, 2) left and right boundary estimation and top and bottom curves extraction, and 3) document rectification. As an additional post-processing step, the participants used their programs to remove graphics and images from the processed pages. The results thus produced are referred to as CTM2.

3 Proposed Solution

The major issue while preparing a document image for processing is to produce quality document image for image analysis. In this design technique an image de-warping algorithm has been presented that removes the distortion and can enhance picture quality, help to improve all subsequent processing steps. The work presented in our approach focuses on the problem of bending of text-lines observed in document images. A method is developed for estimation and removal of line bending deformations introduced in document images during the process of scanning. Our method consists of three stages. In the first stage, a decision methodology is proposed to locate the site of deformation and the direction of deformation. A line drawing approximation model is derived to estimate the amount of deformation in the second stage. Finally, a transformation process brings out the correction. The method has been tested on varieties of printed document images containing the bent text-lines at page borders.

In our approach, the technique that we used, performs a line-by-line de-warping of the observed paper surface. First, we divide the book into two equal parts because in the case of curled pages and books, it is more complicated to correct these distortions. To reconstruct an image without distortions it is required to separate the book into two parts. This means finding the middle of a book and separating the book image into two pages before applying any dewarping process. This method is needed to make it convenient to de-warp them when we separate the images into pages. To achieve more accurate results we will divide the book into two parts.

Now, we have to detect the side and direction of text lines bending. Here we will observe that the lines in document are clearly separated, bending of lines occur only at the borders of the document, bending is either upward or downward in a line. It is evident from number of samples that bending of text-lines is towards borders and the mid region of any document image is always free from bending deformation.

Deformation normally occurs only in $1/3^{rd}$ part, near left or right border of the document image, and middle of the document image is always free from deformation. First the process of warp detection is applied on 1st and 3rd part of the document image as 2nd part (middle) is always free of warp as per observation.

A digital differential analyzer (DDA) line drawing algorithm draws a line between any two specified points. An imaginary elliptical arc and an imaginary line based on the side and direction of deformation from the position of deformation to edge of document helps in further processing. A series of elliptical arcs and lines are drawn until a suitable arc and a line are encountered such that the arc and line enclose the deformed region.

After detecting the area of bend the next step is to perform transformation for straightening the bends. A spatial domain transformation is performed on image to remove the bending of text-line in the deformed region. The correction of estimated bending deformation adopts a point processing technique of shift in y-direction. In this transformation the deformed part of document is shifted in y-direction based on the estimation of bending of text-lines.

Transformation process can be represented as,

$$g'(x,y) = T[g(x,y)]$$

Where g is the original deformed image in spatial coordinate's g' is the deformation corrected image in the same coordinate system and T is the point processing transformation function for deformation correction.

The proposed algorithm consists of three main steps given below:

1. Determine the side of deformation (Left or right)
2. Determine the direction of deformation (left upward, left downward, right upward or right downward)
3. Perform Correction
      for each line in the document image
          A. fix up an imaginary arc or line.
          B. for each point p(x, y) on arc or line
               i) d=distance between p(x, y) and q(x, y') on top text line
               ii) Shift all points at x up by a value d

4 Experimental Results

The proposed method is tested using more than 50 samples of English and Gurmukhi document images containing text-lines bending deformations. Tests were conducted on all four types of text-line bending deformed document images. The test samples were considered with different font size and style, different line spacing and different scanning resolution. Fig 2-5 shows the different input images and corrected output images as result.

The proposed algorithm has some limitations:

- It does not work if the document image is skewed or distorted.
- If there are large spaces between words, the line tracker sometimes tries to make two lines out of one; this results in visually strong distortions.

Fig. 2(a). image with left-upward deformation

Fig. 2(b). image with left-upward deformation (corrected i.e. leftup1)

Fig. 3(a). image with left-downward deformation

Fig. 3(b). image with left-downward deformation (corrected i.e. leftdown1)

Fig. 4(a). image with right-upward deformation

Fig. 4(b). image with right-upward deformation (corrected)

Fig. 5(a). image with right-downward deformation

Fig. 5(b). image with right-downward deformation (corrected)

6   Conclusion

The developed method for estimation and removal of line bending deformation does not require any special arrangements to acquire the document image, the image obtained by flat -bed scanner is sufficient. The experimental results are fairly close to the results of the method proposed by Breuel and Zhang which requires set up for acquiring stereo vision image and Complex mathematical models for construction of 3D image vision and interpolation techniques. The developed model is very sensitive to noise and skew. Since this is an initial effort on the problem further refinement/enhancement on the work could be attempted like investigation of different approaches to fix a suitable elliptical arc that can show consistent performance of bending deformation correction, investigation of generic models to correct multiple types of bending deformations in a single document. A proposal is under investigation to extend the work to decompress the compressed characters in the bend region while carrying out transformation process.

References

[1] Vasudev, T., Hemanthakumar, G., Nagabhushan, P.: An Elliptical Approximation Model for Removal of Text-line Bending Deformations at Page Borders in a Document Image. In: Proceedings of the International Conference on Cognition and Recognition, pp. 645–654 (September 2005)

[2] Gatos, B., Ntirogiannis, K.: Restoration of arbitrarily warped document images based on text line and word detection. In: Proceedings of Fourth IASTED Int. Conf. on Signal Processing, Pattern Recognition, and Applications, pp. 203–208 (February 2007)

[3] Yin, X.-C., Sun, J., Naoi, S.: Perspective rectification for mobile phone camera-based documents using a hybrid approach to vanishing point detection. In: Proceedings of the Second International Workshop on Camera Based Document Analysis and Rectification, pp. 37–44 (2007)

[4] Fu, B., Wu, M., Li, R., Li, W., Xu, Z., Yang, C.: A Model based Book Dewarping Method Using Text Line Detection. In: Proceedings of 2nd International Workshop on Camera Based Document Analysis and Recognition, pp. 63–70 (2006)

Developing Oriya Morphological Analyzer Using Lt-Toolbox

Itisree Jena, Sriram Chaudhury, Himani Chaudhry, and Dipti M. Sharma

International Institute of Information Technology, Hyderabad
itisree@research.iiit.ac.in, sriram_c@research.iiit.ac.in,
himani@research.iiit.ac.in, dipti@iiit.ac.in

Abstract. In this paper we present the work done on developing a Morphological Analyzer (MA) for Oriya language, following the paradigm approach. A paradigm defines all the word forms of a given stem, and also provides a feature structure associated with every word. It consists of various paradigms under which nouns, adjectives, indeclinables (avyaya) and finite verbs of Oriya are classified. Further, we discuss the construction of paradigms and the thought process that goes into their construction. The paradigms have been created using an XML based morphological dictionary from the Lt-toolbox package.

Keywords: Oriya, Morph, Agglutinative, Apertium, Lt-toolbox, Morphological analyzer.

1 Introduction

Oriya is an Indo-Aryan language spoken by about 31 million people mainly in the Indian state of Orissa. "Oriya is a syntactically head final and morphologically agglutinative language." [9] "The subject agrees with the verb in person, number and honorificity." [5] It has natural gender and it doesn't affect other grammatical categories like pronoun and verb. Oriya has four tenses, present, past, future and conditional. The last one is a mood but it behaves like a tense. Nouns in Oriya are generally characterized by the presence of inflectional categories like number, gender and vibhakti. They can take classifiers like -ti, -tA. These classifiers can come with singular nouns, like bALakati[1] where -ti is the classifier. All the nouns in Oriya have the third person features. Oriya doesn't have a grammatical gender. However, there exist a number of nouns indicating natural gender distinction at lexical level and there is no reflection of this distinction in the agreement with the verb. Only a few attributive adjectives referring to physical properties show gender agreement with their head noun.

Since Oriya is an agglutinative language a word by itself is a bundle of linguistics information. Ex: let us take the noun 'bALakaku'. Here, 'bALaka' is noun and 'ku' is the vibhakti. In Oriya, the Vibhakti attaches itself with the noun, unlike Hindi, where it is two different words. Also, unlike Hindi, Oriya nouns do not change their root. Eg:

---

[1] All examples have been given in the WX notation.
http://sanskrit.inria.fr/DATA/wx.html

C. Singh et al. (Eds.): ICISIL 2011, CCIS 139, pp. 124–129, 2011.
© Springer-Verlag Berlin Heidelberg 2011

Table 1. Agglutinative feature of Oriya

Oriya	Hindi
bALaka**ku** PaLa xia	ladake **ko** Pal xo
bALaka**ti** PaLa KAuCi	LadakA Pal KA rahA hE

In Hindi the form of "ladakA" (boy) changes as seen in table 1 above, whereas in Oriya, in both of the sentences in Table 1 the form of bALaka doesn't change. Only the vibhakti attached with the noun changes.

In Oriya some suffixes are added only to a selected number of noun stems. Eg. '-mAne' is the suffix, that adds only to animate nouns.

Table 2. Suffixes attached with number in Oriya

Oriya		English	
singular	**plural**	**singular**	**plural**
maNiRa	maNiRa+mAne	man	men
pua	Pua + mAne	son	sons

But this suffix can not be added to either human proper nouns or inanimate nouns. Eg. we can not say paWara+ mAne (stones). The inflectional plural suffix '-mAne' can only be added to animate nouns, as seen in table 2 above.

Further, in Oriya finite verb always agrees with the subject noun and is reflected by an agreement marker. The agreement suffix follows the tense suffix as a compulsory constituent of the finite verb construction.

Non-finite verb forms don't take agreement markers. "A non-finite verb is created when the non-finite verb inflections add to the verb stem." [4]

Eg: muz BAwa KAi skulaku gali.

I rice eating school go (past)

Here, the non-finite suffix is '-i'. It shows the completion of an action and forms a perfect particle. Eg:

mowe miTA KAibAku Bala lAge.

I (Dat) sweet eat particle good feel

"I like to eat sweet"

Some non-finite constructions are formed by adjoining postpositions/avyaya. They don't take any grammatical information like nouns and verbs do. Postpositions come under this category.

eg: apekRA (comparatively), anusAre (according to), xbArA (by), CadA (without), TAru (from).

2 Related Works

Bharati et al. (2004) proposed a paradigm based algorithm for morphological analysis of Hindi, an inflecting language. It works with paradigms based on add-delete strings. All the word forms of a stem along with their corresponding feature structures are given. Bharti et. al. (1995, Chapter 3) discusses that this model works well for moderately rich inflectional forms [10]. Since Marathi has a well defined paradigm-based system of inflection, Bapat et. al. (2010) [2] developed their morphological analyzer based on the paradigm approach. However, since Marathi is highly inflectional and also shows derivation to a high degree, they combined the paradigm-based inflectional system with finite state machines (FSMs) to handle its morphotactics. As also seen in our work, Vaidya et. al. (2009)[1] too make use of the Lt-toolbox package to incorporate handling of derivational morphology of Marathi. Their work builds on the paradigm based MA for Marathi by Bharti et. al. (1998)[11].

3 Design

"The morphological analyzer was developed as a part of a new language pair for Apertium, hence the analyzer data conforms to Apertium's dix format." [8] The data for the analyzer is contained in an XML file, that is, the language paradigm dictionary (Oriya here). "This dictionary has two principal parts, the first one is pardef section, which contains the inflection rules for a particular type of word, the other is entry section, which contains the list of words stating which pardef they belong to." [8] "The dictionary shows correspondences between surface forms and lexical forms. [1]". Surface forms are the inflected forms of the words found in texts, whereas lexical forms are the base forms of those words. For example, 'KAe' (eat) is a surface form and its lexical form is "KA" (eat) with features like verb, singular, first person.

"The benefit of using Apertium is its robust architecture." [8] Given an XML dictionary it can create a compiled dictionary which is generally faster than a normal database based dictionary.

4 Paradigm Approach

A paradigm defines all the word forms of a given stem, and also provides a feature structure associated with each word form [3]. At first we focused on the open category words like nouns, verbs, and adjectives and later on, closed category words like postpositions and conjunctions. We classified words from a given corpus into different grammatical categories. Then we took all possible forms of each word, derived their root by detaching the suffix/es and then assigned each form a feature structure with grammatical information such as gender, number, person for nouns and tense, aspect, modality (TAM) for verbs. The root dictionary and paradigms for roots and their word forms were made using Lt-toolbox. To avoid redundancy, paradigms that have identical grammatical information are grouped. However, all words with similar endings/suffixes may not follow the same paradigm. Eg. the two verbs 'KA' (eat) and 'gA' (sing) follow the same paradigm. But the verb 'yA' falls in a different paradigm though it has the same ending.

Table 3. Current Dictionary Size

Categories	Entry in dictionary
Nouns	2449
Verbs	200
Adjectives	373
Postpositions	44
Conjunctions	21
Total	**3087**

4.1 Current Dictionary

The present work has been carried out on a limited number of words ranging from closed categories to open categories. And the task of updating the figures mentioned in table 3 above, is on going.

4.2 Source for Database

The Oriya dictionary named Shabdatatwabodha Abhidhana [6] forms the main source for creating the database, along with other resources like the online Hindi dictionary Shabdanjali, created at LTRC, IIIT-H. The existing Hindi morph analyser based on the same approach, also served as a guide regarding the creation and analysis of the data. At present we have 14 paradigms for verbs, 15 for nouns, and 11 for adjectives. In "Oriya animate and inanimate nouns take different suffixes" [7] and paradigms for nouns are made on the basis of the suffixes.

5 Experiment and Results

5.1 Coverage of Verbs

We currently have the CIIL corpus of 85000 words. Of which we randomly extracted 10,000 words to experiment the coverage of verbs--since our preliminary focus was on verbs. We extracted words at the end of each sentence, since Oriya is a verb final language and a most constructions in Oriya end in verbs. As seen in Table 4, out of the 10,000 extracted words there are 595 unique verbs.

Table 4. Results

Total words	10000
Unique verbs	595
Recognized verbs	448
Unrecognized verbs	147

5.2 Total Coverage

We took another set of data consisting of 32,000 words from the CIIL Oriya corpus to run our analyzer to find out its total coverage of all the parts of speech which our morph analyzer handles. The results can be seen in table 5 below.

Table 5. Results

Total words	32000
Total unique words	9836
Total recognized words	3807
Total unrecognized words	6029

6 Error Analysis

The words that remain unrecognized by the morph can be easily accounted for. The unanalyzed words mainly include the words that are not present in the dictionary used. This is a shortcoming of the dictionary based method for morph analysis that the words not present in the dictionary are not recognized and analyzed. The main problem encountered with the recognition of the verbs is that they do not fall into any of the existing verb paradigms. This can either be ascribed to the less robustness of the paradigms, or to the need for separate paradigms for these verbs. Further, since causative verbs and verbal complexes are currently not being handled, these remain unanalyzed too. Also, it needs a mention here, that more data/paradigms need to be added to the database for better performance.

7 Conclusion and Future Work

In the present work we have talked about developing a Morph analyzer using Lt-toolbox. At present this morph analyzer handles only inflectional morphology, and we are working on nouns, verbs, indeclinables and adjectives. In the future this work can be extended to the remaining grammatical categories. Also, as a next step it can be extended to handle compound words and derivational morphology. Further, the dictionary of paradigms being currently used gives information about suffixes. Since this is an ongoing work, constant additions to the database are underway. The same work can be done to handle prefixes. It needs a mention here, that currently the morph analyzer is in its preliminary stage and there is scope for its improvement such as addition of remaining categories, which will lead to an expansion of its current coverage, and improve its performance.

A "Morph analyzer forms the foundation for applications like Part of Speech (POS) tagging, Chunking and Machine Translation." [2] Developing an Oriya morph analyzer will help build these applications and also, in machine translation of Oriya to other languages and vice versa. It is thus a useful resource for the language.

References

1. Vaidya, A., Sharma, D.M.: Using Paradigms For Certain Morphological Phenomena In Marathi. In: ICON 2009, pp. 132–139 (2009)
2. Bapat, M., Gune, H., Bhattacharyya, P.: A Paradigm-Based Finite State Morphological Analyzer For Marathi. In: 23rd International Conference on Computational Linguistics, Coling 2010, pp. 26–34 (2010)
3. Jayan, J.P., Rajeev, R.R., Rajendran, S.: Morphological Analyser for Malayalam - A Comparison of Different Approaches. IJCSIT 2(2), 155–160 (2009)
4. Dash, G.N.: Descriptive Morphology Of Oriya. Visva-Bharati. Research Publications Committee, Santiniketan (1982)
5. Singh, B.C.: A Morphological Generator for Oriya. M. Phil. dissertation. University of Hyderabad, Hyderabad (2004)
6. Sharma, G.N.: Shabdatatwabodha Abhidhana. Bi. Si. Dāśa (1962)
7. Mahapatra, B.P.: Prachalita OdiA Bhashara eka Byakarana. Vidyapuri, Cuttack (2007)
8. Faridee, A.Z.M., Tyers, F.M.: Development of a morphological analyser for Bengali. Universidad de Alicante. Departamento de Lenguajes y Sistemas Informaticos, pp. 43–50 (2009)
9. Mohapatra, R., Hembram, L.: Morph-Synthesizer for Oriya Language-A Computational Approach. Language In India 10, 205–211 (2010)
10. Akshar, B., Chaitanya, V., Sangal, R.: Natural Language Processing: A Paninian Perspective. Prentice Hall of India, New Delhi (1995)
11. Bharti, A., Kulkarni, A., Chaitanya, V.: Challenges in Developing a Word Analyzer for Indian Languages. In: The workshop on Morphology, CIEFL, Hyderabad (1998)

Durational Characteristics of Indian Phonemes for Language Discrimination

B. Lakshmi Kanth, Venkatesh Keri, and Kishore S. Prahallad

International Institute of Information Technology, Hyderabad, India – 500032
{lakshmikanth.b,venkateshk}@research.iiit.ac.in,
kishore@iiit.ac.in

Abstract. Speech is the most important and common means of communication. Human beings identify a language by looking at the acoustics and the letter to sound rules (LTS) that govern the language. But pronunciation is governed by the person's exposure to his/her native language. This is a major issue while considering words, especially nouns in Indian languages. In this paper, a new methodology of analyzing phoneme durations for language discrimination is presented. The work has been carried out on a database built with words, mostly nouns, common to Hindi, Tamil and Telugu languages. Durational analysis of phonemes has been carried out on the collected database. Our results show that phoneme durations play a significant role in differentiating Hindi, Telugu and Tamil languages with regard to stop sounds, vowels and nasals.

Keywords: Phoneme, Duration and Statistical Significance.

1 Introduction

Analysis of language discrimination is important in a country like India with many languages. The differences between languages can be used in language identification, speech recognition, speaker recognition and also for building text-to-speech systems. In the past, the analysis on languages has been carried mostly using prosodic and spectral information. Much of the research on languages so far has been on spectral information, mainly using the phonemic features and their alignment. Such systems may perform well in similar acoustic conditions [4]. Indian languages such as Hindi, Tamil and Telugu share a similar phoneme set in the production of speech sounds. In the case of words that are common to Indian languages, the discrimination rate decreases as the phonemic alignment and their sound almost remains constant for a particular phoneme. Therefore, looking at prosodic features like duration, pitch, intensity is a significant key for language discrimination.

In this paper we are looking into durational characteristics of phonemes for language discrimination. Klatt [1] studied the segmental duration of English language and showed that phoneme durations have potential cue in carrying language information. Samudravijaya [2] stated that durational cues are useful in discriminating Hindi language phonemes. Y.K.Muthusamy [3] stated that phonemic transcriptions

C. Singh et al. (Eds.): ICISIL 2011, CCIS 139, pp. 130–135, 2011.
© Springer-Verlag Berlin Heidelberg 2011

are advantageous than arriving at accurate phonetic transcriptions. The study of durational characteristics of languages with similar phoneme set is therefore an interesting task.

In section 2, building the phonemically balanced database is discussed. It includes collection of words, speech recording and manual labeling of speech. In section 3, analysis of language discrimination is discussed with the final conclusion in section 4.

2 Database and Labeling

In performing the phonemic durational analysis, we have jointly prepared a phonemically balanced word database. The words are nouns e.g., person names, place name, etc. of Indian origin and are common to the three Indian languages Hindi, Tamil and Telugu. A total of 440 words have been collected and ensured that each phoneme occurs at least 5 times in the beginning, middle and end of a word in the database.

The speech database was collected with 25 speakers with an average age group of 18 to 25 years having 13 males and 12 females in each language. All the speakers are bilingual in the sense that they have educational background in native language at least upto senior schooling. Speakers were provided with words in their native script and were instructed to speak each word in their respective language with normal rate and intonation into the microphone. The problem with the recording phase is that each speaker of a particular nativity has his own way of pronouncing the words. This will effect the phonemic coverage in each language. In order to avoid the above problem, we made sure that each word has been recorded with same pronunciation in three languages. All these utterances were recorded using Edirol R09 speech recorder at 48 kHz frequency. Also, the high quality speech is digitized at 16 kHz with 16bits per sample in a quite environment to avoid echo effect.

Table 1. The Indian Language Transliteration (ITRANS) symbols of phonemes common to three languages Hindi, Tamil and Telugu

Semi Vowels					Long Vowels					Dipthongs	
a	i	u	e	o	aa	ii	uu	ei	oo	ai	au

Unvoiced Unaspirated	Unvoiced Aspirated	Voiced Unaspirated	Voiced Aspirated	Nasals
k	kh	g	gh	-
ch	chh	j	jh	nj~
t:	t:h	d:	d:h	nd~
t	th	d	dh	n
p	-	b	bh	m

Semi Vowels				Fricatives			
y	r	l	v	sh	shh	s	h

Manual labeling of the recorded data has been carried out for phoneme durational analysis. Figure 1 illustrates the waveform and spectrogram with phoneme labels of word "auku". A total of 43 phonemes were considered for the study as tabulated in Table 1. In the manually labeled speech database, special symbols were used to signify the geminate consonants, and closure and release durations of stop consonants separately which were tabulated in Table 2. For e.g., the word "auku" in Fig 1 is labeled into phonemes au, kcl, k, and u. The singleton stops have same closure and release stop phonemes, the cluster stops have dissimilar closure and release stop phonemes and gemination refers to repetition of same phoneme.

Table 2. Rules followed in manual labeling

	Consonant	Rule
Singleton stop	k	kcl k
Cluster stop	tk	tcl k
Gemination Stop	kk	k1cl k1
Gemination	nn	n1

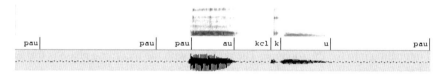

Fig. 1. Waveform and spectrogram of word "auku" spoken by a Hindi speaker

3 Analysis

The speech database has words that were intentionally pronounced to have similar sound words. The mean values and standard deviation values of phoneme durations were computed and used to study the durational characteristics in differentiating languages. As shown in Table 3, the broad-phonetic classifications are categorized into short vowels, long vowels, semivowels, diphthongs, nasals and fricatives, and stop sounds. The stop sounds are separately labeled into closure and release phonemes. The following section provides an analysis of the duration of phonemes in the collected word database.

Student's t-test is used in analyzing the significance of durational differences between phonemes. This test is used to calculate the statistical difference as a function of the difference between means relative to the variability [5]. Each of the other columns in Table 3 contains the results of comparison of phoneme durations of any two languages. This was conducted at 99.95% confidence level.

In stop sounds, release durations have more practical significance than closure durations. Stop sounds shows that release durations have statistical significance in the case of unvoiced unaspirated, unvoiced aspirated and voiced aspirated phonemes. Their durational statistics are tabulated in Table 4. In Hindi, the durations of unvoiced aspirated are twice the durations of unvoiced unaspirated. In Tamil language, durations of unvoiced aspirated are equal to durations of unvoiced unaspirated, this may be because of having single alphabet for the stop sound in the place of articulation. In Telugu, durations of voiced unaspiration are equal to that of voiced aspiration and durations of unvoiced aspiration are twice the durations of unvoiced unaspiration.

The vowels show significant difference between Telugu and Hindi in short vowels and between Tamil and Telugu in long vowels. The vowels helps in differentiating Telugu language, because Hindi and Tamil speakers pronounce both short and long vowels almost at equal duration rates. We found that Hindi has more duration and Tamil has less duration than Telugu in short and long vowels respectively and are tabulated in Table 5.

In the case of nasals, Hindi language has a significant durational characteristic. Hindi phonemes m, n, and n1, have more duration and nd~, nj~ have lesser duration compared to other two languages. This is more of a practical significance in differentiating Hindi language and are tabulated in Table 6.

From the above analysis, Telugu can be discriminated using vowels and Hindi can be discriminated using nasals. Along with the help of singleton stop phoneme durations, the three languages can be discriminated.

Table 3. Table shows the phoneme durations that are statistically significant at 99.95% confidence level between languages Hindi, Tamil and Telugu. "S" represents the significance between two languages.

Broad Phonetic Classification		Hindi-Tamil	Tamil-Telugu	Telugu-Hindi
Short Vowels				S
Long Vowels			S	
Semi Vowels				
Unvoiced Unaspirated	Closure			
	Release	S	S	S
Unvoiced Aspirated	Closure			
	Release	S	S	S
Voiced Unaspirated	Closure			
	Release			
Voiced Aspirated	Closure			
	Release	S	S	S
Diphthongs				S
Nasals		S		S
Fricatives			S	

Table 4. Mean and standard deviations, std (in msec) of release durations of singleton stop sounds in three languages Hindi, Tamil and Telugu languages

		Hindi		Tamil		Telugu	
	Singleton Stops	Mean	Std	Mean	Std	Mean	Std
Unvoiced Unaspirated	k	34.60	11.25	37.72	11.99	39.36	11.91
	t:	20.27	6.02	21.58	5.73	20.52	5.96
	t	25.89	7.90	28.27	8.67	27.98	8.62
	p	23.00	8.32	26.60	9.74	26.69	10.46
Unvoiced Aspirated	kh	76.67	22.22	37.80	10.28	54.94	22.70
	t:h	69.40	24.73	25.26	8.04	43.57	29.83
	th	60.77	20.78	30.36	7.72	40.89	19.45
Voiced Unaspirated	g	60.73	26.21	60.84	26.60	54.93	20.88
	d:	46.46	31.80	43.27	28.88	37.60	25.24
	d	63.73	32.60	58.87	32.70	55.23	23.77
	b	63.18	27.59	65.44	26.49	59.75	21.25
Voiced Aspirated	gh	62.97	30.92	63.94	35.00	53.97	20.01
	d:h	59.35	33.59	64.16	34.53	56.23	28.18
	dh	60.89	28.60	57.05	27.25	52.11	21.96
	bh	60.69	22.70	69.77	21.95	61.85	37.35

Table 5. Mean and standard deviations,std (in msec) of short and long vowels between Hindi and Telugu and between Telugu and Tamil languages respectively

	Hindi		Telugu	
Phone	Mean	Std	Mean	Std
a	79.080	31.137	76.776	23.936
e	109.439	43.779	92.286	38.739
i	106.740	59.646	99.948	53.136
o	103.772	29.205	75.068	22.171
u	85.850	41.814	78.676	39.406

	Telugu		Tamil	
Phone	Mean	Std	Mean	Std
aa	156.671	40.531	164.670	42.208
ei	148.157	37.818	156.548	37.861
ii	122.819	37.657	131.766	35.809
oo	135.856	33.975	145.210	33.652
uu	125.987	42.452	137.889	41.624

Table 6. Mean and standard deviations,std (in msec) of nasal sounds

Phone	Hindi		Telugu		Tamil	
	Mean	Std	Mean	Std	Mean	Std
m	82.59	27.94	77.70	24.10	78.69	27.24
n	82.44	31.82	76.10	31.03	76.59	31.82
n1	130.89	27.37	124.98	23.87	117.37	18.78
nd~	67.62	38.06	81.32	34.34	80.70	35.20
nj~	74.11	29.35	84.34	35.56	79.90	31.23

4 Conclusion

A study on the analysis of language discrimination on phonemically balanced speech database is presented. With intentionally spelled words having same pronunciation, we find major differences in the prosodic level. From the analysis, Telugu can be discriminated using vowels and Hindi can be discriminated using nasals. Along with the help of singleton stop phoneme durations, the three languages can be discriminated. Hence, phoneme duration holds key information for language discrimination.

References

1. Klatt, D.H.: Linguistic uses of segmental duration in English: Acoustic and perceptual evidence. J. Acoust. Soc. Am. 59(5), 1208–1221 (1976)
2. Samudravijaya, K.: Durational characteristics of Hindi phonemes in Continuous Speech. Technical Report, TIFR (April 2003)
3. Muthuswamy, Y.K.: A segmental approach to Automatic Language Identification, Ph.D. Thesis, Oregon Graduate Institute (1993)
4. Zissman, M.A.: Comparison of four approaches to automatic language identification of telephone speech. IEEE Transactions on Speech and Audio Processing 4(1), 31–44 (1996)
5. Douby, S., Weardon, S., Chilko, D.: Statistics for research. John Wiley and Sons, Inc., Hoboken (2004)

A Transliteration Based Word Segmentation System for Shahmukhi Script

Gurpreet Singh Lehal and Tejinder Singh Saini

Punjabi University, Patiala 147 002 Punjab, India
{gslehal,tej74i}@gmail.com

Abstract. Word Segmentation is an important prerequisite for almost all Natural Language Processing (NLP) applications. Since word is a fundamental unit of any language, almost every NLP system first needs to segment input text into a sequence of words before further processing. In this paper, Shahmukhi word segmentation has been discussed in detail. The presented word segmentation module is part of Shahmukhi-Gurmukhi transliteration system. Shahmukhi script is usually written without short vowels leading to ambiguity. Therefore, we have designed a novel approach for Shahmukhi word segmentation in which we used target Gurmukhi script lexical resources instead of Shahmukhi resources. We employ a combination of techniques to investigate an effective algorithm by applying syntactical analysis process using Shahmukhi Gurmukhi dictionary, writing system rules and statistical methods based on n-grams models.

Keywords: Shahmukhi, Gurmukhi, Word Segmentation, Transliteration.

1 Introduction

Segmentation of a sentence into words is one of the necessary preprocessing tasks of NLP. Word segmentation can be split into two main processes: word candidate generation and word candidate selection. The first process aims at constructing all possible word candidates from a given input text. While, the latter process aims at choosing the most suitable candidate. For languages like English, French, and Spanish etc. tokenization is considered trivial because the white space or punctuation marks between words is a good approximation of where a word boundary is. Whilst many Asian languages like Urdu, Persian, Arabic, Chinese, Dzongkha, Lao and Thai have no explicit word boundaries [5-7]. Therefore, one must resort to higher levels of information such as: information of morphology, syntax, and statistical analysis to reconstruct the word boundary information [1-4]. In general the problem of segmenting word can be classified into dictionary based and statistical based methods. Statistical methods are considered to be very effective to solve segmentation ambiguities. Durrani [5] and Durrani and Hussain [6] have discussed in detail the various Urdu word segmentation issues. A word segmentation system for handling space insertion problem in Urdu script has been presented by Lehal [9].

In this paper, Shahmukhi word boundary issues have been discussed in detail. The word segmentation module is part of Shahmukhi-Gurmukhi transliteration system and

C. Singh et al. (Eds.): ICISIL 2011, CCIS 139, pp. 136–143, 2011.
© Springer-Verlag Berlin Heidelberg 2011

the novel approach presented in this paper, mainly uses target script lexical resources instead of Shahmukhi resources because Shahmukhi script is usually written without short vowels leading to potential ambiguity. We employ a combination of techniques to investigate an effective algorithm by applying syntactical analysis process using Shahmukhi Gurmukhi dictionary, writing system rules and statistical methods, including n-grams to solve word segmentation.

1.1 Shahmukhi Script

Shahmukhi is a local variant of cursive Urdu script used to record the Punjabi language in Pakistan. It is based on right to left Nastalique style of the Persian and Arabic script. Shahmukhi script has thirty eight letters, including four long vowel signs Alif ا[ə], Vao و[v], Choti-ye ی[j] and Badi-ye ے[j]. Shahmukhi script in general has thirty seven simple consonants and eleven frequently used aspirated consonants. There are three nasal consonants (ݨ[ɳ], ن[n], م[m]) and one additional nasalization sign, called Noon-ghunna ں [ɲ]. In addition to this, there are three shot vowel signs called Zer ِ[ɪ], Pesh ُ[ʊ] & Zabar َ[ə] and some other diacritical marks or symbols like hamza ء [ɪ], Shad ّ, Khari-Zabar ٰ[ə], do-Zabar ً[ən], do-Zer ٍ[ɪn] etc.

Shahmukhi characters change their shapes depending upon neighboring context. But generally they acquire one of these four shapes, namely isolated, initial, medial and final. Arabic orthography does not provide full vocalization of the text, and the reader is expected to infer short vowels from the context of the sentence. Any machine transliteration or text to speech synthesis system has to automatically guess and insert these missing symbols. This is a non-trivial problem and requires an in depth statistical analysis [6].

2 Word Boundary Issues in Shahmukhi Text

Shahmukhi is written in cursive Urdu script. The concept of space as a word boundary marker is not present in Urdu script but with the increasing usage of computer it is now being used, both to generate correct shaping and also to separate words [6]. The word boundary identification for Shahmukhi text is not simple. Due to cursive script and irregular use of space, Shahmukhi word segmentation has both space omission and space insertion problems as discussed below. Space insertion refers to insertion of extra spaces in a word, while space omission refers to deletion of spaces between adjacent words.

2.1 Space Insertion Problem

There are two basic reasons for space insertion in a Shahmukhi word.

- The space within a word is also used to generate correct shaping while writing Shahmukhi words. Therefore, space is introduced as a tool to control the correct letter shaping and not to consistently separate words. For Example consider a word واد ات /att vād/ and گنجل دار /guñjhal dār/ having a space to generate the correct shape of ت [t] and ل [l] respectively. Without space both are having visually incorrect forms as اتواد /attvād/ and گـنـجـلـد ا ر /guñjhaldār/ respectively. Presence of this type of space in Shahmukhi text leads to space insertion problem in Shahmukhi word which needs to be handled accordingly while processing the Shahmukhi text.

- Many Shahmukhi words which are written as combination of two words are written as single word in Gurmukhi script. So if the two words are as such transliterated to Gurmukhi, they cannot be read properly and in some cases their meaning also gets changed. For example, if the Shahmukhi word ذمےواری /zimmē vārī/ is as such transliterated to Gurmukhi, then it will be read as ਜ਼ਿੰਮੇ ਵਾਰੀ /zimmē vārī/

while it should be written as single word ਜ਼ਿੰਮੇਵਾਰੀ/zimmēvārī/. Thus, the two Shahmukhi words had to be combined before transliteration so that the correct Gurmukhi word is generated. Similarly the city names like حیدر آباد /haidar ābād/, جیکب آباد /jaikab ābād/, جعفر آباد /jāfar ābād/ after transliteration produce unacceptable names in Gurmukhi script as ਹੈਦਰ ਆਬਾਦ /haidar ābād/, ਜੈਕਬ ਆਬਾਦ/jaikab ābād/, ਜਾਫ਼ਰ ਆਬਾਦ/jāfar ābād/. To produce correct transliteration the extra space between the names should be removed to combine them as a single word as ਹੈਦਰਾਬਾਦ /haidrābād/, ਜੈਕਬਾਬਾਦ /jaikbābād/, ਜਾਫ਼ਰਾਬਾਦ /jāfrābād/.

2.2 Space Omission Problem

While writing in Urdu/Arabic script a common user finds that it is unnecessary to insert space between the two Urdu words because the correct shape is produced automatically when the first word ends with a non-joiner Urdu character [6]. The same case is observed in Shahmukhi text that many times the user omits word boundary space between the consecutive words where the first word ends with a non-joiner character. This is because the absence of space after non-joiner character has no visible implication and do not affect the readability of the Shahmukhi text. But during computational processing where space is used as a word boundary delimiter, these two or more words are found to be merged together. This gives rise to space omission problem in Shahmukhi text.

Table 1. Space Omission Problem with Multiple Merged Words

Word					Romanized			
w	w4	w3	w2	w1	w1	w2	w3	w4
انسپیکٹرمہمدخان		خان	مہمد	انسپیکٹر	imspaikṭar	muhmmad	k͟hān	
رشتےدےمقام		مقام	دے	رشتے	rishtē	dē	mukām	
داہےایہدےوچ	وچ	ایہدے	ہے	دا	dā	hai	Ihdē	vic

For example, consider the following Shahmukhi words آ گیا /ā giā/ and ہو سکدا /hō sakdā/ having the first word token ends with a non-joiner character. We can see that they will retain same shape after deleting word boundary space as آگیا /āgiā/ and ہوسکدا /hōsakdā/. Therefore, user can easily skip word boundary space because it does not affect the readability of the Shahmukhi words. More examples of Shahmukhi words having space omission problem with multiple merged words is shown in table 1.

3 Algorithm for Handling Space Insertion Problem

Rule based techniques like longest matching, maximum matching and statistical methods including n-grams have been extensively used for word segmentation. We employ a combination of both rule based and statistical n-gram techniques for Shahmukhi word segmentation, as proposed by Lehal [9] for Urdu space insertion problem. Based on the idea presented by Lehal [9] we have divided the whole process into two stage architecture as shown in fig.1. In the first stage, writing system rules have been applied to decide if the adjacent Shahmukhi words have to be joined. The rule based analyzer is incorporated based on the knowledge of the writing system specific information for instance some characters such as ں [ɲ] and ً come at the end of a word only, certain characters such as (ۇ, ئ ء ً ٓ ں and ۂ), cannot come at the beginning of a word and the presence or absence of *hamza*(ء) before the second vowel gives a indication of joining or not joining of words. Along with these rules there are some typical words in Shahmukhi for example یا /yā/, یاں /yāṃ/ and نہ /nā/ which need special care while processing.

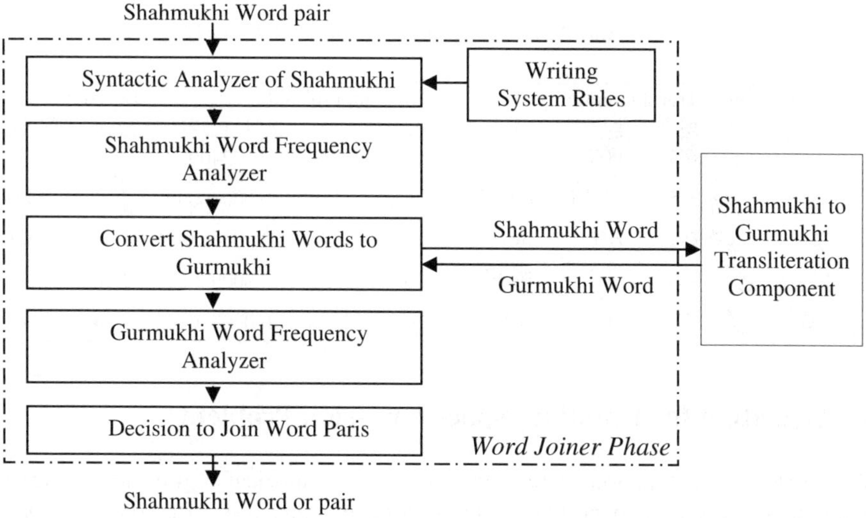

Fig. 1. Word Joiner Phase of Transliteration

In case these rules give a definite answer, then we do not move to the second stage. Otherwise, after rule based analyzer the word pairs are analyzed for statistical analysis. In this stage, we have made use of Gurmukhi corpus resources to make the final decision. We use Shahmukhi resources only if the Gurmukhi resources are not sufficient to make a decision for example in case of out-of-vocabulary words (OOV) and unknown cases where the corresponding Gurmukhi transliteration is not present. The algorithm of the statistical analysis is as follows:

Step 1: We have to first transliterate the individual (w1, w2) Shahmukhi tokens and their joined form (w1 concatenated with w2) into Gurmukhi say g1, g2 and g3 respectively and then look for the probability of occurrence in Gurmukhi corpus p(g1),p(g2) and p(g3).

Step 2: If the probability of occurrence of Joined Gurmukhi form p(g3) is greater than the individual Gurmukhi tokens then the words are joined else not.

Step 3: If the joining decision at step2 is to join the word tokens then we additionally look for the existence of the bigram (g1, g2) in Gurmukhi corpus. If the bigram is present, then the two Shahmukhi words are not joined. This is to overcome the situation when the product of probabilities p(g1).p(g2) becomes much more small. As a result many times step2 give the decision to join the words even though they were not to be joined.

Consider the five outputs provided in table 2 to understand the detailed processing of statistical analysis. The system evaluated the unigram probabilities and found that at step 2 the condition is true for all the cases except the first case and the decision is to join them. But at step 3 system found that the last two cases are not joined because the corresponding bigrams (ਚੰਨ/cann/, ਵਲੀ/valī/) and (ਗੁਣ/guṇ/, ਗਾ/gā/) are present in the bigram lexicon.

Table 2. Processing Steps of Statistical Analysis

Input tokens		Transliteration			Unigram Probability		Decision	
w2	w1	g1	g2	g3	p(g3)	p(g1).p(g2)	Step2	Step3
اج	کول	ਕੇਲ	ਅੱਜ	ਕੇਲਾਜ	0.00003919	0.00240909	No	-
شائن	سن	ਸਨ	ਸ਼ਾਇਨ	ਸਨਸ਼ਾਇਨ	0.00001120	0.00000039	Join	Join
سلو	ہن	ਹਨ	ਸਲੂ	ਹੰਸਲੇ	0.00004478	0.00000387	Join	Join
ولی	چن	ਚੰਨ	ਵਲੀ	ਚਨੈਲੀ	0.00003639	0.00000060	Join	No
گا	گݨ	ਗੁਣ	ਗਾ	ਗੰਗਾ	0.00172694	0.00001642	Join	No

4   Algorithm for Handling Space Omission Problem

We employ a combination of both rule based and statistical n-gram techniques for handling space omission problem. This is a challenging task to predict the correct combination of words from the merged word string. Firstly, Input multi-word has to be broken up into character combinations (CC) as per defined rules. The position of non-joiner characters in the multi-word and the position of ں, ے[e] and ٔ characters is a good broken point with in a multi-word. Then each adjacent CC's are combined to form a list of the purposed Shahmukhi words. After which, each CC in all the purposed words is transliterated using the transliteration component. Next, we have to design a strategy to select the most probable correct segmentation from the purposed word list. In this stage, the Shahmukhi and Gurmukhi lexical resources are used to make the final decision. For example consider the merged token تیلاتیلاایکٹھاکرکے /tīlātīlāikṭṭhākarkē/ which is broken into کے ، کر ، اکٹھا، تیلا، تیلا five CCs using the CC rules. Then each pair of adjacent CC's are combined to form a list of 16 purposed Shahmukhi words. After transliteration and statistical analysis of all the purposed words, the best probable word is selected as an output by the system. To handle over segmentation of out-of-vocabulary (OOV) or unknown words we have imposed the condition that the system will accept only those purposed word combinations which contain at least one character combination of length greater than three or at least one valid bigram character combination exist. For example, consider the Shahmukhi word خانسامیاں /khānsāmīāṃ/ which is out-of-vocabulary and it can be broken down into three valid Gurmukhi CCs ਖਾ/khā/, ਨੱਸਾ/nassā/ and ਮੀਆਂ/mīāṃ/ by this algorithm.

Clearly, these CCs qualify the first condition but they do not have existence of valid bigram. Hence, this word will not be broken down by the system due to imposed condition and transliterated into Gurmukhi script as ਖਾਨਸਾਮੀਆਂ /khānsāmīāṃ/ which is correct transliteration. The system architecture is shown in fig. 2.

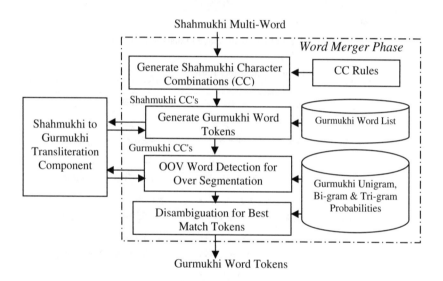

Fig. 2. Word Merger Phase of Transliteration

5   Experiments and Results

A study of segmentation analysis of Shahmukhi text is conducted on a Shahmukhi corpus of size 3 million words. This corpus is a collection of data like news, articles, short stories, books, novels, poetry etc. collected from Pakistan and downloaded from popular Shahmukhi Unicode website http://www.wichaar.com. It is observed that the Shahmukhi corpus has 1.49% words with space omission and 1.05% of words with space insertion problem. The algorithm for space insertion problem was tested on this corpus and after manual evaluation we found that this algorithm works at 95.23% of accuracy. The system has shown good performance except some over joining cases are also observed. The main cases for consideration and improvement are those Shahmukhi tokens having no bi-gram in Gurmukhi lexicon as a result they are over joined. This type of situation can be improved by increasing the size of lexicon.

Table 3 shows the observed occurrence of space omission cases which are broken up with respect to number of merged words. It is observed that the maximum number of merged words in a multi-word ligature is five and their occurrence in the corpus is 0.037%. The percentage of occurrence of four merged words is observed to be 0.23%

Table 3. Occurrence of Merged Words in Shahmukhi Corpus

Number of Merged words (n)	Occurrence (%)	Segmentation Accuracy (%)
n=5	0.036778	75
n=4	0.229864	77.5
n=3	3.83413	76.11
n=2	96.99338	93.77

Table 4. Failure Cases of Space Omission Algorithm

SN	Merged words	Error Type	Incorrect Form	Correct Form	Romanized
1	تے فراق	OOV	ਤੇ ਫ਼ਰ ਇਕ / تے فر اق	ਤੇ ਫ਼ਿਰਾਕ / تے فراق	tē firāk
2	اورکٹ	OOV	ਔਰ ਕੱਟ / اور کٹ	ਔਰਕੁਟ / اورکٹ	aurkuṭ
3	وینزیلاوچ	OOV	ਵੇਨਜ਼ ਯੂਲਾ ਵਿਚ / وینز یلا وچ	ਵੇਜ਼ੁਏਲਾ ਵਿਚ / وینزیلا وچ	vēñjuēlā vic
4	آسٹرولوجی	OOV	ਆਸਟਰ ਵੱਲੋ ਜੀ / آسٹر ولو جی	ਆਸਟ੍ਰੋਲੋਜੀ / آسٹرولوجی	āṣṭraulōjī
5	ناصرخان	Prob.	ਨਾ ਸਿਰ ਖ਼ਾਨ / نا صر خان	ਨਾਸਿਰ ਖ਼ਾਨ / ناصر خان	nāsir khān
6	پرتانوالی	Prob.	ਪਰ ਤਾਂ ਵਾਲ਼ੀ / پر تاں والی	ਪਰਤਾਂ ਵਾਲ਼ੀ / پرتاں والی	partāṃ vāl̤ī
7	ونڈداربیا	Prob.	ਵੰਡ ਦਾ ਰਿਹਾ / ونڈ دا ربیا	ਵੰਡਦਾ ਰਿਹਾ / ونڈدا ربیا	vaṇddā rihā
8	خدانخواسطہ	Izafat	ਖ਼ੁਦ ਅੲਖਵਾ ਸੱਤਾ / خد انخوا سطہ	ਖ਼ੁਦਾ-ਨ-ਖ਼ਾਸਤਾ / خدانخواسطہ	khudā-na-khāstā
9	دورفاروقی	Izafat	ਦੋਰ ਫ਼ਾਰੂਕੀ / دور فاروقی	ਦੌਰ-ਏ-ਫ਼ਾਰੂਕੀ / دورِفاروقی	daur-ē-fārūkī
10	سیدمحمودالحسن	Izafat	ਸੱਯਦ ਮਹਿਮੁਦ ਅਲਹਸਨ / سیِد محمود الحسن	ਸੱਯਦ ਮਹਿਮੂਦ-ਉਲ-ਹਸਨ / سیِد محمودالحسن	sayyad mahimūd-ul-hasan

which is also very less in number. After that, relatively high occurrence 3.83% of three merged words is observed. The most frequent space omission cases are two merged words having maximum coverage 96.99% of the corpus.

The overall segmentation accuracy of space omission algorithm is 92.97%. The system has shown highest accuracy 93.77% when two merged words are found in the multi-word ligatures. The accuracy of the system decreases when the number of merged word is more that two.

The analysis of system errors shows that there are three types of errors that the system had made with the current input. As shown in table 4 first type of words are those which are out of vocabulary and system performed over segmentation. The second type of error words are those in which the joined word ligature (unigram) has less probability then the probability of individual word tokens (bi-gram) e.g. the unigram ਪਰਤਾਂ/partām/ has very less probability of occurrence where as the probability of bi-gram ਪਰ/par/ and ਤਾਂ/tām/ is much more. The third type of error words are special unknown Izafat or compound words from Urdu domain which need to be handled. We can produce better results in the future with the scope to increase the size of the training corpus.

References

1. Papageorgiou, C.P.: Japanese Word segmentation by hidden Markov model. In: Proceedings of the Workshop on Human Language Technology, pp. 283–288 (1994)
2. Nie, J.Y., Hannan, M.L., Jin, W.: Combining dictionary, Rules and Statistical Information in Segmentation of Chinese. In: Computer Processing of Chinese and Oriental Languages, vol. 9, pp. 125–143 (1995)
3. Wang, X., Fu, G., Yeung, D.S., Liu, J.N.K., Luk, R.: Models and Algorithms of Chinese Word Segmentation. In: Proceedings of the International Conference on Artificial Intelligence (IC-AI 2000), Las Vegas, Nevada, USA, pp. 1279–1284 (2000)
4. Xu, J., Matusov, E., Zens, R., Ney, H.: Integrated Chinese Word Segmentation in Statistical Machine Translation. In: Proceedings of the International Workshop on Spoken Language Translation, Pittsburgh, PA, pp. 141–147 (2005)
5. Durrani, N.: Typology of Word and Automatic Word Segmentation in Urdu Text Corpus. MS Thesis, National University of Computer and Emerging Sciences, Lahore, Pakistan (2007)
6. Durrani, N., Hussain, S.: Urdu Word Segmentation. In: The 2010 Annual Conference of the North American Chapter of the ACL, Los Angeles, California, pp. 528–536 (2010)
7. Akram, M., Hussain, S.: Word Segmentation for Urdu OCR System. In: Proceedings of the 8th Workshop on Asian Language Resources, Beijing, China, pp. 88–94 (2010)
8. Sproat, R., Shi, C., Gale, W., Chang, N.: A stochastic finite state word segmentation algorithm for Chinese. Computational Linguistics 22, 377–404 (1996)
9. Lehal, G.S.: A Two Stage Word Segmentation System for Handling Space Insertion Problem in Urdu Script. World Academy of Science, Engineering and Technology 60, 321–324 (2009)
10. Lehal, G.S.: A Word Segmentation System for Handling Space Omission Problem in Urdu Script. In: 1st Workshop on South and Southeast Asian Natural Language Processing (WSSANLP) 23rd COLING, Beijing, China, pp. 43–50 (2010)

Optimizing Character Class Count for Devanagari Optical Character Recognition

Jasbir Singh and Gurpreet Singh Lehal

Department of Computer Science, Punjabi University, Patiala, India
jbs.5@rediffmail.com, gslehal@gmail.com

Abstract. Optical character recognition is a widely used technique for generating digital counterpart of printed or handwritten text. A lot of work has been done in the field of character recognition of Devanagari script. Devanagari script consists of several basic characters, half form of characters, vowel-modifiers and diacritics. From character recognition point of view only 78 character classes are sufficient for the identification of these characters. But in Devanagari the characters fuse with each other, which result in segmentation errors. Therefore to avoid such errors we shall consider such compound characters as separate recognizable unit. We have identified 864 such compound characters that make a total of 942 recognizable units. But it is very difficult to handle such a large number of classes; therefore we have optimized the character class count. We have found that the first 100 classes can contribute to 98.0898% of the overall recognition.

Keywords: Conjuncts, Segmentation, Recognizable unit.

1 Introduction

Optical character recognition belongs to the family of techniques performing automatic identification. It deals with the problem of recognizing optically processed characters. The character recognition work on Devanagari script started in 70's when Sinha and Mahabala [1] presented a syntactic pattern analysis system with an embedded picture language. Sinha and Bansal [2] have discussed the use of various knowledge sources at all levels in Devanagari document processing system. Chaudhuri and Pal [3] have suggested primary grouping of characters, where each character is assigned to one of the three groups namely basic, modifier and compound character group before going for actual recognition process. Bansal and Sinha [4-5] have presented method for segmentation and decomposition of Devanagari composite characters into their constituent symbols. Kompalli et al [6] have discussed the wide range of challenges in Devanagari script that are not seen in Latin based scripts. They also mentioned that half consonants have different shapes from full-consonants; therefore the use of post-processing techniques or half-consonant classifiers for the left part can improve conjunct recognition. In our work we shall consider each compound character as separate class so as to reduce the errors introduced due to over and under segmentation of such characters. Therefore in this paper we have presented the analysis and optimization of characters classes that would be sufficient to get the desired recognition rate.

C. Singh et al. (Eds.): ICISIL 2011, CCIS 139, pp. 144–149, 2011.
© Springer-Verlag Berlin Heidelberg 2011

2 Problems with Segmentation (Need of Multiple Classes)

One of the significant phases in any optical character recognition system, upon which the performance of the overall system depends is the *segmentation phase*. Segmentation phase consists of the line segmentation, word segmentation and character segmentation. Out of these, character segmentation is most critical one, as the most of the recognition errors in optical recognition system are due to the character segmentation. In Devanagari the constituent characters may join horizontally or vertically to form new characters. When the constituent characters (*consonants*) fuse horizontally (*laterally*), they result in the formation of conjuncts (Fig. 1).

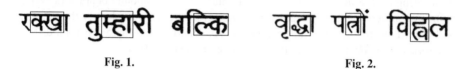

Fig. 1.   **Fig. 2.**

Although in the resulting conjuncts, the constituent characters are at adjacent positions, but they fuse laterally in such a way that there is no vertical space between them and hence it becomes very difficult to separate them during segmentation phase. In certain cases the constituent characters may combine in such a way which leads to the formation of new single character in which the constituent characters does not appear at adjacent positions i.e. they merge with each other (Fig. 2).

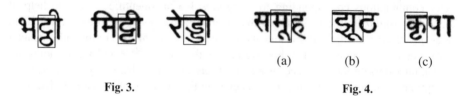

                                                              (a)   (b)   (c)

Fig. 3.   **Fig. 4.**

Similarly the constituent components may combine vertically, which result in the formation of characters having more height than the normal character height (Fig. 3).There is another class of characters in which height of the character itself or when combined with descenders (lower zone vowel modifiers) is such that it results in the problems in segmentation. In some cases the height of the primitive character is so large as compared to the adjacent consonant-descender combination that it leads to under segmentation (Fig. 4-a). Similarly the height of consonant itself may be so large that it will lead to the over segmentation of consonant-descender combination (Fig. 4-b). In few cases the consonant-descender combination itself is small in height as compared to adjacent character that it lead to the under segmentation (Fig. 4-c). From above it is clear that it is quite difficult to separate the constituent characters from these compound characters. If we try to segment them, it will lead to over or under segmentation and hence segmentation errors. Therefore treating such type of compound characters as a separate recognizable unit can decrease the segmentation problems. In the same way all the consonant-descender combinations will be treated as separate classes.

3 Identification/Optimization of the Classes

Before getting into the optical recognition process one must know the character classes that are going to be used for the underlying script. Any word in Devanagari can be divided into three zones: middle, upper and lower zone. The table shown below (Table 1.) provides some of the recognizable units corresponding to all three zones. The recognizable unit is the smallest possible unit that can be recognized by character recognition process. For example न contribute to both ¬ and ा, which are two recognizable units that will be separated during segmentation phase if one removes the headline. Some other examples are given after Table 1. Apart from these basic recognizable units there exist a large number of compound characters that can be treated as separate recognizable units. Therefore we have to identify them, as they will contribute to the recognition process. As stated above there are four different categories of the character combination that are to be identified:

Category-1: The combination in which consonants fuse to create conjuncts (Fig-1).
Category-2: The combination in which the adjacent consonants fuse to form a composite character having more height as compared to other (Fig-3).
Category-3: The characters may join which result in a single character (Fig-2).
Category-4: All of the consonant with the lower vowel modifier (Fig-4).

In order to identify the possible character classes a corpus of approximately 3 million words has been used. The corpus comprises of Unicode data, therefore most of character combinations corresponding to categories 1-3 are identified with the help of diacritic ् (*halant*). It is to be noted that in many cases even the presence of halant does not cause the adjacent consonants to combine to form the compound character corresponding to categories 1-3. For example in the word ट्वेंटी, ट and व does not result in conjunct despite the presence of halant. Therefore such characters along with the exceptions, which do not form conjuncts even in the presence of halant are identified. For example ट will not form the compound characters corresponding to categories 1-3 with any consonant except with ट ठ य र. For the category 4, the presence of descender is checked after the consonant. Again there are certain exceptions to it, for example in the word अद्भुत even though द and भ can lead to conjunct द्भ, but the presence of ु will result in separate consonant with descenders द् and भु. Therefore care is taken to count them separately.

Table 1. First 10 recognizable units (out of 78) along with their frequency of occurrence

S.No	Recognizable unit	Overall % occurrence	S.No.	Recognizable unit	Overall % occurrence
1.	ा	20.7631	6.	न	3.7870
2.	े	7.4461	7.	स	3.4365
3.	क	5.2938	8.	ह	3.3931
4.	र	5.2631	9.	म	3.2991
5.	ि	3.7936	10.	ं	3.2023

In the above table आ, आँ, ओ, ओ, औ, ि, ी, ों, ो, ो, ों, ग, ण, श, ग़ contribute to ा. Similarly "॥ Contribute twice to ।", "ऐ, ओ, ो contribute to ॊ", "औ, ो contribute to ॏ", "ऍ, ऑ, ॉ contribute to ॓" and "ॊ, ऎ, ॏ, ॉ contribute to ॓".

The following table (Table 2.) depicts first few recognizable units obtained by lateral fusion of the characters along with their constituent characters, their occurrence in middle zone and their overall contribution. A total of 655(unique) such recognizable units corresponding to category-1 and category-2 have been identified.

Table 2. Some recognizable units obtained by lateral fusion of the characters

Horizontally fused consonants	Character Combination	% Occurrence with in all middle zone characters	Overall % Occurrence
प्र	प ् र	0.3766	0.2900
त्र	त ् र	0.1606	0.1237
स्त	स ् त	0.1601	0.1233
क्ष	क ् ष	0.1332	0.1026
न्ह	न ् ह	0.1198	0.0923
रु	र ु	0.1121	0.0863
क्य	क ् य	0.1052	0.0810
न्द	न ् द	0.0949	0.0731

The table shown below (Table 3.) provides the overall contribution of these recognizable units if we select a definite number of these units. From this table (Table 3.) it is evident that if we select first 200 such units they will contribute to 2.9040%, and they all will contribute to 2.9349% toward overall recognition.

Table 3. Overall contribution of laterally fused recognizable units

Horizontally fused consonants selected out of 655	% Contribution toward middle zone	Overall % contribution
10	1.4286	1.1003
20	2.0878	1.6080
50	3.0632	2.3593
100	3.5570	2.7396
150	3.7094	2.8570
200	3.7705	2.9040
250	3.7915	2.9202
350	3.8043	2.9301
400	3.8066	2.9318
450	3.8079	2.9329
500	3.8089	2.9336
655	3.8106	2.9349

Similarly 209 (unique) and total of 317980 recognizable units corresponding to category-3 and category-4 have been identified. The table (Table 4.) depicts first few such units along with the frequency of their occurrence in middle zone and their overall occurrence.

Table 4. Overall occurrence of consonants with descender/vertically overlapped consonants

Consonants with descender/vertically overlapped consonants	% Occurrence with in all middle zone characters	Overall % occurrence
कु	0.2879	0.2217
हु	0.2744	0.2113
मु	0.2734	0.2105
सु	0.2301	0.1772
गु	0.1380	0.1063
पु	0.1142	0.0879
पू	0.1141	0.0879

The following data provides the overall contribution of these recognizable units if we select a fix number of these units. From this data it is clear that if we select all such units they will contribute to 2.6206% toward overall recognition.

Table 5. Percentage contribution of consonants with descender/vertically overlapped conso-nants selected out of 209 recognizable units

Consonants with descender/vertically overlapped consonants out of 209	% Contribution toward middle zone	Overall % contribution
10	1.6988	1.3084
20	2.3778	1.8314
50	3.1215	2.4042
100	3.3695	2.5952
150	3.4000	2.6187
209	3.4025	2.6206

The total number of recognizable units counts to 942. These units are so large in number that it is very difficult to handle all these as separate classes, therefore we optimize the class count by considering all possible recognizable units and then evaluating their overall contribution. The table (Table 6.) depicts first few recogniz-able units and their contribution with in all 942 classes.

Table 6. Percentage occurrence of few recognizable units out of 942 recognizable units

S.No.	Recognizable unit	Overall % Occurrence	S.No.	Recognizable unit	Overall % Occurrence
1.	ा	22.9577	6.	न	3.7131
2.	्ा	8.2331	7.	ि	3.5408
3.	क	5.1207	8.	ह्	3.3730
4.	र	4.9624	9.	ं	3.3140
5.	ा	4.1946	10.	म	3.1325

From data given below (Table 7.) we find that if we chose first 100 recognizable units they will contribute to 98.0898%. Similarly the selection of first 600 units will result in the contribution of 99.9951%.

Table 7. Percentage contribution of recognizable units

Recognizable units	% contribution	Recognizable units	% contribution
20	82.0185	300	99.8817
30	90.1112	400	99.9672
40	93.4336	500	99.9883
50	95.0826	600	99.9951
70	96.6985	700	99.9976
100	98.0898	800	99.9989
150	99.1658	942	100.0000
200	99.5661		

4 Conclusion

In Devanagari script each zone contributes to the different recognizable units. From character recognition point of view there are 78 distinct recognizable units in Devanagari script. But in Devanagari script the characters may fuse (Fig.1-4) resulting in compound characters. These characters may be very difficult to separate, and hence contribute to segmentation errors. So to avoid such errors we shall consider all such compound characters as separate recognizable unit. We have identified 864 such unique units. The overall recognizable units are obtained by adding basic 78 recognizable units and the 864 (Table 2, 4.) compound recognizable units, which result in a total of 942 recognizable units. As it would be very difficult to handle such a large number of classes, so the contribution of each class is evaluated to find how many classes would be sufficient to get desired recognition rate. From above table it is clear that if we consider first 100 classes they will contribute to 98.0898%, similarly first 150 classes will contribute to 99.1658%. It has also been found that the single recognizable unit 'ोा' contributes 22.9577% toward the overall accuracy.

References

1. Sinha, R.M.K., Mahabala, H.N.: Machine recognition of Devanagari script. IEEE Transactions on Systems, Man and Cybernetics SMC-9, 435–441 (1979)
2. Sinha, R.M.K., Bansal, V.: On Devanagari Document Processing. In: IEEE International Conference on Systems, Man and Cybernetics, vol. 2, pp. 1621–1626 (1995)
3. Chaudhuri, B.B., Pal, U.: An ocr system to read two Indian language scripts: Bangla and Devanagari (Hindi). In: Proceedings of the 4th International Conference on Document Analysis and Recognition, Germany, vol. 2, pp. 1011–1015 (1997)
4. Bansal, V., Sinha, R.M.K.: Integrating Knowledge Sources in Devanagari Text Recognition. IEEE Transactions on Systems, Man and Cybernetics-part A: Systems and Humans 30, 500–505 (2000)
5. Bansal, V., Sinha, R.M.K.: Segmentation of touching and fused Devanagari characters. Pattern Recognition 35, 875–893 (2002)
6. Kompalli, S., Nayak, S., Setlur, S.: Challenges in OCR of Devanagari Documents. In: Proceedings of the 8th International Conference on Document Analysis and Recognition (ICDAR), vol. 1, pp. 327–331 (2005)

Multifont Oriya Character Recognition
Using Curvelet Transform

Swati Nigam and Ashish Khare

Department of Electronics and Communication
University of Allahabad, Allahabad, India
swatinigam.au@gmail.com, ashishkhare@hotmail.com

Abstract. In this paper, we have proposed a new character recognition method for Oriya script which is based on curvelet transform. Multi font Oriya character recognition has not been attempted previously. Ten popular Oriya fonts have been used for the purpose of character recognition. The wavelet transform has widely been used for character recognition purpose, but it cannot well describe curve discontinuities. We have used curvelet transform for recognition which is done using curvelet coefficients. This method is suitable for Oriya character recognition as well as various other scripts' recognition purpose also. The proposed method is simple and extracts effectively the features in target region, which characterizes better and represents more robustly the characters. The experimental results validate that the proposed method improves greatly the recognition accuracy and efficiency than other traditional methods.

Keywords: Optical Character Recognition, Oriya Characters, Curvelet Transform, Morphological Operations, SVM Classifier.

1 Introduction

The optical character recognition (OCR) in image sequences is a very popular problem in the field of image processing today. It is one of the most common techniques using these days among face recognition, signature recognition, text recognition and fingerprint recognition, etc. High accuracy character recognition is a challenging task for scripts of languages. Various Indic scripts are used in the Indian mainland. Some of them are Bengali, Devanagari, Gujarati, Gurmukhi, Kannada, Malayalam, Oriya, Tamil, and Telugu. The character recognition techniques for the English script have been exploited very much. Even commercial software is also available for this purpose. However, for the major part of other scripts such as Arabic [1] and Indian [2], OCR is still an active domain of research. For English and Kanji scripts, good progress has been made towards the recognition of printed scripts, and the focus nowadays is on the recognition of handwritten characters [3].

OCR research for different Indian languages is still at a nascent stage. There has been limited research on recognition of Oriya [4, 5]. The proposed system [4] was developed by combining traditional and modern proposed techniques. In [5], the focus was on a bilingual OCR for recognition of the printed English and Oriya texts.

C. Singh et al. (Eds.): ICISIL 2011, CCIS 139, pp. 150–156, 2011.
© Springer-Verlag Berlin Heidelberg 2011

Although some work has been done for optical character recognition over the years, but almost all existing work makes an important implicit assumption that the script or language of the document to be processed is known. In practice, it means that human intervention is necessary in identifying the script or language of each document. In our growingly interconnected world, this is clearly an inefficient and undesirable task.

To minimize the human intervention, we implemented a curvelet transform based technique. In this paper, we have presented an innovative multiresolution character recognition technique based on the curvelet transform given by Candes and Donoho [6]. The wavelet transform does not process edge discontinuities optimally, and discontinuities across a simple edge affect all the wavelets coefficients on the edge. The Curvelet transform was introduced to overcome the weakness of wavelets in higher dimensions. The new tight frame of curvelet is an effective nonadaptive representation for objects with edges [6]. It is well capable of handling two dimensional singularities also. Curvelet Transform has been successfully applied for image de-noising [7] and object tracking [8] purposes, however, we have not found much application of curvelets in character recognition [9]. Different characters and various fonts of Oriya script are shown in Fig. 1 and 2 respectively.

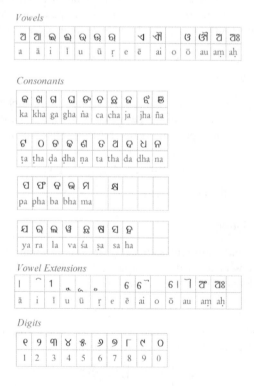

Fig. 1. Different Oriya Characters

Ori 1 Uni	RaghuOriya	SamyakSans	utkal

Fig. 2. Variability of Oriya fonts

For the enhancement of the recognition accuracy, in this paper, the morphological thinning and thickening of the original image (containing characters) was done. After that, the curvelet coefficients of these images were used for training of separate classifiers. At last, the outputs of these classifiers were used to get results.

The rest of the paper is organized as follows. Section 2 describes the basic curvelet transform. Section 3 briefly discusses the proposed method. In Section 4, the implementation along with the experimental setup and the results obtained are discussed. Finally in Section 5, conclusions are made.

2 The Curvelet Transform

Curvelet transform is a new transform which is based upon parabolic scaling law width $\approx$ length². It was given by Candès and Donoho in 1999. Curvelet transform is capable to handle curve discontinuities well as compared to wavelet and ridgelet transform. There are four steps in the implementation of curvelet transform. These are:

2.1 Sub-band Decomposition

This is the first step. An image is decomposed into different subbands by using energies of different layers. Let f is an image which is needed to be decomposed and P_0 and Δ_s are low pass and high pass filters respectively. Then this image can be decomposed as

$$f \mapsto (P_0 f, \Delta_1 f, \Delta_2 f, \ldots \ldots \Delta_k f) . \tag{1}$$

where $1 \leq s \leq k$ and k is the last subband.

2.2 Smooth Partitioning

After decomposing into different subbands, we get different small images. These small images are now centerised along a grid of dyadic squares, so that we can get each image in the domain of dyadic squares. We get this by:

$$h_Q = w_Q . \Delta_s f . \tag{2}$$

where h_Q is the output of this step and w_Q is small windowing function applied on each small image. The grid of dyadic squares is defined as

$$Q = \left[\frac{k_1}{2^s}, \frac{k_1+1}{2^s}\right] \times \left[\frac{k_2}{2^s}, \frac{k_2+1}{2^s}\right] \in Q_s . \tag{3}$$

where Q_s Є all the dyadic squares of the grid, k_1 corresponds to first highpass filter and k_2 corresponds to last highpass filter.

2.3 Renormalization

After smooth partitioning, Renormalization of each subband is done. In this step, each subband is converted from dyadic square to unit square [0,1] x [0,1]. This is done to make the further process easy and to reduce the computational complexity.

For renormalization, an operator T_Q is defined for each Q as

$$\left(T_Q f\right)(x_1, x_2) = 2^s \, f\left(2^s x_1 - k_1, 2^s x_2 - k_2\right). \tag{4}$$

where x_1, x_2 are ridge lines along first and last high pass filters. This operation is used for renormalization of each subband. The renormalization of each subband is done as

$$g_Q = T_Q^{-1} h_Q. \tag{5}$$

Hence we get g_Q as renormalized subband image.

2.4 Ridgelet Analysis

The last and important step of curvelet transform is doing ridgelet analysis of different renormalized subbands. In this step, each subband is processed in ridgelet domain. For field $L^2(R^2)$, the basis elements are defined as ρ_λ and we get,

$$\alpha_{(Q\lambda)} = \langle g_Q, \rho_\lambda \rangle. \tag{6}$$

The curvelet transform is useful for character recognition due to its following properties:

(1) The curvelet coefficients are directly calculated in the Fourier space. In the context of the curvelet transform, this allows avoiding the computation of the 1D inverse Fourier transform along each radial line.

(2) Each subband is sampled above the Nyquist rate, hence, avoiding aliasing – a phenomenon typically encountered by critically sampled curvelet transform.

(3) The reconstruction is trivial. The curvelet coefficients simply need to be co-added to reconstruct the input signal at any given point. In our application, this implies that the curvelet coefficients simply need to be co-added to re-construct Fourier coefficients.

(4) In curvelet domain, the most essential information in the image is com-pressed into relatively few large coefficients, which coincides with the area of major spatial activity.

3 The Proposed Method

In this paper, we have presented an innovative character recognition method. This method is based on the curvelet transform of morphologically transformed versions of an original character. In previous years, Mohammad and Husain [10] have been used morphology for the character recognition task.

Curvelets provide a very good representation of curves in an image. Thinning or thickening of this image changes the position of the curves. With the use of curvelet transform, the changes in position of curves were encoded in the transformed domain.

In the proposed method, 2-level thinning and 2-level thickening of the image was done. This was performed one by one. First a thinning process was performed followed by thickening. After that again a thinning process was performed followed by thickening again. After that the original image and the morphologically changed images, both were used. According to definition of curvelet transform, there were five sets of curvelet coefficients, which were used to train five SVM classifiers. If a character was not recognized from original image, then it could be recognized from transformed image. Also, in the process of recognition of test characters, five versions of the input image were created. The curvelet coefficients of these five versions were classified using the corresponding SVM classifiers. The block diagram of the classification scheme is shown in Fig. 3.

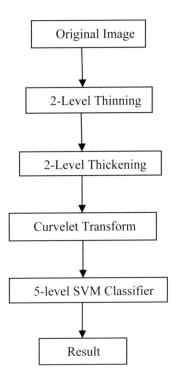

Fig. 3. Classification Scheme based on Curvelet Transform

4 Experiments and Results

Ten Oriya fonts were used in the proposed method. For each of the fonts 10 different font sizes viz., 8, 10, 12, 14, 16, 18, 20, 22, 24 and 28 points were taken. The experimental part of recognition of Oriya characters was divided into two parts. In the first part of the experiments, 8 different font sizes of all the 10 fonts were used as the prototype for the SVM classifier. The remaining 2 sizes were used for testing of experiments. For each of the fonts, the two different sizes that need to be tested were selected randomly, i.e. say for the Ori 1 Uni font the testing set comprised of sizes 12 and 20 while for RaghuOriya the font sizes for testing were randomly selected to be 18 and 28 points. For the other different fonts the selected sizes were still different.

For every font, the remaining eight font sizes behaved as the prototype for the SVM. In the second part, the 2 fonts were randomly selected for testing and the remaining 8 served as the prototypes for the SVM classifiers. Here division was not done on the basis of the font sizes. For the testing all the sizes for the 2 fonts were considered, the same being true for the 8 prototypes used by the SVM. Two examples of original images, thinned images and thickened images are shown in Fig. 4.

Original Image   Thinned Image   Thickened Image

Fig. 4.

As described earlier, 2-level thinning and 2-level and thickening was performed on the original image to obtain morphologically transformed version of the original image. The curvelet transform was performed on these original and transformed images of the same character at a single scale. The curvelet coefficients served as the feature set for training SVM classifiers. The final result was made on the basis of these classifiers. The output from classifier with the highest recognition accuracy was chosen to be the final result. Both the experiments were repeated 4 times. The results of the experiments are presented in the Table 1 and Table 2 below.

Table 1. Results of testing with different font sizes

Experiment	1	2	3	4
Accuracy(%)	94.75	94.40	95.30	95.15

Table 2. Results of testing with different fonts

Experiment	1	2	3	4
Accuracy(%)	93.70	94.40	94.15	94.70

While testing for different font sizes, the best results were achieved during experiment No. 3 and for testing with different fonts, the best results were obtained for experiment No. 4. The detailed results for these two experiments are shown in Table 3 and Table 4.

Table 3. Detailed results for experiment No. 3 for testing different font sizes

Classifier trained with	Recognition Accuracy (%)
Original image	92.80
Thinning	92.45
Thickening	91.70
Total accuracy	95.25

Table 4. Detailed results for experiment No. 4 for testing different fonts

Classifier trained with	Recognition Accuracy (%)
Original image	92.00
Thinning	91.70
Thickening	90.80
Total accuracy	94.70

5 Conclusions

In this paper, we have developed and demonstrated a new algorithm for character recognition of different fonts of Oriya language that exploits new tight frames of curvelet and provides a sparse expansion for typical images having smooth contours. We use curvelet coefficients for character recognition and feature extraction of the object in the image. The curvelet transform provides near-ideal sparsity of representation for both smooth objects and objects with edges. It is clear that the proposed method performs well.

Acknowledgment. This work was supported in part by the University Grants Commission, New Delhi, India under Grant No. F.No.36-246/2008(SR).

References

1. Hassin, A.H., Tang, X.L., Liu, J.F., Zhao, W.: Printed Arabic character recognition using HMM. Journal of Computer Science and Technology 19(4), 538–543 (2004)
2. Bansal, V., Sinha, R.M.K.: A complete OCR for printed Hindi text in Devanagari script. In: 6th Int. Conf. Document Analysis and Recognition (ICDAR 2001), pp. 800–804 (2001)
3. Hanmandlu, M., Murali, M.K.R., Kumar, H.: Neural based handwritten character recognition. In: 5th Int. Conf. Document Analysis and Recognition, pp. 241–244 (1999)
4. Chaudhuri, Pal, U., Mitra, M.: Automatic recognition of printed Oriya script. In: 6th Int. Conf. Document Analysis and Recognition (ICDAR 2001), pp. 795–799 (2001)
5. Mohanty, S., Dasbebartta, H.N., Behera, T.K.: An Efficient Bilingual Optical Character Recognition (English-Oriya) System for Printed Documents. In: Seventh International Conference on Advances in Pattern Recognition, ICAPR 2009, pp. 398–401 (2009)
6. Candès, E.J., Donoho, D.L.: Curvelets-a surprisingly effective nonadaptive representation for objects with edges. In: Schumaker, L.L., et al. (eds.) Curves and Surfaces. Vanderbilt University Press, Nashville (1999)
7. Binh, N.T., Khare, A.: Multilevel Threshold based Image Denoising in Curvelet Domain. Journal of Computer Science and Technology 25(3), 633–641 (2010)
8. Nigam, S., Khare, A.: Curvelet Transform based Object Tracking. In: IEEE International Conference on Computer and Communication Technology, pp. 230–235 (2010)
9. Majumdar, A.: Bangla Basic Character Recognition Using Digital Curvelet Transform. Journal of Pattern Recognition Research, 17–26 (2007)
10. Mohammad, F., Husain, S.A.: Character Recognition Using Mathematical Morphology. In: International Conference on Electrical Engineering, pp. 1–5 (2007)

Exploiting Ontology for Concept Based Information Retrieval

Aditi Sharan, Manju Lata Joshi, and Anupama Pandey

School of Computer & Systems Sciences, Jawaharlal Nehru University
New Delhi-110067 India
aditisharan@gmail.com,
manjulatajoshi@gmail.com,
pandey.anupama008@gmail.com

Abstract. Traditional approaches for information retrieval from textual documents are based on keyword based similarity. A key limitation of these approaches is that they do not take care of meaning and semantic relationship between words. Recently some work has been done on concept based information retrieval(CBIR), which allows to capture semantic relations between words in order to identify importance of a word. These semantic relations can be explored by using ontology. Most of the work for CBIR has been done in English language. In this paper we explore the use of Hindi Wordnet ontology for CBIR from Hindi text documents. Our work is significant because very limited amount of work has been done on CBIR for Hindi documents. Basic motivation of this paper is to provide an efficient structure for representing concept clusters and develop an algorithm for identifying concept clusters. Further we suggest a way of assigning weights to words based on their semantic importance in the document.

Keywords: Information retrieval, semantic relations, concept cluster, vector space model, Hindi Wordnet ontology, Concept based information retrieval.

1 Introduction

Information retrieval is devoted to finding relevant documents from a collection of documents in response to a query provided by the user [12]. Vector Space Model(VSM) is the most popular method for information retrieval. In traditional VSM a document is represented by providing weights to the keywords, mostly based on tf-idf measure. Similarity between query and document depends on terms, which are common to both query and document. Some of the popularly used similarity measures in VSM are: Cosine measure, Jaccard's coefficient, dice coefficient etc. [12, 13].

Considering subjectivity in NLP (Natural language processing), it is well known fact that the keywords (important terms present in the document) are not sufficient to provide information about content of the document. Therefore, we say that **VSM does not take into account the content of a document. The content of a document can be extracted by finding semantically important terms** [2, 3, 7]. This can be explained by an example. Let us consider a document D_1 with following content:

C. Singh et al. (Eds.): ICISIL 2011, CCIS 139, pp. 157–164, 2011.

D₁: भारतीय समाज में नारी के स्थान की बात हो तो दुर्गा, शक्ति से ले कर विज्ञान, तकनीकी, जैसे क्षेत्रों में काम करने वाली स्त्रियों, इंदिरा गाँधी जैसी नेताओं के उदाहरण दे कर हम अपना गर्व व्यक्त करते हैं. हर क्षेत्र में नारियाँ पुरुषों से कदम से कदम मिला कर बढ रही हैं।

Reading document D1, one can make out that नारी,   दुर्गा and स्त्री are more important than other terms such as पुरुष, विज्ञान and तकनीकी. However using VSM with traditional TF measure weight of the term 'स्त्री is 1, which is same as that of terms पुरुष, विज्ञान, and तकनीकी. Therefore, 'स्त्री' and पुरुष, विज्ञान, and तकनीकी will have equal importance. This leads to the need of Concept Based Information Retrieval (CBIR), which captures semantic relation between terms in order to identify importance of a term. There can be several approaches for developing concept based information retrieval systems. Most of these approaches use ontology for extracting concepts from documents. Various attempts have been made to capture semantic similarities by exploiting linguistic ontology such as WordNet [4, 6, 10].

Ontology can be used to build up a lexical chain. Two words are in a lexical chain if they are related by a relation [1, 5]. There are many relations among words such as - identity, synonymy, hypernymy-hyponymy and meronymy-holonymy. These relations link related terms in a document to represent the lexical cohesion structure of the document. Thus presence of a lexical chain identifies semantically important terms. Now if we again consider the earlier example (document D₁) we observe that नारी, दुर्गा and स्त्री form a lexical chain. Therefore, these words can be identified as semantically important terms.

We suggest use of ontology for representing a document as a collection of concept clusters. When a document is associated with multiple concepts, there will be multiple clusters and each cluster may represent single concept.

2   Ontology Based Model for Concept Based Information Retrieval

Kang et. al [3] have provided a method for exploiting ontology for CBIR. However, we found that this method has certain limitations. We are suggesting a modified method which is more efficient and overcomes these limitations. In this section we present a brief review of the model provided by Kang et. al [3] followed by presenting the modified model. Using Kang's approach a document can be represented as a collection of concept clusters. A concept cluster is a weighted lexical chain that represents one aspect of the meaning of a document and expresses the degree of relatedness among the semantic terms within a document. Concept clusters have been defined in several ways (refer to Kang et.al [3] for details).

Now again consider document D₁. Figure1(a) shows the relationship between some of the nouns (shown as nodes) extracted from document D₁ and links (shown as edges) denoting relations between terms:

If we observe representation of concept clusters (figure 1), we find some ambiguities in the representation:-

1 Firstly, there can be different possible representations for the same concept cluster. For instance for the example discussed in D₁, two possible representations are shown in Figure 1(a) and 1(b)

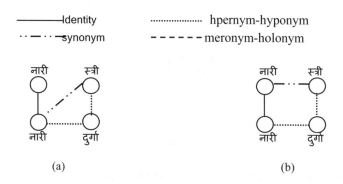

Fig. 1. A representation of concept cluster for D_1

This problem arises because there are multiple (two) nodes representing नारी. Therefore the nodes attached to the node नारी can be linked to any one of the above nodes.

2 This problem is a consequent of first problem, where we cannot develop a well defined algorithm for constructing concept cluster due to ambiguity in structure itself.

3 There is a redundancy in the resulting concept clusters as same node can appear multiple times.

A   Modified Representation for Concept Clusters

Keeping in view the above discussed problems in the generation of concept clusters, we are suggesting a modified structure, which captures the same information, but avoids ambiguities. In this model each term has a unique representation with one attribute giving frequency of the term. This frequency actually captures identity relationship among the terms. According to our model a concept cluster can be defined as follows:

Definition. Let $T=\{(t_1,f_1),(t_2,f_2),\ldots,(t_m,f_m)\}$ be the set of terms and their frequency in a concept cluster, where t_i is term and f_i is frequency of ith term. Let $R = \{$identity, synonyms, hypernym-hyponym, meronym-holonym $\}$ be the set of lexical relations. Let $M(r_k, t_g)$ be the sum of frequency of all the terms linked to term $t_g \in T$ through relation $r_k \in R$ and let $W(r_k)$ be the weight of relation r_k. Then the score $S_{Term}(t_g)$ of term t_g in a concept cluster is defined as:

$$S_{Term}(t_g) = (f_g - 1) * W(r_1) + f_g * \sum_{k=2}^{4} M(r_k, t_g) * W(r_k) \quad 1 \le g \le m \quad (1)$$

Where r_1, r_2, r_3, r_4 are identity, synonym, hypernym -hyponym and meronym-holonym relations respectively. Concept cluster in Figure 1(a) can now be constructed as:

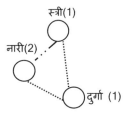

Fig. 2. Modified representation of figure 1

Based on above structure we suggest an algorithm for developing concept clusters.

3   Algorithm for Identifying Concept Clusters

Before giving the algorithm for identifying concept clusters we start with defining concept clusters in terms of a graph. Each document can be represented as graph where nodes represent terms and edges represent relations between the terms. Each concept clusters then represents a connected component of the graph. (Refer to Figure 1 and 2).

Now we present simplified algorithm for concept based IR.

1   Initialize weights $W (r_k)$ of semantic relations r_k.
2   For a document D find T, where T represents the set of all nouns in document D (use Part of Speech(POS) tagger [9]).
3   Repeat

    3(a)   For document D identify a concept cluster represented as a graph C(V, E)
        3.1   Start with a term $x \in$ T and add x as vertex V to cluster C.
        3.2   Assign f_x as frequency of x in D
        3.3   Assign T = T - x
        3.3   For each term $y \in$ T related to any $x \in$ V through any relation r_k

      (a)   Assign f_y as frequency of y in D
      (b)   Use y to extend the graph C by adding y to V as vertex and relation r_k as edge between x and y.
      (c)   Assign T = T-y.

  Until T=φ

4   Find weight of each term x using concept cluster

      4.1   Find the concept cluster to which term x belongs.
      4.2   Repeat following steps within concept cluster.

      4.2.1   For node x, identify the relations using the edges connected to x and calculate weight of x using formula given in equation (1).

5   Represent document by the concept clusters identified along with weights assigned by concept based model.

Now we present a small example that explains working of algorithm and highlights importance of CBIR in comparison to traditional VSM. In this example we consider 5 documents as shown in appendix 1 and a query "ग्रह". Assigning similarity manually (considering actual relevance), we can say that for query "ग्रह", the ranked relevant documents are: Doc 1, Doc 2, Doc 4, Doc 3 and Doc 5.

From our algorithm (step 3) concept clusters identified for document Doc 1 are:

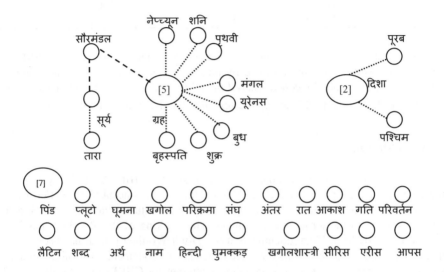

Fig. 3. Concept clusters for Doc 1

Inside the oval, the numeral shows the frequency of the term in document.

Table 1 shows weight assigned to query using tf-idf technique and Table 2 shows semantic weight assigned to same query by our algorithm (step 4).

After constructing concept clusters we can calculate weight of each term using equation (1). Accordingly weight of term 'ग्रह' is equal to ((4*0.7) (5*8*.3) (0.1*5)) = 15.3.

Table 1. Weights assigned to terms using TF-IDF technique

Doc/Term	"ग्रह"
Doc 1	0.4845
Doc 2	0.2907
Doc 3	0.1938
Doc 4	0.6783
Doc 5	0.0

Table 2. Weights assigned to terms using concept based model

Doc/Term	"ग्रह "
Doc 1	15.3
Doc 2	7.4
Doc 3	0.7
Doc 4	1.8
Doc 5	0.0

Table 3. Ranked relevant documents retrieved by VSM, concept based technique and the expected results

Query/Techniques	VSM	CBIR	Expected Results
	Doc 4	Doc 1	Doc 1
	Doc 1	Doc 2	Doc 2
ग्रह	Doc 2	Doc 4	Doc 4
	Doc 3	Doc 3	Doc 3
	Doc 5	Doc 5	Doc 5

Table 3 shows the ranked result of relevant documents retrieved by VSM and CBIR. The result shows that returned by concept based model is more accurate. In this case result obtained by concept based model is same as actual result. This is quite obvious, as this model is able to capture semantic similarity between the terms.

4 Conclusion

In this paper, we have suggested the use of ontologies for CBIR. Specifically, we have presented a way to represent a document as a collection of concept clusters. Further we have developed an algorithm for finding concept based clusters and for assigning semantic weights to the terms in a document. The working of the algorithm has been shown for CBIR from Hindi documents using Hindi Wordnet [8]. It has been shown that concept based model can increase the efficiency of document retrieval in comparison to traditional VSM. This is a preliminary attempt for CBIR using Wordnet ontology in Hindi, which is definitely better than traditional VSM. In future the results need to be further investigated on larger data set to see the extent of improvement. Further our next attempt is to give a more computationally efficient algorithm.

References

1. Ercan, G., Cickeli, I.: Using Lexical Chains for Keyword extraction. Information Processing and Management 43, 1705–1714 (2007)
2. Kang, B., Kim, V., Lee, S.: Exploiting Concept Clusters for Content-based Information Retrieval. Information Sciences 179(2-4), 443–462 (2005)
3. Kang, B., Kim, D., Kim, H.: Fuzzy Information Retrieval Indexed by Concept Identification. In: Matoušek, V., Mautner, P., Pavelka, T. (eds.) TSD 2005. LNCS (LNAI), vol. 3658, pp. 179–186. Springer, Heidelberg (2005)
4. Hirst, G., St-Onge, D.: Lexical Chains as Representations of Context for the Detection and Correction of Malapropisms, WordNet: An electronic Lexical Database. The MIT Press, Cambridge (1998)
5. Galley, M., McKeown, K.: Improving word sense disambiguation in lexical chaining. In: Proceedings of the 18th International Joint Conference on Artificial Intelligence, pp. 1486–1488 (2003)
6. Stokes, N., Carthy, J.: Combining Semantic and Syntactic Document Classifiers to Improve First Story Detection. In: Proceedings of the 24th ACM SIGIR Conference, New Orleans, pp. 424–425 (2001)

7. Sharan, A., Joshi, M.L.: An insight into semantic similarity aspects using Wordnet. International Journal of Information Communication and Technology 2(4), 331–341 (2010)
8. Hindi WordNet, http://www.cfilt.iitb.ac.in/wordnet/webhwn/
9. Society for Natural Language Technology Research, http://nltr.org/snltr-software/
10. Budanitsky, A.: Lexical Semantic Relatedness and its Application in Natural Language Processing, Technical Report CSRG-390, University of Toronto (1999)
11. Parsons, K., Mc Cormac, A., Butavicius, M., Dennins, S., Ferguson, L.: The Use of Context-Based Information Retrieval Technique. In: Command control, Communications and Intelligence Division DSTO Defence Science & Technology Organisation, Australia (2009)
12. Grossman, David, Frieder .: Information Retrieval: Algorithm and Heuristic. Kulwer Academic Press, London (2004)
13. Meadow, C.T., Boyce, B.R., Kraft, D.H., Barry, C.L.: Text Information Retrieval Systems. Academic Press, London (2007)

Appendix

Document 1

सूर्य या किसी अन्य तारे के चारों ओर परिक्रमा करने वाले खगोल पिण्डों को ग्रह कहते हैं। अंतर्राष्ट्रीय खगोल संघ के अनुसार हमारे सौरमण्डल में आठ ग्रह हैं - बुध, शुक्र, पृथ्वी, मंगल, बृहस्पति, शनि, युरेनस और नेप्चून. इनके अतिरिक्त तीन बौने ग्रह और हैं - सीरिस, प्लूटो और एरीस। प्राचीन खगोलशास्त्रियों ने तारों और ग्रहों के बीच में अन्तर इस तरह किया- रात में आकाश में चमकने वाले अधिकतर पिण्ड हमेशा पूरब की दिशा से उठते हैं, एक निश्चित गति प्राप्त करते हैं और पश्चिम की दिशा में अस्त होते हैं। इन पिण्डों का आपस में एक दूसरे के सापेक्ष भी कोई परिवर्तन नहीं होता है। इन पिण्डों को तारा कहा गया। पर कुछ ऐसे भी पिण्ड हैं जो बाकी पिण्डों के सापेक्ष में कभी आगे जाते थे और कभी पीछे - यानी कि वे घुमक्कड़ थे। Planet एक लैटिन का शब्द है, जिसका अर्थ होता है इधर-उधर घूमने वाला। इसलिये इन पिण्डों का नाम Planet और हिन्दी में ग्रह रख दिया गया।

Document 2

मंगल, सौरमंडल में सूर्य से चौथा ग्रह है। पृथ्वी से देखने पर, इसको इसकी रक्तिम आभा के कारण " लाल ग्रह " के रूप मे भी जाना जाता है। मंगल ग्रह का धरातल स्थलीय और के वातावरण विरल है। इसकी सतह देखने पर चंद्रमा के गर्त और पृथ्वी के ज्वालामुखियों, घाटियों, रेगिस्तान और ध्रुवीय बर्फीली चोटियों की याद दिलाती है। यह स्थान है ओलम्पस मॉस का जो हमारे सौरमंडल का सबसे अधिक ऊँचा पर्वत है साथ ही विशालतम कैन्यन वैलेस मैरीनेरिस भी यहीं पर स्थित है। अपनी भौगोलिक विशेषताओं के अलावा, मंगल का घूर्णन काल और मौसमी चक्र पृथ्वी के समान हैं।

Document 3

ग्रह पीड़ा निवारक टोटके-
सूर्य
१. सूर्य ग्रह को बली बनाने के लिए व्यक्ति को प्रातःकाल सूर्योदय के समय उठकर लाल पुष्प वाले पौधों एवं वृक्षों को जल से सींचना चाहिए।
२. रात्रि में ताँबे के पात्र में जल भरकर सिरहाने रख दें तथा दूसरे दिन प्रातःकाल उसे पीना चाहिए।
३. ताँबे का कड़ा दाहिने हाथ में धारण किया जा सकता है।
४. लाल गाय को रविवार के दिन दोपहर के समय दोनों हाथों में गेहूँ भरकर खिलाने चाहिए। गेहूँ को जमीन पर नहीं डालना चाहिए।

Document 4

कुछ दिनों पहले धरती जैसा एक ग्रह मिलने की खबर से वैज्ञानिक जगत का रोमांच कम भी न हुआ था कि अचानक पता चला है कि उन दूर अंधेरों में तो कुछ और ग्रह भटक रहे हैं जिनसे सृष्टि की उत्पत्ति के रहस्य को समझने में बड़ी मदद मिलेगी. अमेरिकी स्पेस एजेंसी नासा की महादूरबीन केपलर ने पांच नए ग्रहों का पता लगाया है. नए तलाशे गए ग्रहों के ज़खीरे में ये पांच ग्रह हैं और बहुत गरम हैं. यानी जीवन की कोई संभावना नहीं लगती. लेकिन एक संभावना इस बात की है कि ये ग्रह उन अध्ययनों में बड़ी कारगर भूमिका निभा सकते हैं जो इस बात को समझने के लिए दशकों से जारी हैं कि ब्रह्मांड कैसे बना, सृष्टि कैसे अस्तित्व में आई और ग्रहों नक्षत्रों का निर्माण किस तरह गैस और धूल से हुआ

Document 5

इस वर्ष की थीम का पर्याय "अनेकता में एकता" वाला है, समन्वय, पारस्परिक सदभाव, प्रेम व सहिष्णुता की बात है, जो खुद को औरों को अमन-चैन से जीने का मौका देने का सन्देश है!- क्योंकि वास्तविक स्वरूप में हम सब एक धरती पर और एक जैसे तत्वों की सरंचना मात्र ही तो हैं। और एक ही पर्यावरण का हिस्सा भी. हम शक्ल व सूरत में जुदा-जुदा होने के बावजूद भी एक धरती के बासिन्दें है और हम सब का भविष्य भी एक है।

Parsing of Kumauni Language Sentences after Modifying Earley's Algorithm

Rakesh Pandey[1], Nihar Ranjan Pande[1], and H.S. Dhami[2]

[1] Amrapali Institute of Technology and Sciences, Haldwani
Uttarakhand, India, 263139
[2] Department of Mathematics
Kumaun University, SSJ Campus, Almora Uttrakhand, India, 263601
rakeshpandeyaits@gmail.com

Abstract. Kumauni language is one of the relatively understudied regional languages of India. Here, we have attempted to develop a parsing tool for use in Kumauni language studies, with the eventual aim of developing a technique for checking grammatical structures of sentences in Kumauni language. For this purpose, we have taken a set of pre-existing Kumauni sentences and derived rules of grammar from them, which have been converted to a mathematical model using Earley's algorithm, suitably modified by us. The Mathematical model so developed has been verified by testing it on a separate set of pre-existing Kumauni language sentences. This mathematical model can be used for parsing new Kumauni language sentences, thus providing researchers a new parsing tool.

Keywords: Kumauni language, Context Free Grammar, Earley's Algorithm, Natural Language Processing, Parsing.

1 Introduction

The first stage in parsing is *Token Generation* or lexical analysis, by which the input character stream is split into meaningful symbols defined by a grammar of regular expression. The next stage is *Parsing* or syntactic analysis, which involves checking that the tokens form an allowable expression. This is usually done with reference to a Context Free Grammar (CFG) that recursively defines components, which can make up an expression and the order in which they must appear. The final phase is *Semantic Parsing* or analysis, which requires working out the implications of the expression just validated and taking the appropriate action. In the case of a calculator or interpreter, the action is to evaluate the expression or program; a compiler, on the other hand, generates some kind of code. Attribute grammars can also be used to define these actions. Brian Roark (2001) presents a lexicalized probabilistic top-down parser, which performs very well both in terms of accuracy of returned parses and in terms of the efficiency with which they are found, relative to the best broad-coverage statistical parsers.

C. Singh et al. (Eds.): ICISIL 2011, CCIS 139, pp. 165–173, 2011.
© Springer-Verlag Berlin Heidelberg 2011

Top-down backtracking language processors have some advantages compared to other methods, i.e.

1) They are general and can be used to implement ambiguous grammars.
2) They are easy to implement in any language that supports recursion.
3) They are highly modular, i.e. the structure of the code is closely related to the structure of the grammar of the language to be processed.
4) Associating semantic rules with the recursive functions that implement the syntactic productions rules of the grammar is straightforward in functional programming,.

Languages that cannot be described by CFG are called Context Sensitive Languages. Tanaka (1993) has developed an algorithm for CFG. An informal description of a new top-down parsing algorithm has been developed by Richard A. Frost et al (2006) that accommodates ambiguity and left recursion in polynomial time. Shiel (1976) noticed the relationship between top-down and chart parsing and developed an approach in which procedures corresponding to non-terminals are called with an extra parameter, indicating how many terminals they should read from the input. Fujisaki Tetsunosuke (1984) has tested a corpus to parse it using Stochastic Context Free Grammar and probability theory to make the parse tree. R. Frost et al (2007) presented a method by which parsers can be built as modular and efficient executable specifications of ambiguous grammars containing unconstrained left recursion. In 2008, the same authors, R. Frost et al (2008), described a parser combinator as a tool that can be used to execute specifications of ambiguous grammar with constraints left recursion, which execute polynomial time and which generate compact polynomial sized representation of the potentiality.

Devdatta Sharma (1985), a leading linguist, was the first to study Kumauni language linguistically. Carrying forward his initiative, we have taken Kumauni language for information processing, i.e. to check the grammars of input sentences. Parsing process makes use of two components; a parser, which is a procedural component and a grammar, which is declarative. The grammar changes depending on the language to be parsed while the parser remains unchanged. Thus, by simply changing the grammar, a system would parse a different language. We have taken Earley's Parsing Algorithm for parsing Kumauni sentences according to a grammar that we have defined for Kumauni language, using a set of pre-existing Kumauni sentences.

2   Earley's Parsing Algorithm

The task of the parser is essentially to determine if and how the grammar of a pre-existing sentence can be determined. This can be done in two ways, Top-down Parsing and Bottom-up parsing.

Earley's algorithm is a top-down dynamic programming algorithm. We use Earley's dot notation: given a production $X \rightarrow xy$, the notation $X \rightarrow x \bullet y$ represents a condition in which x has already been parsed and y is expected.

The state set at input position k is called S(k). The parser is seeded with S(0), consisting of only the top-level rule. The parser then iteratively operates in three stages; prediction, scanning, and completion.

- Prediction: For every state in S (k) of the form (X → x • Y y, j) (where j is the origin position as above), add (Y → • z, k) to S (k) for every production in the grammar with Y on the left-hand side (Y → z).
- Scanning: If a is the next symbol in the input stream, for every state in S(k) of the form (X → x • a y, j), add (X → x a • y, j) to S(k+1).
- Completion: For every state in S(k) of the form (X → z •, j), find states in S(j) of the form (Y → x • X y, i) and add (Y → x X • y, i) to S(k)

For example, let us take the input sentence, "You eat the food in the restaurant". The following numeric key can be supplied to the words of this sentence: "0 *You* 1 *eat* 2 *the* 3 *food* 4 *in* 5 the 6 *restaurant* 7", where the numbers appearing between words are called position numbers. For CFG rule S →NP VP, we will have three types of dotted items:

• [S→ .NP VP, 0, 0]   • [S→ NP.VP, 0, 1]   • [S→ NP VP., 0, 4]

Here,   S → Starting Symbol   NP → Noun Phrase   VP → Verb Phrase

i. The first item indicates that the input sentence shall be parsed applying the rule S → NP VP from position 0.
ii. The second item indicates the portion of the input sentence from the position number 0 to 1 that has been parsed as NP and the remainder left to be satisfied as VP.
iii. The third item indicates that the portion of input sentence from position number 0 to 4 has been parsed as NP VP and thus S is accomplished.

3   Derivation of Kumauni Language Grammar and Modification of Earley's Algorithm

In this section, we have attempted to develop a grammar of Kumauni language. We have taken some pre-existing Kumauni language sentences randomly and tried to derive rules of grammar from them, as it is next to impossible to collect all types of sentences of any language.

We see that a sentence can be written in different forms that have the same meaning, i.e. positions of tags are not fixed. The grammar rules derived here may not apply to all the sentences in Kumauni language, since we have not considered all types of sentences possible in Kumauni language. Some sentences that have been used to make the rules of grammar for Kumauni language are given below:

Let K be the set of all parts of speeches in Kumauni language K = (NP, PN, VP, ADV, ADJ, PP, ART, IND), where

NP → Noun   PN → Pronoun   VP → Verb   ADV → Adverb

ADJ → Adjective PP →   Preposition   ART → Article   IND → Indeclinable

Table 1.

In Kumauni	In English	Grammar
kAn Je re?	Where are you going?	PP –VP
Sab thEk Chan	They all are fine	NP- ADJ- VP
Ook byAh pichal sAl haigou	he got married last year	PN - ADJ- NP - VP
theek cha pein ItvAr din milOn	Well, see you on Sunday.	ADVP- PP – NP - VP
mein itvAr din Onake koshish karou	I will try to come on Sunday	PN- NP- ADV- VP
main pushp vihar sAketak paas roo(n)chou	I live in Pushp Vihar near Saket	PN- NP- PP- VP
Par jAno pein	Good Night	VP
myAr bAbu fauj me naukari kareni	My father is serving in Indian army	NP- NP- PP- VP
jaduk AshA, utuk haber jyAdA hainch	It is more than expected	NP- ADJ- VP
champAwat bahute bhal jAg chuu	Champawat is a very beautiful place	NP- ADJ-VP
makai wanki ligi bahut door chaln pado	I have to go far for that place	NP- PP- ADJ- VP
ter much to nai buwAr jas chamakano	Your face is shining like a new bride.	NP - PP- VP
ab mee jaa	Now I am going	ADVP- PN- VP

Formation of vector space for a language

Using English language – since it has 8 parts of speeches – we can form a matrix (called connection matrix) of the order 8 x 8, where rows and columns are represented by parts of speeches. This matrix pertains to the FOLLOW relation.

> PREV (x) = {Set of all lexical categories that can precede x in a sentence}
>   = {y: (Row y, Column x) is 1}
> FOLLOW (x) = {y: (Row x, Column y) is 1}

For example, upon parsing the sentence *"John is looking very smart"* in parts of speech, it becomes "NP VP ADV NP". Its connection matrix representation is depicted as:

	NP	PN	VP	ADV	ADJ	PP	ART	IND
NP	0	0	1	0	0	0	0	0
PN	0	0	0	0	0	0	0	0
VP	0	0	0	2	0	0	0	0
ADV	3	0	0	0	0	0	0	0
ADJ	0	0	0	0	0	0	0	0
PP	0	0	0	0	0	0	0	0
ART	0	0	0	0	0	0	0	0
IND	0	0	0	0	0	0	0	0

Using a text document, we get several sentences and each sentence can be represented by a connection matrix of the order 8 x 8. Thus, a set of all matrices of the order 8x8 forms a vector space V of dimension 64 over the field of integers under addition and usual multiplication. Therefore, in a text document each sentence is an element of this vector space. As each sentence has several parts of speech, it can be a subspace of the vector space generated for language. Similarly, parts of a sentence will also be a subspace of the sentence.

To carry this argument further, we propose some linear transformations of subspaces of Kumauni language sentence. In the following sequence:

- T is linear transformation of Sentence subspace in Kumauni.
- U is linear transformation of the Proposition-phase subspace in Kumauni.
- W is the linear transformation of the Noun-phase subspace in Kumauni.

Additionally, Identity transformation has also been used.

Table 2.

T: (S)	U: (PP)	W: (NP)
T_1: (S)= PP VP	U_1: (PP)= PN NP	W_1: (NP)= NP PP
T_2: (S) = PP	U_2: (PP)= NP PN	W_2: (NP)= PP NP
	U_3: (PP)= ADJ NP	W_3: (NP)= ADV NP
	U_4: (PP)= NP ADJ	W_4: (NP)= PP
	U_5: (PP)= NP	W_5: (NP)= ART NP
	U_6: (PP)= ADJ	W_6: (NP)= NP ART
	U_7: (PP)= IND NP	W_7: (NP)= IND PN
	U_8: (PP)= PN	W_8: (NP)= PN IND
	U_9: (PP)= ADV NP	W_9: (NP)= VP
	U_{10}: (PP)= ADV	

4   Modification of Earley's Algorithm for Kumauni Text Parsing

We know that Earley's algorithm uses three operations, Predictor, Scanner and Completer. We add Predictor and Completer in one phase and Scanner operation in another phase.

Let x, y, z, PP, VP are sequence of terminal or non-terminal symbols and S, B are non terminal symbols:

Phase 1: (Predictor + Completer)
For an item of the form [S →x .By, i, j], create [S →x.zy, i, j] for each production of the [B→z]. Mathematically in phase 1 we apply the transformations suggested earlier.

Phase 2: (Scanner)
For an item of the form [S→x.wy, i, j] create [S→xw.y, i, j+1], if w is a terminal symbol appeared in the input sentence between j and j+1. When the transformation is successfully applied, it allows us to move in to next position or transformation.

Our Algorithm

Input: Tagged Kumauni Sentence
Output: Parse Tree or Error message
Step 1: If Verb is present in the sentence then $[T: S \rightarrow .PP\ VP, 0, 0]$ then we use trans-
formation T_1. Else $[T: S \rightarrow .PP, 0, 0]$ then we use transformation T_2.
Step 2: Use the transformation U and W and do the following steps in a loop until
there is a success or error
Step 3: For each item of the form of $[S \rightarrow x.By, i, j]$, and we use transformations T_i, U_i,
W_i.
Step 4: For each item of the form of $[S \rightarrow .xwy, i, j]$, apply phase 2
Step 5: If we find an item of the form $[S \rightarrow x., 0, n]$ $i.e$ the transformations work successfully, then we accept the sentence as success else error message, where n is the
length of input sentence. And then come out from the loop.
Step 6: Generate the parse trees for the successful sentences according to the used
transformations. A transformation is said to be successful if it same as any member
of table 1.

5   Parsing Kumauni, Using Proposed Grammar and Algorithm

Let us take a Kumauni sentence, मी त्यार दगड़ बजार जू (Mee tyar dagad bazAr
joo). In English it means, "I will go to the market with you".

Here, the position number for words is assigned based on the sequence of their
parsing.

0 *Mee* 1 *tyar* 2 *dagad* 3 *bazaar* 4 *joo* 5
In our sentence: 1. PN → "mee"   2. PN → "tyar"   3. PP → "dagad"
4. NP→"bazaar"   5. VP → "joo"

Now we use the transformation defined earlier (Table- 2).
The parsing process will proceed as follows:

Table 3.

Sr. No	Rule	Phase applied
1	[S → .PP VP , 0, 0] by T1	Apply Phase 1
2	[S → .NP VP, 0 , 0] by U5	Apply Phase 1
3	[S → .PP NP VP, 0, 0] by W2	Apply Phase 1
4	[S → .PN NP NP VP, 0, 0] by U1	Apply Phase 1
5	[S →. "mee" NP NP VP, 0, 0]	Apply Phase 2
6	[S →. "mee" .NP NP VP, 0, 1] by identity transformation	Apply Phase 1
7	[S → "mee" .PP NP VP, 0, 1] by W4	
8	[S → "mee" PN NP NP VP, 0, 1] by U1	Apply Phase 1
9	[S → "mee". "tyar" NP NP VP, 0, 1]	Apply Phase 2
10	[S → "mee". "tyar" .NP NP VP, 0, 1] by identity transformation	Apply Phase 1

Table 3. *(continued)*

11	[S → "mai" "tyar" .PP NP VP, 0, 2] by W4	Apply Phase 1
12	[S → "mee" "tyar"."dagad" NP VP, 0, 2]	Apply Phase 2
13	[S → "mee" "tyar" "dagad" .NP VP, 0, 3] by identity transformation	Apply Phase 1
14	[S → "mee" "tyar" "dagad".."bazaar" VP, 0, 3]	Apply Phase 2
15	[S → "mee" "tyar" "dagad".."bazaar" .VP, 0, 4] by identity transformation	Apply Phase 1
16	[S → "mee" "tyar" "dagad".."bazaar" . "joo", 0, 4]	Apply Phase 2
17	[S → "mee" "tyar" "dagad".."bazaar" . "joo", 0, 5]	Complete

In the above example, we have shown only the steps which lead to the goal. The other steps are ignored.

6 Stages of The Model

In the model there are 3 stages:

• Lexical Analysis   •   Syntax Analysis   • Tree Generation

In the Lexical Analysis stage, the programme finds the correct tag for each word in the sentence by searching the database. There are seven databases (NP, PN, VP, ADJ, ADV, PP, ART, IND) for tagging the words.

In the Syntax Analysis stage, the program tries to analyze whether the given sentence is grammatically correct or not.

In the Tree Generation stage, the programme finds all the production rules that lead to success and generates parse tree for those rules. If there are more then one paths to success, this stage can generate more than one parse trees. It also displays the words of the sentences with proper tags. The following figure shows a parse tree generated by the model. The original parse tree for the above sentence is depicted in figure (1).

7 Verification of Program

After implementation of Earley's algorithm using our proposed grammar, it has been seen that the algorithm can easily generate parse tree for a sentence if the sentence structure satisfies the grammar rules. For example, we take the following Kumauni sentence, मेर नाम कमल छ (Mer nAma Kamal chh). The structure of the above sentence is NP-NP-VP. This is a correct sentence according to Kumauni literature. According to our proposed grammar, a possible top down derivation for the above sentence is:

```
1. S [Handle]
2. >>PP VP              [T₁: S→PP VP]
3. >> NP VP             [U: PP→NP]
4. >>NP PP VP           [W: NP→NP PP]
5. >>NP NP VP           [U: PP → NP]
```

6. >>mer nAma NP VP   [W: NP → mer nAma]
7. >>mer nAma kamalVP   [W: NP→ Kamal]
8. >>mer nAma kamal chh   [VP → chh]

From the above derivation it is clear that the sentence analysed by the model is correct according to the proposed grammar, thus proving that our parsing model generates a parse tree successfully. The actual programme shall be as follows and the figurative representation is shown in figure (2)-

Input sentence- Mer nAma Kamal chh.
Sentence recognized
Tree ---->

1. S   2. [S --->(PP VP)]   3. [PP --->(NP)] VP   4. [NP --->(NP PP)]NP
5. [NP ---> (np :Mer nAma)]PP VP   6. [PP]VP   7. [PP ---> (NP)]VP
8. [NP ---> (np : Kamal)]VP   9. VP   10. [VP ---> (vp :chh)]

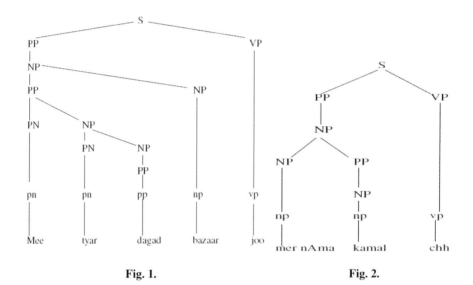

Fig. 1.   Fig. 2.

This model tests only the sentence structure according to the proposed grammar rules. So, if the sentence structure satisfies the grammar rules and follows Earley's algorithm, then the model will recognize the sentence as a correct sentence and generate a parse tree. Otherwise, it gives an error output.

8 Conclusion and Future Work

The project explained above has been conducted using everyday sentences, in order to make the result applicable widely. We have developed a context free grammar (CFG) for simple Kumauni sentences, studied the issues that arise in parsing Kumauni sentences and produced an algorithm suitable for those issues. This algorithm is a modification of Earley's Algorithm, which has proved to be simple and effective. In the traditional Earley's algorithm there are many more steps in parsing than in our model. Thus, our model reduces the length of parsing. It has an added feature in the sense that whereas Earley's algorithm contains three stages, our model works only in two steps. In this work, we have considered a limited number of Kumauni sentences for deriving the grammar rules. We have also considered only the seven main tags. In future work(s) related to the field of study covered in this paper, an attempt can be made to consider many more Kumauni sentences and more tags, for developing a more comprehensive set of grammar rules.

References

1. Brian, R.: Probabilistic top-down parsing and language modeling. Computational Linguistics 27(2), 249–276 (2001)
2. Frost, R., Hafiz, R., Callaghan, P.: Modular and Efficient Top-Down Parsing for Ambiguous Left-Recursive Grammars. In: 10th International Workshop on Parsing Technologies (IWPT), ACL-SIGPARSE, Prague, pp. 109–120 (June 2007)
3. Frost, R., Hafiz, R., Callaghan, P.: Parser Combinators for Ambiguous Left-Recursive Grammars. In: Hudak, P., Warren, D.S. (eds.) PADL 2008. LNCS, vol. 4902, pp. 167–181. Springer, Heidelberg (2008)
4. Fujisaki, T.: A Stochastic approach to sentence parsing. In: Annual Meeting of the ACL Proceedings of the 10th International Conference on Computational Linguistics and 22nd Annual Meeting on Association for Computational Linguistics, pp. 16–19 (1984)
5. Jay, E.: An efficient context free parsing algorithm. Communications of the ACM 13(2) (February 1970)
6. Richard, A., Frost, R., Hafiz: A New Top-Down Parsing algorithm to Accommodate Ambiguity and Left Recursion in Polynomial Time. ACM SIGPLAN Notices 41(5), 45–54 (2006)
7. Shiel, B.A.: Observations on context-free parsing. Technical Report TR 12–76, Center for Research in Computing Technology, Aiken Computational Laboratory, Harvard University (1976)
8. Sarma, D.: The formation of Kumauni language. SILL: series in Indian languages and linguistics. Bahri Publications (1985)
9. Tanaka, H.: Current trends on parsing - a survey, TITCS TR93-00031 (1993),
   http://www.citeseerx.ist.psu.edu/viewdoc/
   summary?doi=10.1.1.56.83

Comparative Analysis of Gabor and Discriminating Feature Extraction Techniques for Script Identification

Rajneesh Rani[1], Renu Dhir[1], and G.S. Lehal[2]

[1] Department of CSE, NIT Jalandhar,
Punjab, India
ranir@nitj.ac.in, dhirr@nitj.ac.in
[2] Department of CSE, Punjabi University, Patiala,
Punjab, India
gslehal@gmail.com

Abstract. A considerable amount of success has been achieved in developing monolingual OCR systems for Indian Scripts. But in a country like India, where many languages and scripts exist, it is more common that a single document contain words from more than one script. Therefore a script identification system is required to select the appropriate OCR. This paper presents a comparative analysis of two different feature extraction techniques for script identification of each word. In this work, for script identification discriminating and Gabor filter based features are computed of Punjabi words and English numerals. Extracted feature are simulated with Knn and SVM classifiers to identify the script and then recognition rates are compared. It has been observed that by selecting the appropriate value of k and appropriate kernel function with appropriate combination of feature extraction and classification scheme, there is significant drop in error rate.

Keywords: Script Identification, Gabor Features, Discriminating Features, Support Vector Machines, Knn.

1   Introduction

For a multilingual country like India where the documents contain more than one language, to develop an OCR is a great challenge. Mostly, two different kinds of techniques can be used to develop this type of system. One technique is combined database approach [1].That is the database of reference characters has alphabets from all of its languages in which the document is printed. So database is larger at the recognition level of individual character. The second technique is based on the identification of the script of each character before taking the characters for recognition. This helps in reduced search in the database at the cost of script recognition task. A number of techniques for determining the script of printed/handwritten documents can be typically classified into four categories [2, 3]: a) connected component based Script Identification b)Script Identification at text block level c) Script Identification at text line level d) Word level Script Identification.

C. Singh et al. (Eds.): ICISIL 2011, CCIS 139, pp. 174–179, 2011.

Feature Extraction is an important phase for script identification system of a word. Feature Extraction has been defined as "Extracting from the raw data the information which most relevant for classification purposes, in the sense of minimizing the within-class pattern variability while enhancing the between-class pattern variability" [4].There are a number of techniques available for feature extraction for script identi-fication [5-15]. Selection of a feature extraction technique is the single most important factor in achieving high performance of script identification systems. Gabor filters [10-12] can be used as a directional feature extractor. Other types of features are dis-criminating features [13-15] which means that every language can be identified based on its distinct visual appearance. These features can be extracted by using morpho-logical reconstruction of an image. This paper presents a comparison of these two methods for identification of Punjabi words and English numerals.

The paper is organized as follows. The theory of Gabor filters and feature extrac-tion using these is discussed in Section 2. Discriminating features of Punjabi words and English numerals have been described in Section 3. Section 4 deals with different classification techniques and finally Section 5 contains the experimental results and conclusion.

2   Gabor Filters

A Gabor Filter is a linear filter whose impulse response is defined by a harmonic func-tion multiplied by a Gaussian function.

$$h(x, y) = g(x, y)s(x, y) \tag{1}$$

Where $s(x, y)$ is a complex sinusoid, known as carrier and $g(x, y)$ is a Gaussian shaped function, known as envelope. Thus the 2-D Gabor filter can be written as

$$h_{x, y, \theta, f} = e^{-\frac{1}{2}\left(\frac{x'^2}{\sigma_x^2} + \frac{y'^2}{\sigma_y^2}\right)} . e^{j 2 \pi f x} \tag{2}$$

Where σ_x and σ_y explain the spatial spread and are the standard deviations of the Gaussian envelope along x and y directions. x' and y' are the x and y co-ordinates in the rotated rectangular co-ordinate system given as

$$x' = x\cos\theta + y\sin\theta \tag{3}$$

$$y' = y\cos\theta - x\sin\theta \tag{4}$$

Any combination of θ and f, involves two filters, one corresponding to sine function and other corresponding to cosine function in exponential term in Equation 2. The co-sine filter, also known as the real part of the filter function, is an even symmetric filter and acts like a low pass filter, while the sine part being odd-symmetric acts like a high pass filter.

In the present work, multi-bank Gabor filters having five different values for Spa-tial frequency (f = 0.0625, 0.125, 0.25, 0.5, 1.0) and six different values for orientation

$(\theta = 0°, 30°, 60°, 90°, 120°, 150°, 180°)$ are chosen to give a total of 70 Gabor filters with a combination of 35 even and 35 odd filters. From the output of each Gabor filter mean and standard deviation are computed, which serves as Gabor features. Thus for each word we get a feature vector of 140 values given by

$$F = [\mu_1, \sigma_1, \mu_1, \sigma_1, \mu_1, \sigma_1 .... \mu_{70}, \sigma_{70}]$$

3 Discriminating Features of Punjabi Words and English Numerals

Punjabi words and English numerals have a distinct visual appearance as shown in Fig. 1.

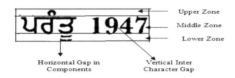

Fig. 1. Sample image of Punjabi Word and English Numeral showing different Zones

After a careful study of shapes of Punjabi words and English numerals, nine features for automatic classification of English numerals and Punjabi words' script are:

F1: Average Aspect ratio (AAR): The average aspect ratio (AAR) is defined as:

$$AAR = \frac{1}{N} \sum_{i=1}^{N} \frac{height(component_i)}{width(component_i)} \qquad (5)$$

Here N is the number of connected components of input word image.

F2: Average Eccentricity (AE): The average eccentricity (AE) is defined as

$$AE = \frac{1}{N} \sum_{i=1}^{N} \frac{len\_maj\_axis(component_i)}{len\_min\_axis(component_i)} \qquad (6)$$

Here N is the number of connected components of input word image.

F3-F6: Based on Stroke Density in a Direction (SD): Features F3, F4, F5 and F6 are based on stroke densities in vertical, horizontal, left diagonal and right diagonal directions. The stroke density in a direction is computed as:

$$SD = \frac{\sum_{i=1}^{N} no\_onpixels\_instroke_i}{size\_of\_word} \qquad (7)$$

Here N is the number of strokes in that direction.

To extract the stroke density in a direction, we have performed the morphological opening operation on the input binary word/numeral image with line structuring element having length=k × Mean (Connected_components_Height) and angle depending on the direction.

F7: Pixel Ratio after Filling Holes (PRFH): For fill holes, we choose the marker image, f_m to be 0 everywhere except on the image border, where it is set to 1-f. Here f is the original image.

$$f_m(x\,y) = \begin{cases} 1\text{-}f(x,y) \ if\, f(x,y) is\ on\ the\ border\ of\, f \\ otherwise\ \ 0. \end{cases}$$

(8)

PRFH is computed as:

$$PRFH = \frac{Sum\,of\_onpixels\_afterfillhole}{size\_of\_word}$$

(9)

F8: Vertical Inter Character Gap (VICG): To extract the value of this feature, vertical projection histogram is taken of the image. If   any vertical projection profile value is equal to zero then that means there is a gap between two characters and the value of this feature is set to1 otherwise is set to 0.

F9: Horizontal Break in Components (HBIC): To extract the value of this feature, horizontal projection histogram is taken of the image. If   any horizontal projection profile value is equal to zero then that means there is a gap between components of a word and the value of this feature is set to1 otherwise is set to 0.

4   Classification

The objective of classification is to identify the script of words taken form the test set. Features extracted from the words are sent to the Classifier.

KNN (k nearest neighbor) Classification

The k- nearest neighbor (k-nn) approach attempts to compute a classification function by examining the labeled training point sin n dimensional space. Then the Euclidean distance is calculated between the test point and all reference points q in order to find k nearest neighbors. A test sample is labeled with the same class label as the label of the majority of its K nearest neighbors. Nearest Neighbor is a special case of k-nn, where k=1.

SVM (Support Vector Machines) Classification

SVM is a kind of learning machine whose fundamental is statistics learning theory. For these, it finds the optimal hyper-plane which maximizes the distance, the margin, between the nearest examples of both classes, named support vectors (SVs). If the data is nonlinear, there arises the need of mapping the data to higher dimensional feature space by function ϕ. So the linear classifier is extended to nonlinear classifier by computing the dot product in the input space rather than in the feature space via constructing a kernel function. Variant learning machines are constructed according to different kernel functions and thus construct different hyper planes in the feature space.

Different types of kernel functions used in the reported work are: Linear, RBF, Polynomial and Sigmoid.

5 Experimental Results and Discussion

The experiments are done in Matlab 7.4(R2007a). In order to investigate the effectiveness of each method, data set of 4505 words has been created form various documents. Documents are created in different fonts and printed from a laser printer. Then these documents are scanned. Fonts used are AnmolLipi and Anmol Kalmi for Punjabi words and Times New Roman and Calibri for English Numerals. So from all these documents 4505 words are segmented, out of which 1900 and 2605 are English Numerals and Punjabi words.

Fivefold defines the data set of 4505 words into five disjoint subsets each having 901 words. Here, four subsets are used for training and one is used for testing. So this process is repeated five times leaving one different subset for evaluation each time. Then the average accuracy is calculated.

Table 1 provides the details of recognition results for   different subsets with different kernel functions using SVM.

Table 1. Script Identification Results Using SVM with Discriminating Features and Gabor Features

Input	Classification Accuracy with Different Kernel Functions in %			
	Linear Kernel	Polynomial Kernel	RBF Kernel	Sigmoid Kernel
Discriminating Features	97.23	94.96	95.53	93.85
Gabor Features	99.75	99.82	96.67	57.82

Table 2 provides the details of recognition results for different subsets with Knn with different values of K.

Table 2. Script Identification Results Using KNN with Discriminating Features and Gabor Features

Input	Classification Accuracy with Different Values of K			
	K=1	K=-3	K=5	K=7
Discriminating Features	99.02	99.13	98.98	98.93
Gabor Features	97.62	97.11	96.91	96.40

It has been observed that for discriminating features, KNN Classifier gives the better results and for Gabor features, SVM Classifier gives the better results. Again it has been observed that different kernel functions and different values of K, for each of features, give better results. However error rate is more for increasing the value of K beyond 7. None gives 100% accuracy. So a combination of these classifiers and these feature extraction techniques can be used to get more accurate results.

References

1. Dhanya, D., Ramakrishnan, A.G.: Simultaneous Recognition of Tamil and Roman Scripts. In: The Proc. Tamil Internet, Kuala Lumpur, pp. 64–68 (2001)
2. Rani, R., Dhir, R.: A Survey: Recognition of Scripts in Bi-Lingual/Multi-Lingual Indian Documents. National Journal of PIMT Journal of Research 2(1), 55–60 (2009)
3. Abirami, S., Manjula, D.: A Survey of Script Identification Techniques for Multi-Script Document Images. international journal of Recent trends in Engineering 1(2), 246–249 (2009)
4. Devijver, P.A., Kittler, J.: Pattern Recognition: A statistical Approach. Prentice –Hall, London (1982)
5. Wood, S., Yao, X., Krishnamurthi, K., Dang, L.: Language identification from for printrd trxt independent od fsegmentation. In: Proc of International Conference on Image Processing, pp. 428–431 (1995)
6. Dhanya, D., Ramakrishnan, A.G., Pati, P.B.: Script identification in printed bilingual documents. Sadhana 27(part 1), 73–82 (2002)
7. Pal, U., Sinha, S., Chaudhuri, B.B.: Word-wise Script identification from a document containing English,Devnagari and Telgu Text. In: The Proc. of NCDAR, pp. 213–220 (2003)
8. Padma, M.C., Vijya, P.A.: Language Identification of Kannada, Hindi and English Text Words through Visual Discriminating features. The International Journal of Computational Intelligence Systems 1(2), 116–126 (2008)
9. Dhir, R., Singh, C., Lehal, G.S.: A Structural Feature Based Approach for Script Identification of Gurmukhi and Roman Character and Words. In: The proc. of 39th Annual National Convention of Computer Society of India (CSI) held at Mumbai, India (2004)
10. Pati, P.B., Raju, S.S., Pati, N., Ramakrishnan, A.G.: Gabor filters for document analysis in Indian Bilingual Documents. In: The Proc. Of ICISIP, pp. 123–126 (2004)
11. Pati, P.B., Ramakrishnan, A.G.: HVS inspired system for Script Identification in Indian Multi-Script Documents. In: Proc. of 7th International Workshop on Document Analysis System, Nelson Newland, pp. 380–389 (2006)
12. Pati, P.B., Ramakrishnan, A.G.: Word level multi-script identification. The Pattern Recognition Letters 29, 1218–1219 (2008)
13. Dhandra, B.V., Mallikarjun, H., Hegadi, R., Malemath, V.S.: Word-wise Script Identification from Bilingual Documents based on Morphological Reconstruction. In: The Proc. of First IEEE International Conference on Digital Information Management, pp. 389–394 (2006)
14. Dhandra, B.V., Mallikarjun, H., Hegadi, R., Malemath, V.S.: Word–wise Script Identification based on Morphological Reconstruction in Printed Bilingual Documents. In: The Proc. of IET International Conference on Vision Information Engineering VIE, Bangalore, pp. 389–393 (2006)
15. Dhandra, B.V., Hangarge, M.: On Separation of English Numerals from Multilingual Document Images. The Journal of Multimedia 2(6), 26–33 (2007)

Automatic Word Aligning Algorithm for Hindi-Punjabi Parallel Text

Karuna Jindal[1,4], Vishal Goyal[2,4], and Shikha Jindal[3,5]

[1] Student, [2] Assistant Professor, [3] Student
Department of Computer Science
[4] Punjabi University, Patiala, [5] PEC University of Technology, Chandigarh
jindal.karuna@yahoo.com, vishal.pup@gmail.com,
er.shikhagoel@gamil.com

Abstract. In this paper, an automatic alignment system for Hindi-Punjabi parallel texts at the word level has been described. Automatic word alignment means that without the human interaction the parallel corpus should be aligned word by word with the machine accurately. Boundary-detection and minimum distance function approaches have been used to deal with multi-words. In the existing algorithm, only 1:1 partial word alignment had been done with very less accuracy. But for the multi-words alignment, no work had been implemented. For removing this limitation in the existing system, Different techniques like Boundary-detection, Dictionary lookup and Scoring based Minimum distance function for word alignment has been used in the present system. After implementing above mentioned techniques, the present system accuracy was found to be 99% for one-to-one word alignment and 83% accuracy for multi-word alignment.

Keywords: Automatic word alignment, Automatic Hindi-Punjabi Dictionary generation, scoring, dictionary lookup, boundary-detection.

1 Introduction

A corpus is a collection of spoken or written utterances of natural language usually accessible in electronic form. A parallel corpus is a text in one language together with its translation in another language. Our research aim is to automatically align Hindi-Punjabi parallel text word-wise. The words will be aligned such that the pair will be consisted of Hindi word and the corresponding translated Punjabi word. For this task, we need Hindi-Punjabi parallel corpus. Word alignment of Parallel corpus is the identification of the corresponding words in both halves of the parallel text. Thus, by using this algorithm, Hindi to Punjabi dictionary can be generated automatically for machine translation.

2 Word Alignment

Word alignment means deciding which pairs of words can be the translation of each other in source and target language.

C. Singh et al. (Eds.): ICISIL 2011, CCIS 139, pp. 180–184, 2011.

For Example: हमने वह निशान दिखाया।   ਅਸੀਂ ਉਹ ਨਿਸ਼ਾਨ ਵਖਾਇਆ ।

(Humne veh nishaan dikhaya)   **(Asi oh nishaan dikhaya)**

Alignment algorithm will align every Hindi word with the corresponding Punjabi word.

Hindi_Words	Punjabi_Words
हमने	ਅਸੀਂ
वह	ਉਹ
निशान	ਨਿਸ਼ਾਨ
दिखाया	ਵਖਾਇਆ

3   Related Work

Most of the researchers have worked on non-Indian languages but very little work has been done for Indian languages. Gale and Church (1993) used estimate translation probabilities and used these probabilities to search for most probable word alignment. D.Wu, (1994) used Gale's method for Chinese and English parallel text applying the length-based approach and further extended it to adding lexical cues. Mukda Suktara-chan et al. (1997) used statistical alignment approach. Somboonphol et al. (2002) used Gale's method along with threshold function for estimating word correspondences. Aswani et al. (2005) used simple sentence length approach for sentence alignment, dictionary lookup and nearest aligned neighbours approach to deal with many-to-many word alignment for Hindi-English parallel text.

4   Alignment Algorithm

The algorithm works on the principle that a shorter sentence tends to translate into shorter sentence and a longer sentence tends to translate into a longer sentence. For this, we need the parallel corpus in which we have the source text and target text.

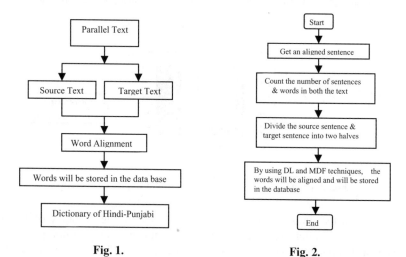

Fig. 1.   Fig. 2.

4.1 Algorithm

- Load sentence aligned Hindi and Punjabi parallel text files.
- These files are processed through the align function that counts the number of sentences and words in the files.
- Split the source sentences and target sentences into word order and create arrays of words in each sentence.
- Then one by one, each sentence of source and target files is taken and words of these sentences are matched. The matching is based on Boundary-Detection [11], improved dictionary lookup and the scoring given by the minimum distance function [11]. It considers only 1:1, 1:2, 2:1, 1:3 and 3:1 type of words alignments.
- The words are aligned and stored into the database.

4.1.1 Improved Dictionary Lookup Approach

In this approach, first and last word of the sentence are matched with dictionary (Hindi-Punjabi) of most common words and stored in database. In case, they not matched, matching of middle words of Hindi-Punjabi sentence is done. There may be a case the above matching results comes out to be false. Then last word of first half and first word of second half is checked. If last word of first half and first word of second half of Hindi-Punjabi are found in the dictionary then they will be stored in the data base and the remaining unaligned words will be align according to minimum distance with scoring techniques. If they do not match with the dictionary, then directly we will apply the minimum distance technique.

5 Comparison

The existing [11] and improved algorithms are implemented using VB.NET with ASP.NET. The results of both algorithms are compared with accuracy of word alignment. It has been found that existing system is unable to align the multi-words like Hindi word 'माता पिता' with Punjabi word 'ਮਾਪੇ'. This problem of alignment was resolved in improved algorithm.

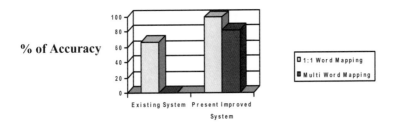

Fig. 3. Comparison of Existing and Present Improved System

As we see in the graph, according to existing system, the accuracy of 1:1 word mapping is 67% and after implementing above techniques in present improved system explained in this paper, system accuracy was found to be 99% for one-to-one word alignment. For the multi word mapping, no work has been done in the previous system. But by using present improved algorithm, the accuracy for multi words is found to be 83%.

6   Evaluation and Results

The accuracy of the system is calculated by the following formula:

Accuracy percentage = (No. of correctly aligned words/Total number of words)*100
Hindi-Punjabi parallel corpus has been collected from resources like EMILLE corpus and others for the word alignment. Parallel Corpus was also developed using the existing Hindi to Punjabi Machine Translation System available online at website http://h2p.learnpunjabi.org. The Hindi text was downloaded from number of online Hindi newspaper websites like BBC Hindi, Dianik Jagram, Bhaskar etc. Hindi-Punjabi parallel corpus used comprised of 50K sentences of variable lengths. An evaluation was performed based on parallel corpus from different fields. We obtained 99% accuracy for one-to-one word alignment and 83% accuracy for multi word alignment. We see that Hindi and Punjabi are close languages in term of length of sentences and also in words. The same approaches can be applied to other languages which are closely related by doing little modification. The model was motivated by the observation that the longer regions of text tend to have longer translations, and that the shorter regions of text tend to have shorter translations.

7   Conclusion

Most of the researchers have worked on non-Indian languages but very little work has been done for Indian languages. In the existing algorithm, only 1:1 partial word alignment had been done with very less accuracy. But for the multi-words alignment, no work had been implemented. For removing this limitation in the existing system, Different techniques like Boundary-detection, Dictionary lookup and Scoring based Minimum distance function for word alignment has been used in the present system. After implementing above mentioned techniques, the present system accuracy was found to be 99% for one-to-one word alignment and 83% accuracy for multi-word alignment.

References

1. Dagan, I., Church, K., Gale, W.: Robust Bilingual Word Alignment for Machine Translation. In: Proceedings of the Workshop on Very Large Corpora (1993)
2. Wu, D.: Aligning a Parallel English-Chinese Corpus Statistically with Lexical Criteria. In: Proc. of the 32nd Annual Conference of the ACL, Las Cruces, NM, pp. 80–87 (1994)
3. Kawtrakul, A., Thumkanon, C., Oovorawan, Y., Varasrai, P., Suktarachan, M.: Automatic Thaiunknown word recognition. In: Proceedings of the Natural Language Processing Pacific Rim Symposium, pages, Thail, pp. 341–348 (1997)

4. Le, S., Youbing, J., Lin, D., Yufang, S.: Word Alignment of English-Chinese Bilingual Corpus Based on Chunk. In: Proceedings of the Joint SIGDAT conference on Empirical methods in NLP and very large corpora: held in conjunction with the 38th Annual Meeting of the ACL, Hong Kong, vol. 13, pp. 110–116 (2000)
5. Somboonphol, N., Sornlertlamvanich, V.: Statistical Technique for Estimating Word correspondence for Bilingual Dictionary Development. In: Proceedings of SNLP-Oriental COCOSDA (2002)
6. Zhao, B., Vogel, S.: Word Alignment Based on Bilingual Bracketing. In: HLT-NAACL Workshop: Building and Using Parallel Texts: Data Driven MT and Beyond (2003)
7. McInnes, B., Pedersen, T.: The Duluth Word Alignment System. Participated in the HLT-NAACL Workshop on Parallel Text (2003)
8. Tufis, D., Ion, R., Deasu, A., Stefanescu, D.: Combined Word Alignment. In: Proceedings of the ACL Workshop on Building and using Parallel Corpora, Data-driven Machine Translation and Beyond, Ann Arbor (2005)
9. Aswani, N., Gaizauskas, R.: Aligning words in English-Hindi parallel corpora. In: Proceeding of the ACL Workshop on Bilingual & Using Parallel Texts, Ann Arbor, pp. 115–118 (June 2005)
10. Gaizauskas, R., Aswani, N.: A hybrid approach to align sentences and words. In: Proceeding of the ACL Workshop on Bilingual & Using Parallel Texts, Ann Arbor, pp. 57–64 (June 2005)
11. Goyal, V., Garcha, L.: Automatic Word Alignment Algorithm for Bilingual Hindi-Punjabi Parallel Text. In: Proceeding of the IACC at Thapar University, Patiala (2009)

Making Machine Translations Polite: The Problematic Speech Acts

Ritesh Kumar

Centre for Linguistics
Jawaharlal Nehru University, New Delhi, India
riteshkrjnu@gmail.com

Abstract. In this paper, a study of politeness in a translated parallel corpus of Hindi and English is done. It presents how politeness in a Hindi text is translated into English. A theoretical model (consisting of different situations that may arise while translating politeness from one language to another and different consequences of these situations) has been developed to compare the politeness value in the source and the translated text. The polite speech acts of Hindi which are most likely to be translated improperly into English are described. Based on this description, such rules will be developed which could be fed into the MT systems so that the problematic polite speech acts could be handled effectively and efficiently by the machine while translating.

Keywords: Politeness, impoliteness, speech acts, structural model, machine translation.

1 Introduction

Polite (or, politic) behaviour has been defined as "socioculturally determined behaviour directed towards the goal of establishing and/or maintaining in a state of equilibrium the personal relationships between the individuals of a social group". [1] Till now there has been no study related to the politeness divergence in translation as far as I know, even though it is very much expected because of 'sociocultutrally determined' nature of politeness. It is a first study of its kind to study a pragmatic aspect of the language like politeness from the point of view of machine translation.

The 'Structural Model of Politeness' [2, 3] is being used for the formal description and classification of the data. In this paper, the kind of illocutionary speech acts that have been improperly or badly translated across different situations, are described, with an aim to incorporate the final results in the machine translation systems. The data for the present study is taken from the parallel translation corpus of 50,000 tagged sentences in 12 Indian languages currently under preparation by the Special Centre for Sanskrit Studies, Jawaharlal Nehru University, New Delhi under the project titled Indian Language Corpora Initiative (ILCI) [4,5]. For the purposes of translation, the rule of structural equivalence or structural parallelism (which implies that the translated texts are intended to have correspondence at lexical [one word is translated into one word and not multiple words and vice-versa] phrasal [one phrase is

C. Singh et al. (Eds.): ICISIL 2011, CCIS 139, pp. 185–190, 2011.
© Springer-Verlag Berlin Heidelberg 2011

translated into one phrase and not more than one and vice-versa] as well as clausal level [one clause being translated into only one clause, not more than one and vice-versa]) is given the prime importance in this project. This has a lot of implications for the present study.

2   The Corpus and the Situations

In the study done in Kumar and Jha [3] (taking into consideration 4000 sentences), the sentences in the corpus are divided into different kinds of speech acts based on the taxonomy of speech acts given by Searle [6]. Since the data in the corpus is taken from written texts, it has only two kinds of speech acts – the assertives and the directives. There is a possibility for other speech acts but they are yet not found.

The above study shows that the maximum share is covered by the situation 2 (the structure of the source language (SL) is translated into the target language (TL) but not the politeness value) while 41% sentences are properly translated (situation1, where both the structure and the politeness value of SL is properly carried over into TL, leading to 'consequence 1', which is the 'proper translation'). There are only three examples of situation 3 (politeness value of the SL is properly translated into TL but the structure is not preserved) and there are only 4% cases where situation 4 (neither the structure nor the politeness value of the SL is carried across in the TL) arises. Moreover there are only two instances of consequence 3 (polite or non-polite sentence in SL is translated into an impolite sentence in TL), as of now, but it cannot be ruled out completely.

In the present paper a classification of the speech acts that occur in situations 2 and 4, leading to the consequence 2 (polite sentence in SL is translated into a non-polite sentence in the TL) has been given.

3   Classifying the Problematic Speech Acts

The sentences whose politeness value could not be translated properly are defined as the problematic sentences or speech acts. It should be especially noted that these speech acts, for the present purposes, are problematic because while translating the sentences from Hindi to English (while following the principle of structural equivalence), *only* the politeness value of the source text could not be preserved in the target text; no other criteria whatsoever is taken into consideration while deciding upon the problematic speech acts. The politeness value of a sentence is determined by a survey conducted among native or native-like speakers of Hindi (who speak English as a second language), where they are asked to give a politeness value to the Hindi sentence and whether the translation given here maintains that value or not.

The classification scheme given here is very tentative in the sense that it is based on the study of just around 3000 sentences and it is highly likely to be revised in some ways as more data from the corpus comes in. Moreover there is the need for more fine-grained classification that could be directly related to the structure of the sentence since it is very necessary to identify the general structural cues in order to make the machine well-equipped to handle the situation.

3.1 Problematic Speech Acts in Situation 2

The following illocutionary speech acts have been identified as problematic for situation 2. All the examples, along with their translations, cited in the paper are taken from the parallel corpus mentioned above. Since the corpus is still under preparation, some of the translations may look bad or concocted but they will be revised at a later stage.

The directionals. The speech acts which are used to give direct instructions to someone to do something (as against the suggestives) are termed the directionals. In Hindi, politeness is explicitly expressed by the use of honorific forms of the verb and the second/third person pronouns. However in English the verbs and pronouns lack the information regarding the honorificity. For example,

1. बच्चे को माँ का दूध अवश्य पिलाएँ ।

Certainly make the child drink mother's milk.

The indirectionals. The indirectionals are the opposite of the directionals. Politeness is expressed by the indirectness, use of certain emphatic particles, etc. They are very rare occurrences in the corpus.

2. कोहिमा पहुँचने का सबसे सहज उपाय कोलकाता से हवाई मार्ग से जाना ही है ।

The easiest way to reach Kohima is go by airways from Kolkata.

The suggestives. The speech acts which are used to direct someone to do something in the form of a suggestion are classified as suggestives. They are polite in the sense that instead of telling someone to do something directly, it presents it in the form of a suggestion, which the reader is not obliged to follow. However in English most of the times the modal 'should' is used which suggests some kind of mild compulsion. For example:

3. इस बीच उसे और कोई उपाय दे देना चाहिए ।

Between this she should be given some other alternative.

The conditionals. Conditionals are an extension of the suggestives. A condition of the form of 'if....then' is expressed. The factors producing impoliteness here are similar to that of suggestives. Some of the examples are as follows.

4. यह उस औरत के लिए सही उपाय है जिसे बहुत असरदार उपाय की चाहत हो ।

This is good alternative for that woman who has the desire for very effective way.

5. इससे दिन में 2 गोली खानी पड़ेगी ।

By this two pills have to be eaten in a day.

Negative conditionals. Negative conditionals are an extended classification of conditionals. They are of the forms 'if not....then' or 'if.....then not'. The factors leading to the loss of politeness value in these are same as conditionals. For example

6.   लॉस एंजिल्स में आप ने अगर डिजनीलैंड नहीं देखा तो सच में पछताएँगे ।

If you did not see Disneyland in Los Angeles then you will truly repent.

The explanatives. The explanatives could be related to the indirectionals. Like the indirectionals these are also the indirect statements intended to get some action done. However here an explanation or exposition of the reasons behind carrying out the action is given, unlike the indirectionals where just a statement is given to influence the reader. These are also quite rare in the corpus. Few examples are given here.

7.   इससे नमक में मिलाया गया आयोडीन शरीर के अनुपात में नहीं रहता।

Because of this the iodine mixed in salt does not stay in the proportion to the body.

The possibilatives. The speech acts which show the possibility of some event or achievment are termed as possibilatives. The politeness in Hindi sentence is generated by the fact that the author is not imposing his/her views on the author. It is also an instance of the fuzzy case. For example,

8.   ऑटो - रिक्शा द्वारा सिंहगढ तक आसानी से पहुँचा जा सकता है ।

One can easily reach Singhgarh through auto-rickshaw.

3.2   Problematic Speech Acts in Situation 4

The illocutionary speech acts that are improperly translated in situation 4, as far as politeness is concerned, are similar to those in situation 2. However the kind of structures associated with those speech acts are different from that in situation 4. The basic difference between the two situations is that in situation 2 the structures are translated (but not politeness) but here neither of these is translated.

The directionals. In situation 4 the directionals include the use of causatives (not found in English as a productive strategy) and certain lexical items which could not be translated directly into English. For example,

1.   यौन रोगों से पीड़ित व्यक्ति को कंडोम उपलब्ध कराएँ

Make condom available to the person suffering from sexual diseases.

The conditionals. The conditionals in situation 4 also involve the use of causatives and passives of certain structures which cannot be passivised in English. For example,

1.   यदि अंजान बाजारू महिला से यौन संबंधों से परहेज करें और निरोध का
      इस्तेमाल करें तो एड्स से बचा जा सकता है ।

If you abstain from sexual relation with unknown female sex worker and use nirodh then you can be saved from AIDS.

The explanatives. The explanatives are generally not found in situation 4. It again involves the use of passives in Hindi, which could not be translated. An example is given below.

1. संतरे का नियमित सेवन करने से सामान्य सर्दी , जुकाम , खाँसी , इन्फ्लूएन्जा , रक्तस्राव आदि से बचाव होता है ।

By the regular consumption of orange one can be saved from common cold, running nose, cough, influenza, bleeding etc.

The possibilatives. The possibilatives in situation 4 mainly have passives and lexically non-translatable items. For example,

1. महायान बौद्ध मठ से आप पूरी तवांग घाटी का सुंदर नजारा ले सकते हैं ।

You can get a beautiful view of the whole Tawang valley Mahayana Buddhist monastery.

4 Conclusion and the Way Ahead

The present paper gives an overview of the different kinds of speech acts and the possible structures associated with these speech acts that tend to get mistranslated in terms of politeness value while translating from Hindi to English. It shows that there are limited number of speech acts and limited number of structures associated with those speech acts that tend to get mistranslated in terms of politeness.

Thus the future work would include the formalization of the findings of this study in terms of rules that could be included in the machine translation systems.

Acknowledgement. I would like to thank the project director, ILCI, Dr. Girish Nath Jha and the whole team of ILCI for providing me the parallel corpus and helping me immensely in working and develpoing on this paper. Special thanks to Narayan Choudhary and Devi Priyanka Singh for being a part of the project since its inception and playing a pivotal role in the processes involving the setting up the standards of the corpora and giving it its present form to the actual creation of the corpora which has led to the present work.

References

1. Watts, R.J.: Politeness. Cambridge University Press, Cambridge (2003)
2. Kumar, R.: An Overview of Politeness Strategies in Hindi. Presented at: 4th International Students' Conference of Linguistics in India, SCONLI-4, Mumbai (2010)

3. Kumar, R., Jha, G.N.: Translating politeness across cultures: Case of Hindi and English. In: Proceedings of the Third ACM International Conference on Intercultural Collaboration (ICIC 2010), pp. 175–179 (2010)
4. Jha, G.N.: The TDIL program and the Indian Language Corpora Initiative (ILCI). In: Proceedings of the Seventh conference on International Language Resources and Evaluation (LREC 2010), pp. 982–985 (2010)
5. Jha, G.N.: Indian Language Corpora Initiative (ILCI). Invited talk 4th International Language and Technology Conference (4th LTC), Poznan, Poland (November 6, 2009)
6. Searle, J.R.: Expression and meaning: Studies in the theory of speech acts. Cambridge University Press, Cambridge (1979)

Tagging Sanskrit Corpus Using BIS POS Tagset

Madhav Gopal[1] and Girish Nath Jha[2]

[1] Centre for Linguistics, SLL & CS
[2] Special Centre for Sanskrit Studies
Jawaharlal Nehru University, New Delhi
{mgopalt,girishjha}@gmail.com

Abstract. This paper presents the application of BIS POS tagset for tagging Sanskrit. Traditionally, the number of grammatical categories for Sanskrit varies from one to five [3]. The language has been exhaustively described in the tradition. And this description is still prevalent in today's grammar teaching. In such a situation, the application of this tagset, which is a new paradigm with respect to Sanskrit, is a challenge. In this paper, we explore how this tagset could be used in categorizing/describing the language.

Keywords: POS tagging, tagset, morphology, Sanskrit, corpus, Pāṇinian grammar.

1 Introduction

Sanskrit is one of the well-studied languages of the world, having a sophisticated vocabulary, morphology, literature, research, scholarship and most importantly a rich grammatical tradition. Plenty of literature is available on its morphology and phonology describing it variously. However, its syntax has received least attention of linguists [8]. The language has been exhaustively described in the tradition mostly in terms of morphological categories and rarely in syntactic categories. POS tagging, however, is a mix of morphological and syntactic description of a language and rightly called morpho-syntactic tagging also. The morphological categories of Pāṇinian grammar [7, 8] are, however, very wide and less formal from syntactic point of view. Therefore, they need to be further categorized and described in terms of contemporary morpho-syntactic categories of linguistic description. This paper is an attempt to describe classical Sanskrit in terms of these commonly used categories as they are the label to label words in the current POS tagging scheme.

2 Sanskrit POS Tagging

2.1 Availability of Various POS Tagsets

There are many tagsets available for tagging Sanskrit: JNU-Sanskrit tagset (JPOS), Sanskrit consortium tagset (CPOS), MSRI-Sanskrit tagset (IL-POSTS), IIIT Hyderabad tagset (ILMT POS) and CIIL Mysore tagset for the Linguistic Data

C. Singh et al. (Eds.): ICISIL 2011, CCIS 139, pp. 191–194, 2011.
© Springer-Verlag Berlin Heidelberg 2011

Consortium for Indian Languages (LDCIL) project (LDCPOS) [3]. The first two of these are Sanskrit specific, and the rests are common for all Indian languages. Barring IL-POSTS, all the tagsets are flat. The BIS tagset which is a national standard tagset for Indian languages is a hierarchical one and follows a layered approach for annotating various kinds of linguistic information available in a text. This tagset has 11 categories at the top level (Noun, Pronoun, Demonstrative, Verb, Adjective, Adverb, Postposition, Conjunction, Particle, Quantifier and Residual). The categories at the top level have further subtype level 1 and subtype level 2. In this framework the granularity of the POS has been kept at a coarser level. Most of the categories of this tagset seem to have been adapted either from the MSRI or the ILMT tagset. For morphological analysis it will take help from morphological analyzers, so morpho-syntactic features are not included in the tagset.

The BIS scheme is comprehensive and extensible and can spawn tagsets for Indian languages based on individual applications. It captures appropriate linguistic information, and also ensures the sharing, interchangeability and reusability of linguistic resources. The Sanskrit specific tagsets are not compatible with other Indian languages and with the exception of the IL-POSTS, all other tagsets are flat and brittle and do not capture the various linguistic information. The IL-POSTS, an appreciable framework, captures various linguistic information in one go and this, according to the designers of the BIS tagset, makes the annotation task complex. And from machine learning perspective also it is not a good idea. So, the BIS tagset, as a middle path, is suitable for tagging all Indian languages.

2.2 Tagging Sanskrit Using the BIS Tagset

In the BIS scheme, the top level category of noun has four subtypes at level 1: common, proper, verbal and noun location. The verbal noun is for languages such as Tamil and Malayalam. In our opinion it is equally applicable for Sanskrit also. The *kṛdantas* like *āgamanam* and *hasanam* could be tagged as verbal noun, as they are nominalised in a sentence. The indeclinables like *agré* and *pūrvam* could be labeled as noun location.

The pronoun category is divided in 5 subtypes: personal, reflexive, relative, reciprocal, and wh-word. The nominals like *ātman, sva, svakīya* etc. are tagged as reflexives. Among reciprocal pronouns are *parasparam, itarétaram, mithaḥ* and *anyonyam*.

The next top level category is of demonstrative. Demonstratives have the same form of the pronouns, but distributionally they are different from the pronouns as they are always followed by a noun, adjective or another pronoun. In this category, only deictic, relative and wh-word subtypes fall. Deictics are mainly personal pronouns. Sanskrit doesn't differentiate between demonstrative pronouns and third person pronouns.

The category of verb is somewhat complicated in this framework. It has main and auxiliary divisions under subtype level 1 and finite, non-finite, infinitive and gerund divisions under subtype level 2. Verb main does not seem to be an appropriate tag in case of Sanskrit. However, if anybody insists to use it, it can be utilized in tagging the verbs of present tense when followed by a *sma* and the *kta* and *ktavat pratyayāntas* when followed by an auxiliary, and in doing so the auxiliary verbs and *sma* have to retain their Auxiliary tags.

The *tiṅantas* (inflections of *as*, *ās*, *sthā*, *kṛ*, and *bhū* only) that follow a *kṛdanta* to express its (*kṛdanta's*) aspectual meaning, will be tagged with Auxiliary label and the indeclinable *sma* will also get the same tag when follows a verb in present tense and modifies the meaning of the associated verb.

All the conjugations of the *dhātus* are finite verbs (VF). However, when some of these forms will be used to express the aspectual meaning of the preceding *kṛdanta* will be tagged as auxiliary, as is stated above. In addition, *kta* and *ktavat pratyayāntas* will also be tagged as VF when they are not followed by any auxiliary verb. As we do not have a separate tag for gerundives (like *kāryam*, *karaṇīyam*, *kartavyam*), VF tag could be applied for them as well. *kta* and *ktavat pratyayāntas* will be tagged as verb non-finite (VNF) when followed by an auxiliary and other *kṛidantas* like *śatṛ*, *śānac* and *kānac* will also get the same tag.

Sanskrit infinitives are different from other Indian languages and English. They correspond to the infinitive of purpose in English. Only *tumun pratyayāntas* will be tagged as VINF. In the literature [7, 8] *ktvānta* and *lyabanta* are described as gerund. So, these words will be labeled with verb non-finite gerund (VNG).

Adjectives in Sanskrit are rarely realized as modifiers. Often they occur as substantives. However, there is no dearth of pure adjective usages in the language. When they are used with their modified item, should be tagged as adjectives otherwise as nouns. Only manner adverbs are to be tagged as Adverbs in this framework; thus *uccaiḥ* (loudly), *sukham* (happily) etc. will get the adverb tag.

There is a top level category for Postpositions. Sanskrit does not have postposition as such. But we can tag the *upapada* (like *saha, namaḥ, abhitaḥ*) indeclinables as postpositions as they are indeed ambipositions and cause the assignment of a particular *vibhakti* in the concerned nominal.

Conjunction is a major category in the tagset and has co-ordinator (*ca* etc.), subordinator (*yat* etc.) and quotative (*iti* etc.) as subtypes. We have to first enlist the conjunctions in these subcategories and then tag accordingly.

Particle is a very important category for Sanskrit language, as they have many a role to play. Some of the indeclinables described as *avyayas* in the tradition fall in this category. In the tagset, there are default, classifier, interjection, intensifier and negation subtypes of the Particle category. The classifier tag is not applicable for Sanskrit. Words like *bhṛśam* (very much), *atitarām* (very much) etc. are intensifiers.

The Quantifier category includes general, cardinal, and ordinal subtypes. These terms are equally applicable to both types of quantifiers: written in words (like five, fifth etc.) and in digits (like 5, 5th etc.).

Residual as a major category in this tagset holds foreign word, symbol, punctuation, unknown and echo words as subtypes. In Sanskrit echo words are mainly reduplications of a variety of linguistic items. In this framework a word is considered a foreign one if it is written in a script other than Devanagari. The symbol subtype is for symbols like $, %, # etc. If a word does not fit in any of these categories, will be tagged unknown.

3 Conclusion

In a linguistically rich country like India, having linguistic resource standards like BIS scheme is highly recommendable. Also, where various language technology tasks are based on tagged corpora, a policy of having a unified system can be efficient time and cost wise. The BIS scheme enables us to design tagged corpora which will be cross linguistically compatible, reusable and interoperable. The ILCI (Indian Languages Corpora Initiative) project at Jawaharlal Nehru University is trying to tag 12 Indian languages including English using this scheme.

References

1. Baskaran, S., Bali, K., Bhattacharya, T., Bhattacharyya, P., Choudhury, M., Jha, G.N., Rajendran, S., Saravanan, K., Sobha, L., Subbarao, K.V.: A Common Parts-of-Speech Tagset Framework for Indian Languages. In: LREC, Marrakech, Morocco, pp. 1331–1337 (2008)
2. Chandrashekar, R.: Parts-of-Speech Tagging For Sanskrit. Ph.D. thesis submitted to JNU, New Delhi (2007)
3. Gopal, M., Mishra, D., Singh, D.P.: Evaluating Tagsets for Sanskrit. In: Jha, G.N. (ed.) Sanskrit Computational Linguistics. LNCS, vol. 6465, pp. 150–161. Springer, Heidelberg (2010)
4. Hellwig, O.: SANSKRITTAGGER, A Stochastic Lexical and POS Tagger for Sanskrit. In: Huet, G., Kulkarni, A., Scharf, P. (eds.) Sanskrit Computational Linguistics. LNCS, vol. 5402, pp. 266–277. Springer, Heidelberg (2009)
5. IIIT-Tagset. A Parts-of-Speech tagset for Indian Languages, http://shiva.iiit.ac.in/SPSAL2007/iiit_tagset_guidelines.pdf
6. Jha, G.N., Gopal, M., Mishra, D.: Annotating Sanskrit Corpus: adapting IL-POSTS. In: Vetulani, Z. (ed.) Proceedings of the 4th Language and Technology Conference: Human Language Technologies as a Challenge for Computer Science and Linguistics, pp. 467–471 (2009)
7. Kale, M.R.: A Higher Sanskrit Grammar. MLBD Publishers, New Delhi (1995)
8. Speijer, J.S.: Sanskrit Syntax. Motilal Banarsidass Pvt. Ltd., New Delhi (1886) (Repr. 2006)

Manipuri Transliteration from
Bengali Script to Meitei Mayek: A Rule Based Approach

Kishorjit Nongmeikapam[1], Ningombam Herojit Singh[1],
Sonia Thoudam[1], and Sivaji Bandyopadhyay[2]

[1] Dept. of Computer Science and Engg., Manipur Institute of Technology,
Manipur University, Imphal, India
[2] Dept. of Computer Science and Engg., Jadavpur University, Jadavpur, Kolkata, India
{kishorjit.nongmeikapa,herojitningomba,soniathoudam}@gmail.com,
sivaji_cse_ju@yahoo.com

Abstract. This paper describes about the transliteration of Manipuri from Bengali Script to Meitei Mayek (Meitei script). So far no work of Manipuri transliteration is done and being an Eight Schedule Language of Indian Constitution we felt necessary to start through a rule based. A model and algorithm is being designed for transliterating Manipuri from Bengali script to Meitei Mayek (Meitei Script). Even though the model is a simple rule base approached but to our surprise the algorithm proved to come up with an accuracy of 86.28%.

Keywords: Transliteration, Bengali Script, Meitei Mayek, Iyek.

1 Introduction

Transliteration is the process of mapping a word of a source language script to another target language script.

Manipuri (or Meiteilon) is a Tibeto-Burman (TB) language and also one of the Eight Scheduled languages of Indian Constitution. It is highly agglutinative in nature. Manipuri uses two scripts; the first one is purely of its own origin, *Meitei Mayek* while another one is a borrowed *Bengali Script*. The present design of algorithm is to transliterate Bengali Script to the Meitei Mayek.

Transliteration of Indian language is found in the works of IT3 developed by IISc Bangalore, India and Carnegie Mellon [1]. Other transliteration works for Indian languages can be seen in [2], [3], [4], [5] and [6]. For other foreign languages works of European language in [7] and works on Asian language in [8]. So far upto the best of the authors' knowledge no work of transliteration has been done and this is the first work of transliteration for Manipuri.

The paper is organized with Section 2 giving the details about the Linguistic Transliterating Scheme, the model and algorithm in Section 3, the experimental result and evaluation in Section 4 and the conclusion is drawn in Section 5.

C. Singh et al. (Eds.): ICISIL 2011, CCIS 139, pp. 195–198, 2011.
© Springer-Verlag Berlin Heidelberg 2011

2 Linguistic Transliteration Scheme

Bengali which has 52 consonants and 12 vowels is mapped to Meitei Mayek which has 27 (Twenty seven) alphabets (Iyek Ipee) and its supplements: vowels, Cheitap Iyek, Cheising Iyek and Lonsum Iyek [9] are shown in Tables 1,2,3,4 and 5.

Table 1. Iyek Ipee characters in Meitei Mayek

Iyek Ipee

ক->ꯀ (kok)	স(ছ,শ,ষ)->ꯁ (Sam)	ল->ꯂ (Lai)	ম->ꯃ (Mit)
প->ꯄ (Pa)	ন->ꯅ (Na)	চ->ꯆ (Chil)	ত(ট)->ꯇ (Til)
খ->ꯈ (Khou)	ও->ꯉ (Ngou)	থ(ঠ)->ꯊ(Thou)	ব->ꯋ (Wai)
য(য়)->ꯌ (Yang)	হ->ꯍ (Huk)	উ(ঊ)->ꯎ(Un)	ই(ঈ)->ꯏ(Ee)
ফ->ꯐ (Pham)	অ->ꯑ (Atia)	গ->ꯒ (Gok)	ঝ->ꯓ (Jham)
র->ꯔ (Rai)	ব->ꯚ (Ba)	জ->ꯖ (Jil)	দ(ড)->ꯗ(Dil)
ঘ->ꯘ (Ghou)	ধ(ঢ)->ꯙ(Dhou)	ভ->ꯚ(Bham)	

Table 2. Vowels of Meitei Mayek

Vowel letters

আ->ꯑꯥ(Aa)	এ->ꯑꯦ(Ae)	ঐ->ꯑꯩ(Ei)
ও->ꯑꯣ(o)	ঔ->ꯑꯧ(Ou)	অং->ꯑꯪ(Ang)

Table 3. Cheitap Iyek of Meitei Mayek

Cheitap Iyek

(ো)->ꯣ (ot nap)	ি, ী->ꯤ(inap)	া->ꯥ(aatap)	ে->ꯦ(yetnap)
(ৌ)->ꯧ (sounap)	ু, ূ->ꯨ (unap)	ৈ->ꯩ(cheinap)	ং->ꯪ(nung)

Table 4. Cheising Iyek or numerical figures of Meitei Mayek

Cheising Iyek(Numeral figure)

১->꯱(ama)	২->꯲(ani)	৩->꯳(ahum)	৪->꯴(mari)
৫->꯵(manga)	৬->꯶(taruk)	৭->꯷(taret)	৮->꯸(nipal)
৯->꯹(mapal)	১০->꯰(tara)		

Table 5. Lonsum Iyek of Meitei Mayek

Lonsum Iyek

ক-> ꯛ (kok lonsum)	ল-> ꯜ (lai lonsum)	ম->ꯝ (mit lonsum)	প-> ꯞ(pa lonsum)
ণ, ন-> ꯟ (na lonsum)	ট,ত-> ꯠ (til lonsum)	ও->ꯡ(ngou lonsum)	ই, ঈ->ꯢ(ee lonsum)

Alphabets of Meitei Mayek are repeated uses of the same alphabet for different Bengali alphabet like ৰ, শ, ষ, স in Bengali is transliterated to ꯁ in Meitei Mayek.

In Meitei Mayek, Lonsum Iyek (in Table 5) is used when ক is transliterated to ꯛ, ঙ transliterate to ꯩ, ত transliterate to ꯠ etc. Apart from the above character set Meitei Mayek uses symbols like '꯰' (Cheikhie) for '।' (full stop in Bengali Script). For intonation we use '.' (Lum Iyek) and '_' (Apun Iyek) for *ligature*. Other symbols are as internationally accepted symbols.

3   Model and Algorithm

In our model (Fig. 1) we used two mapped file for Bengali Characters and corresponding Meitei Mayek Characters which are read and stored in the *BArr* and *MMArr* arrays respectively. A test file is used so that it can compare its *index* of mapping in the Bengali Characters List file which later on used to find the corresponding target transliterated Meitei Mayek Characters Combination. The transliterated Meitei Mayek Character Cuombination is stored on an output file.

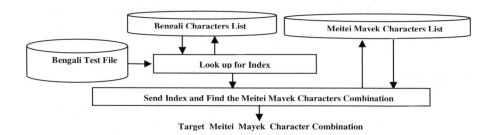

Fig. 1. Model of Transliteration Scheme used in Manipuri Bengali Script to Meitei Mayek

Algorithm use for our model is as follows:

```
Algorithm:transliteration(line, BCC, MMArr[], BArr[])
1.    line : Bengali line read from document
2.    BCC : Total number of Bengali Character
3.    MMArr[] : Bengali Characters List array
4.    BArr[] : Meitei Mayek Character List array
5.    len : Length of line
6.    for m = 0 to len-1 do
7.      tline=line.substring(m,m+1)
8.      if tline equals blank space
9.        Write a white space in the output file
10.   end of if
11.   else
12.    for index=0 to BCC-1
13.      if tline equals BArr[index]
14.        pos = index
```

```
15.      break
16.     end of if
17.     end of for
18.     Write the String MMArr[pos] in the output file
19.     end of else
20.  end of for
```

4 Experiment and Evaluation

Manipuri is a less computerized language and collecting corpus is a hard task. The experiments of the systems are done with the corpus collected from a Manipuri local daily newspaper[1]. A corpus of 20,687 words is collected for testing of the system.

In Evaluation of the result, the system shows an accuracy of 86.28%. Due to use of same character set of the Meitei Mayek relative to Bengali Script as mention in Section 3 we found a lower accuracy.

5 Conclusion

This model being the first model in Manipuri Transliteration it shows a good result. So far other techniques are not yet tried, so plans of implementing other techniques to increment the performance are the future work direction.

References

1. Ganapathiraju, M., Balakrishnan, M., Balakrishnan, N., Reddy, R.: Om.: One Tool for Many (Indian) Languages. In: International conference on Universal Digital Library, vol. 6A(11), pp. 1348–1353. Journal of Zhejiang University SCIENCE, Hangzhou (2005)
2. Surana, H., Singh, A.K.: A More Discerning and Adaptable Multilingual Transliteration Mechanism for Indian Languages. In: IJCNLP 2008, India, pp. 64–71 (2008)
3. Das, A., Ekbal, A., Tapabrata, M., Sivaji, B.: English to Hindi Machine Transliteration at NEWS 2009. In: ACL-IJCNLP 2009, Singapore, pp. 80–83 (2009)
4. Ekbal, A., Naskar, S.K., Bandyopadhyay, S.: A Modified Joint Source-Channel Model for Transliteration. In: COLING/ACL, Morristown, NJ, USA, pp. 191–198 (2006)
5. Ekbal, A., Naskar, S., Bandyopadhyay, S.: Named Entity Transliteration. International Journal of Computer Processing of Oriental Languages (IJCPOL) 20(4), 289–310 (2007)
6. Abbas Malik, M.G.: Punjabi machine transliteration. In: 21st COLING 2006/ACL, Sydney, Australia, pp. 1137–1144 (2006)
7. Marino, J.B., Banchs, R., Crego, J.M., de Gispert, A., Lambert, P., Fonollosa, J.A., Ruiz, M.: Bilingual n-gram Statistical Machine Translation. In: MT-Summit X, pp. 275–282 (2005)
8. Vigra, P., Khudanpur, S.: Transliteration of Proper Names in Cross-Lingual Information Retrieval. In: ACL 2003 Workshop on Multilingual and Mixed-Language Named Entity Recognition, pp. 57–60 (2003)
9. Kangjia Mangang, N.: Revival of a closed account. In: Sanamahi Laining Amasung Punsiron Khupham, pp. 24–29. Imphal (2003)

---

[1] http://www.thesangaiexpress.com/

Online Handwriting Recognition for Malayalam Script

R. Ravindra Kumar, K.G. Sulochana, and T.R. Indhu

Centre for Development of Advanced Computing, Vellayambalam, Trivandrum, Kerala
{ravi,sulochana,indhu}@cdactvm.in

Abstract. Online handwriting recognition refers to machine recognition of handwriting captured in the form of pen trajectories. This paper describes a trainable online handwriting recognition system for Malayalam using elastic matching technique. Each character/stroke is subjected to a feature extraction procedure. The extracted features forms input to a nearest neighborhood classifier which returns the label having the minimum distance. The recognized characters are assigned their corresponding Unicode code points and are displayed using appropriate fonts. With a database containing 8389 handwritten samples, we get an average word recognition rate of 82%.

Keywords: Online Handwriting Recognition, Malayalam, Pre-processing, Feature Extraction.

1 Introduction

Handwriting Recognition can be split into two viz. Offline and Online, depending on the form in which data is presented to the system. In the first system, the handwritten document is converted into an image using a scanner and this is input to the system. In Online handwriting recognition a machine is made to recognize the writing as a user writes on a special digitizer or PDA with a stylus. Here the parameters related to the pen tip like position, velocity, acceleration and sometimes pressure (on the writing surface) are available to the data acquisition system. In both the systems, the ultimate objective is to convert handwritten sentences or phrases in analog form into digital form.

The established procedures to recognize online handwritten characters include data collection & analysis, pre-processing, feature extraction, classification & recognition and post-processing. In this paper we begin with an overview of Malayalam script and the challenges involved. We then present the different procedures that were involved in developing the system. The result of the work done is summarized at the end of the paper.

There are many online character recognizers available in different languages. Works have been reported in Latin, Chinese, English, Arabic, Thai, Urdu, Turkish and Indic scripts namely Gurmukhi, Tamil, Telugu etc. Online Urdu character recognizer makes use of the ligature based approach instead of character based identification [2]. For classifying handwritten Tamil characters DTW and HMM have been reported [3]. Support Vector Machines have been observed to achieve reasonable generalization accuracy, especially in implementations of handwritten digit recognition and character recognition in Roman, Thai, Arabic and Indic scripts such as Devnagari and Telugu [4].

C. Singh et al. (Eds.): ICISIL 2011, CCIS 139, pp. 199–203, 2011.

2 Malayalam Script

Malayalam is one of the four major languages of the Dravidian language family and is spoken by about forty five million people. Like most of the other Indian languages, Malayalam has its origin from the ancient Brahmi script.

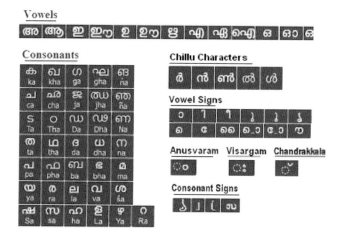

Fig. 1. Malayalam character set

The basic Malayalam character set is given in Figure 1.

3 Challenges

The Online Handwriting recognition for Malayalam script is a greater challenge compared to the recognition of Western scripts because of the following reasons:

- Presence of large number of characters
- Different writing styles & writing speed
- Complexity of the characters & Similarity in character shapes
- Poor reliability of extracted stroke features due to variance in handwriting

4 Data Collection and Analysis

The data was collected from 29 informants, including male and female in the age group 25 to 35, using Wacom Intuos3 tablet with a resolution of 200 lines/inch and with a sampling rate of 200 points/sec. Each and every informant is directed to write a total of 417 Malayalam words (including Latin numbers) which is extracted with help of experts in the Malayalam linguistic domain. Data collected can be used for the analysis of the possible features of each character/stroke and also for the study of the different writing styles of Malayalam characters.

5 Pre-processing

Preprocessing is required to remove irregularities present in the input stroke captured by the digitizing tablet. We performed several pre-processing operations, including dehooking, size normalization, resampling, smoothing and removal of duplicate points.

6 Feature Extraction

A character can be represented by a set of features. The features can be based on the static properties of the characters, dynamic properties, or both. Computational complexity of classification problem can also be reduced if suitable features are selected. For Malayalam, we have considered the following features

- Geometric features - Loop, Intersection, Cusp
- Ink related features - Point density, Stoke Length
- Directional features - Start angle, End angle, Start-End angle
- Global features - Start position, End position, Start-End position
- Local features - Pre-processed X-Y coordinates, quantized slope, dominant points

7 Classification and Recognition

Classification of characters is an efficient way to break a large vocabulary into several smaller groups/classes. We have adopted a two-stage classification scheme using nearest neighbour classifier. The first stage which employs a 3 level classification filters the templates based on the features from the test sample identified during the feature extraction phase and the second stage computes the more expensive Dynamic Time Warping (DTW) [1] distance from the shortlisted templates. The label of the nearest template is assigned to the test sample.

DTW Algorithm: Dynamic Time Warping is a technique that matches two trajectories (handwritten characters) and calculates a distance from this matching. DTW algorithm compares handwritten samples in a way that has results similar to the human handwriting comparing system. It is based on linear matching, but has three conditions or constraints that need to be satisfied. These conditions are as follows

- Continuity condition
- Boundary condition and
- Monotonicity condition

Continuity condition: The continuity condition decides how much the matching is allowed to differ from linear matching. This condition is the core of the Dynamic Time Warping and thus is not optional. If N1 and N2 are the number of points in the first and second curve respectively, the ith points of the first curve and the jth points of the second curve can be matched if (note that the other conditions can bypass this if statement)

$$\frac{N_2}{N_1}i - cN_2 \leq j \leq \frac{N_2}{N_1}i + cN_2 \tag{1}$$

The parameter 'c' determines the amount that the matching is allowed to differ from linear matching.

Boundary condition: The boundary condition, if turned on, forces a match between the first points of the curves and a match between the last points of the curves.

Monotonicity condition: This prevents the matching from going back in time. If at some point in the matching process it is decided that the ith point of the first curve matches with the jth point of the second curve, it is not possible for any point of the first curve with index $> i$ to match with a point of the second curve with index $< j$ and for any point on the first curve with index $< i$ to match with any point on the second curve with index $> j$.

Decision Algorithm: From the selected characters/prototypes, the decision algorithm makes a list of the characters occurring in the top N number of prototypes and decides a weight for each of those characters. Here N is a user specified value that decides how many of the prototypes (counting from the one with the smallest distance to the query) are used by the algorithm. The character with the highest weight is returned as the recognized output.

8 Post Processing

The function of this module is to correct the errors that occurred during the recognition stage, thereby improving the accuracy of the recognized output. The processes performed under post processing were 1) Stroke correction 2) Stroke concatenation 3) Linguistic rules. The post processing stage also makes use of a spellchecker to identify the word written is valid or not. For invalid words, suggestions from the dictionary are provided.

9 Results

The data after analysis was reduced in size by selecting an appropriate subset as prototypes, achieved using the prototype selection method 'Growing of the prototype set' discussed in [5]. With the final database consisting of 8389 character prototypes, DTW- classifier and spellchecker, we have an average recognition rate of 94% at character level and 82% at word level.

Acknowledgements. The authors would like to thank all the people who had helped in any stage in the completion of the work. Special thanks for all who were willing to spend their quality time in the data collection phase and testing.

References

1. Niels, R.: Dynamic Time Warping: An Intuitive Way of Handwriting Recognition? Master's thesis, Radboud University Nijmegen (2004), http://dtw.noviomagum.com/
2. Husain, S.A., Sajjad, A., Anwar, F.: Online Urdu Character Recognition System. In: MVA 2007 IAPR Conference on Machine Vision Applications, Tokyo, Japan, pp. 98–101 (2007)
3. Niels, R., Vuurpijl, L.: Dynamic Time Warping Applied to Tamil Character Recognition. In: Proceedings of 8th ICDAR, pp. 730–734. IEEE, Seoul (2005)
4. Swethalakshmi, H., Jayaram, A., Chakraborty, V.S., Sekhar, C.C.: Online Handwritten Character Recognition of Devanagari and Telugu Characters using Support Vector Machines. In: Proc. 10th IWFHR, pp. 367–372 (2006)
5. Raghavendra, B.S., Narayanan, C.K., Sita, G., Ramakrishnan, A.G., Sriganesh, M.: Prototype Learning Methods for Online Handwriting Recognition. In: 8th ICDAR 2005, Seoul, Korea, August 29-September 1, pp. 287–291 (2005)

Optimized Multi Unit Speech Database for High Quality FESTIVAL TTS

R. Ravindra Kumar, K.G. Sulochana, and T. Sajini

Language Technology Centre, Centre for Development of Advanced Computing,
Vellayambalam, Trivandrum-33, Kerala
{ravi,sulochana,sajini}@cdactvm.in

Abstract. This paper describes the development of optimized multiunit speech database for a high quality concatenative TTS System. The core engine used for the development of Text to speech is the open source FESTIVAL engine and is tested for Malayalam language. The optimal text selection algorithm selects the optimal text, ensuring the maximum coverage of units without discarding entire low frequency units. In this work we created a multiunit database with syllable as the highest unit, ensuring the coverage of all CV-VC units, and phones.

Keywords: Concatenation, Clustering, FESTIVAL, HMM Labeling, Letter to sound rules, Mean opinion score, optimal text selection, Text to speech, Unit selection.

1 Introduction

A Text to speech system is the software which allows the transformation of a string of phonetic and prosodic symbols into synthetic speech. Among the different speech synthesis methods concatenative synthesis is a widely accepted approach because of its good quality output. Higher the units selected for concatenation, better the quality of speech output. Concatenative unit selection requires a large database of a single speaker, which contains multiple realizations of different units with varied prosodic and spectral characteristics to get more natural-sounding synthesized speech.

The quality of a TTS system is measured by the intelligibility and naturalness of the speech output. The factors which affect the quality of synthesized speech are the quality of database, accuracy of labels, text-pronunciation mapping, and units selected for concatenation. Even though there exist good quality TTS for other languages, comparable quality TTS is not available for Malayalam. In this work we developed a Malayalam TTS using unit selection method.

This paper briefs the development of a multi unit database for TTS and the speech quality improvement achieved by using optimized speech database. We have done the work for Malayalam, which is one of the four major Dravidian languages and is the official language of the state of Kerala and the Indian Union territory of Lakshadweep. It is spoken by around 35 million people.. The database is created for female

C. Singh et al. (Eds.): ICISIL 2011, CCIS 139, pp. 204–208, 2011.

voice, with 6 hour data covering 9K syllables. Label files are created using HMM and this improved the accuracy of labels.. The mean opinion score (MOS) for the TTS using this multiunit database is found to be 3.2. The Functional Block Diagram of a multi unit based TTS is given in Fig. 1 below.

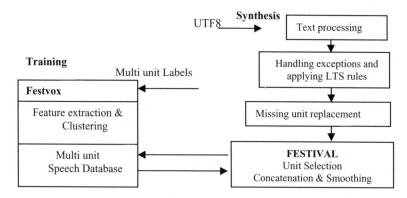

Fig. 1. Functional Block Diagram of TTS

2   Multi Unit Speech Database Creation

The speech database creation for a TTS requires text corpus and speech corpus. This involves collection and selection of text corpus, selection of informant satisfying the requirements and recording of the text in an acoustically treated room. Even though the corpus based TTS requires large database, the sentences/words/phrases selected to train the system must not be too large. It must be the minimum text ensuring the coverage of the units selected for concatenation. This text is recorded, with a good quality voice in a noise free environment by taking care of the syllable rate.

2.1   Corpus Collection and Processing

Text Corpus is collected by crawling web, stories and few contents were manually prepared to ensure the coverage of all syllables.. Text normalization, number suffix pattern handling and sentence wise segmentation is done on the collected corpus. Handling of number suffix pattern is done after normalization. For example '10-നാണ്' is replaced with പത്ത്-നാണ് and then joined with the suffix to get പത്തിനാണ്. 42 number suffix patterns were identified in the collected corpus.

2.2   Letter to Sound Rules (LTS Rules)

Even though Indian languages are phonetic in nature, there exist pronunciation exceptions. For a good quality speech synthesis the use of proper pronunciation is inevitable and this must be considered while selecting text for the database.

The conversion of sentences, to pronunciation is done using the letter to sound rules and dictionary look up for exceptions [3]. The exception list is prepared automatically using the exception patterns. Scoring of sentences is done based on the syllable coverage. Syllable wise segmented sentences and scores are given as the input for optimal text selection.

In addition to the letter to sound rules mentioned in [3], pronunciation variation for /k/ gemination is also incorporated. Taking care of such variations in /k/ gemination in optimal text selection ensures the availability of proper and sufficient units for /k/ gemination in the speech database, and ensures the improvement in the quality of synthesized speech.

2.3 Optimal Text Selection

Selecting text to cover the entire syllables found in the corpus is not possible. Among the 13K syllables, 14% with frequency >=50, 42% with frequency 1, 11% with frequency 2, 6% with frequency 3 and 27% within the frequency range 4-50. Text selection is done to cover maximum high frequent syllables, lower units of the low frequency syllables and the low frequency syllables whose lower units are not available in the corpus. The optimal text selection tool fixes the threshold frequency for the low frequency syllable and checks for the presence of lower syllables and removes all lower syllables for which the next lower syllables are available. Fixing of threshold frequency is done on a trail and error basis and also depends on the planned size of the optimal text. 2 to 3 iterations (changing the threshold value) are done on the collected 0.3 million sentence corpus, to optimize the corpus to 7K sentence covering 9K syllables. The advantage of this method is that it ensures the availability of all high frequency units and low frequency units which cannot be replaced with lower syllable and phone combination. The syllable patterns found in corpus are CVC*, C*V, C*VC, VC, CVC, CV and the pattern C*VC* normally in foreign words.

2.4 Speech Database

The optimal text thus selected is recorded to build the speech corpus. The recording is done with a good quality voice, in a noise free environment at a low speaking rate (5-8 syllables per second). Low syllable rate is enforced to attain clarity of each phone in the utterance. Audio specification of 16bit, 16 KHz mono, in PCM format is used.

2.5 Multi Unit Label for Speech Database

The highest unit selected for concatenation is syllable. The multiunit label is generated from the phone labels. Since labeling using festival is time consuming and the

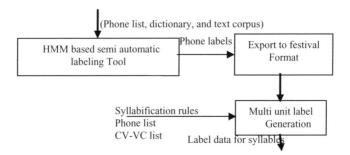

Fig. 2. Semi-automatic tool for multi unit labeling

accuracy depends on the silence duration, we used a HMM [4] based semiautomatic labeling for creating phone level label for the multiunit database.

The multi-units modeled in this database are syllables, all CV-VC not in modeled syllable list and phones. Syllabification rules [2], list of CV-VC units not covered in syllables are given as the input for the multi unit label generator.

3   Building Voice and Synthesis

Festival [1] offers language independent modules for building synthetic voices. The basic steps for building voice are followed. Unit definitions, Letter to Sound rules and syllabification rules were added by modifying the in the required Festival files. The initial, final and medial syllables are clustered separately.

During synthesis, the unit selection algorithm in Festival selects an appropriate decision tree and searches for a suitable realization of the unit which is close to its cluster center and optimizes the cost of joining two adjacent units. At the time of synthesis it uses the NLP module to handle the basic text normalization and UTF8 to phoneme conversion.

The missing syllables are handled by incorporating the missing unit handler. During synthesis, the untrained syllable is replaced with a combination of trained lower syllable and consonant. MOS test gave a score of 3.2, for 50 synthesized sentences with 15 native speakers.

4   Conclusions

Current multi unit database showed a good coverage of syllables for the text input selected from online sources. Remarkable quality improvement in synthesized speech was achieved by incorporating additional rules and exception patterns, HMM labels and by implementing lower unit replacement, which replaces the missing syllables in the synthesis input. In addition to these the HMM based labeling requires less than 25% of the time taken for EHMM based labeling.

Further improvement in quality of speech can be obtained by manually correcting the labels. We also plan to increase the syllable coverage, by covering the missing syllables with words.

Acknowledgements

We would like to thank each and every person who have contributed their valuable time for the project.

References

1. Black, A., Lenzo, K.: Building voices in the Festival speech synthesis system, http://festvox.org/bsv/x740.html
2. Prabodhachandran Nayar, V.R.: Swanavijnanam. Kerala Bhasha Institute (1980)

3. Ravindra Kumar, R., Sulochana Jose Stephen, K.G.: Automatic Generation of Pronunciation Lexicon for Malayalam-A Hybrid Approach: LDC-IL, Creation of Multilingual speech resources academic and Technical Issues.,
   http://www.ldcil.org/up/conferences/speech/presentations/Auto maticGenerationofPronunciationLex- icon_mal%20%5BCompatibility%20Mode%5D.pdf
4. Young, S., Evermann, G., et al.: The HTK Book. Cambridge University Engineering Department (2006), http://htk.eng.cam.ac.uk/docs/docs.shtml

Comparative Analysis of Printed Hindi and Punjabi Text Based on Statistical Parameters

Lalit Goyal

Assistant Professor, DAV College, Jalandhar
goyal_aqua@yahoo.com

Abstract. Statistical analysis of a language is a vital part of natural language processing. In this paper, the statistical analysis of printed Hindi text is performed and then its comparison is done with the analysis already available with printed Punjabi text. Besides analysis of the characters frequency and word length analysis, a more useful unigram, bigram analysis is done. Miscellaneous analysis like Percentage occurrence of various grouped characters and number of distinct words and their coverage in Hindi and Punjabi Corpus is studied.

Keywords: Corpus, Devnagri, Gurmukhi, NLP, Quantitative Analysis.

1 Introduction

Computational Analysis and Processing of Natural Languages are conducted in different parts of Europe and America. In all these languages, a tremendous progress has been made for different applications like Speech synthesis, Lexicography, Handwritten recognition, text summarization, translation between human languages, natural language database and query answering etc. A little research has been done for Indian languages to achieve the goals of these applications. A statistical analysis of languages is must and in this paper, corpus based statistical analysis of the Hindi language (Devnagri) and its comparison with Punjabi language [5] is done.

2 Statistical Analysis

Statistical analysis of different languages is the foremost requirement to have a comprehensive database for all languages. In the present study, quantitative analysis of printed Hindi text has been carried out and then the comparative study with printed Punjabi text is done. The quantitative analysis allows us to discover which phenomena are likely to be genuine reflections of the behavior of a language or variety, and which are merely chance occurrences. Frequency is the main consideration with which the normality and abnormality of a particular phenomena is checked.

A Hindi corpus of size 15.26 MB developed by TDIL, DOE is taken for the research purpose is analyzed statistically to calculate the frequencies and commulative frequencies of different words with different length and characters (unigrams and bigrams).

C. Singh et al. (Eds.): ICISIL 2011, CCIS 139, pp. 209–213, 2011.

3 Results and Discussions

The Corpus contained about 2562577 words and 10501382 characters. A comparison of top 5 most frequently used words in Devnagri and Punjabi corpus[1] is shown in the following table.

Table 1. Comparison of top 5 most frequently used words in Devnagri and Punjabi corpus

S. No.	Hindi Words	Freq.	Comm. Freq.	Punjabi Words	Freq.	Comm. Freq.
1	के	3.877519	3.877519	ਹੈ	2.51	2.51
2	से	3.242188	7.119707	ਦੇ	2.24	4.75
3	में	2.972633	10.092340	ਦੀ	1.85	6.60
4	की	2.534420	12.626760	ਵਿੱਚ	1.82	8.42
5	से	1.810668	14.437428	ਤੇ	1.74	10.16

3.1 Word Length Analysis

Word Length analysis is very useful in the area like information storage and retrieval. The graphs shown below is the comparison of word length frequency of Punjabi and Hindi Text and the cumulative frequency v/s words length of Hindi text.

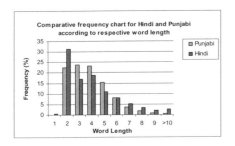
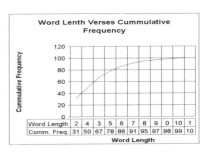

Table 2. Contribution of top 15 words of different lengths in Hindi ad Punjabi Text

Word Length	Comm. Frequency of Top 15 words of Hindi Text	Comm. Frequency of Top 15 words of Punjabi Text
2	73.42	76.75
3	50.27	40.98
4	20.87	22.75
5	10.72	15.19

3.2 Unigram Analysis

Finding the frequency of characters is useful in the areas of cryptography, keyboard design, character recognition etc. The table below gives the frequency and cumulative frequency of top 10 characters in Hindi. When compared, top 10 characters of Hindi Text cover 33.82% irrespective of 31.57% of Punjabi text.

Table 3. The frequency and cumulative frequency of different characters in Hindi

Rank	Character	Freq.	Comm.	Rank	Character	Freq.	Com
1	ा	9.263247	9.26325	6	न	4.244041	38.072
2	क	7.073192	16.3364	7	त	4.083662	42.156
3	र	6.314587	22.6510	8	ि	3.978209	46.134
4	े	5.846335	28.4973	9	स	3.956022	50.090
5	ो	5.331031	33.8283	10	ी	3.894782	53.985

3.2.1 Positional Unigram Analysis

Positional occurrences of top 10 characters (characters that are most commonly used) is analyzed. The most commonly used characters are: ा क र े ो न त ि स ी

Following graphs show top 4 characters according to their position in the word. ा is most frequently used character and its most common position in the word is 2nd.

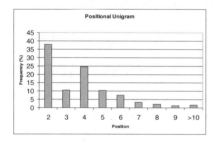

Graph showing positional frequency of ा

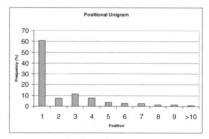

Graph showing positional frequency of क

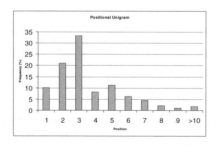

Graph showing positional frequency of र

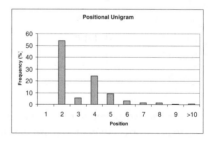

Graph showing positional frequency of े

3.3 Bigram Analysis

Bigram Analysis when compared with Punjabi corpus, top 50 bigrams of Punjabi text covers about 40.65% whereas top 50 bigrams cover 38% of total bigrams in respective languages. Top 10 bigrams of Devnagri Script are shown in the following table:

Table 4. Top 25 bigrams in Devnagri Script

Rank	Bigram	Freq.	Comm. Freq.	Rank	Bigram	Freq.	Comm. Freq.
1	के	1.573613	1.573613	6	ंं	1.262139	8.50083
2	ेर	1.52093	3.094543	7	में	1.175083	9.675913
3	का	1.432667	4.52721	8	या	1.095725	10.77164
4	है	1.430364	5.957574	9	्य	1.091329	11.86297
5	ेर	1.281117	7.238691	10	ने	1.019507	12.88247

3.4 Miscellaneous Analysis

Some more results that are very useful in various applications like Linguistics, speech recognition and optical character recognition, are analyzed here. Various analysis are shown in the table below in terms of percentage of total number of characters.

Table 5. Percentage occurrence of various grouped characters

Character Set	Character Symbol			Percentage
Vowels	अ आ इ ई उ ऊ ओ अनौ ए ऐ ऋ			4.37
Vowel Symbols	अ आ इ ई उ ऊ ओ अनौ ए ऐ ऋ िा ाी ु ू ृ ेा ैा ोा ौा			34.48
Consonants	क ख ग घ ङ च छ ज झ ञ ट ठ ड ढ ण	त थ द ध न प फ ब भ म	य र ल व श ष स ह	33.50
Vowels	अ इ उ ऋ आ ई ऊ ओ औ ए ऐ			4.37
Vowel Symbols	ाा िा ाी ु ू ृ ेा ैा ोा ौा			32.22
Mukta	अ			2.11
Semi Vowel	य र ल व			13.49
Sibilants	श ष स ह			9.27
Halant	्ा			5.33
Nasals	ंं ां			3.42

From the word frequency list coverage can be easily found out. Coverage means how many words covers the particular percentage of the corpus. First 10% of the corpus in Punjabi is covered by 5 words and in Hindi only 3 characters cover the 10% of the whole corpus.

4   Conclusion

The comparative analysis shows that these two languages Hindi and Punjabi are closely related languages. In both the languages top 15 words occupy more than 20% of the whole corpus. Top 15 words of word length 2 occupy more than 70 % of the whole corpus in both the languages. Kanna is the most frequent character covering more than 10% of whole text and its most common position in the word is second. Top 50 bigrams in both Punjabi and a Hindi language covers about 40% of whole corpus. Vowels in Hindi language cover more than half of the corpus whereas in Punjabi the vowels cover only 42% of the whole corpus. In the nutshell, the Hindi and Punjabi languages are closely related languages with most of the similarities.

References

1. Bharti, A., Sangal, R., Bendre, S.M.: Some observation regarding corpora of some Indian languages. In: Proceedings KBCS 1998, pp. 203–213 (1998)
2. Dash, N.S.: A Corpus Based Computational Analysis of the Bangla Language: A Step Towards Natural Language Processing. PhD Thesis submitted to ISI Calcutta (2000)
3. Yannakoudakis, E.J., Tsomokos, I., Hutton, P.J.: N-grams and their implication to natural language understanding. Pattern Recognition 23(5), 509–528 (1990)
4. Church, K.W., Mercer, R.L.: Introduction to the special issue on computational linguistic using large corpora. Computational Linguistic 19(1), 1–23 (1993)
5. Arora, N.: A Statistical Analysis of Gurmukhi Script. M.Tech. Thesis submitted to Punjabi University (2000)

Participles in English to Sanskrit Machine Translation

Vimal Mishra and R.B. Mishra

Department of Computer Engineering, Institute of Technology
Banaras Hindu University, (IT-BHU), Varanasi-221005, U.P., India
vimal.mishra.cse07@itbhu.ac.in, vimal.mishra.upte@gmail.com

Abstract. In this paper, we discuss the participle type of English sentences in our English to Sanskrit machine translation (EST) system. Our EST system is an integrated model of a rule based machine translation (RBMT) with artificial neural network (ANN) model which translates an English sentence into equivalent Sanskrit sentence. We use feed forward ANN for the selection of Sanskrit word like noun, verb, object, adjective etc from English to Sanskrit user data vector (UDV). Our system uses only morphological markings to identify various part of speech (POS) as well as participle type of sentences.

Keywords: Sanskrit, participles, machine translation, English to Sanskrit machine translation, ANN, participles in machine translation, RBMT.

1 Introduction

There have been many MT systems for English to other foreign languages as well as to Indian languages but none for English to Sanskrit MT. Some works on Sanskrit parser and morphological analyzers have done earlier which are as follows. Ramanujan, P. (1992) has developed a Sanskrit parser 'DESIKA', which is Paninian grammar based analysis program. Huet (2003) has developed a grammatical analyzer system, which tags NPs (Noun Phrases) by analyzing sandhi, samasa and sup affixation. Jha et.al (2006) has developed karaka Analyzer, verb analyzer, NP gender agreement, POS tagging of Sanskrit, online Multilingual amarakosa, online Mahabharata indexing and a model of Sanskrit Analysis System.

We have developed a prototype model of English to Sanskrit machine translation (EST) system using ANN model and rule based approach. ANN model gives matching of equivalent Sanskrit word of English word which handles noun and verb. The rule based model generates verb form and noun form for Sanskrit and produce Sanskrit translation of the given input English sentence.

We have divided our work into the following sections. Section 2 presents participles in English and Sanskrit that describe the rules for forming words of participles in Sanskrit which are based on Panini grammar. Section 3 describes the system model of our EST system. Section 4 presents implementation and the result of the translation in GUI form. The conclusions and scope for future work are mentioned in section 5.

C. Singh et al. (Eds.): ICISIL 2011, CCIS 139, pp. 214–217, 2011.
© Springer-Verlag Berlin Heidelberg 2011

2   Participles in English and Sanskrit

In English, the participles are formed from verbs and acts as an adjectives or verbs in a sentence. There are three types of participles in English such as present participles, past participles and future participle. In English, present participles are usually formed by adding "-ing" to a verb. For example, "glowing" and "being" are present participles. Past participles are usually formed by adding "-ed" or "-en" to a verb. For example, "satisfied" and "spoken" are past participles. In Sanskrit, there are many types of participles (called kradanta by Panini) such as present active, present middle, present passive, future active, future middle, future passive (gerundive), past active, past passive, perfect active, perfect middle, gerund and infinitive. In Sanskrit, the participles take krt endings viz. primary nominal endings (Egenes, 2000). In Panini grammar, the rule for past passive particle (PPP) is as follows.

Rule: Kridtind I3I1I93II (Nautiyal, 1997)

In ta (or ita or na), the suffix used for forming the past passive particle from simple verb is ta. As an illustration, consider the following example of English sentence (ES) and their corresponding translation in Sanskrit sentence (SS) that is obtained from our EST system.

ES: Ram went to the forest.
SS: Raamah vanam gatah.

3   System Model of Our EST System

We have developed English to Sanskrit MT (EST) model that comprised the combination of two approaches: rule based model and the dictionary matching by ANN model (Mishra, Vimal and Mishra, R. B., 2010a; 2010b). In this paper, we show the handing of participles in our EST model. In our system, the sentence tokenizer module split the English sentences into tokens (words). The outputs of the sentence tokenizer module are given to POS Tagger module. In POS Tagger module, the part–of-speech (POS) tagging is done on each word in the input English sentence. The output of POS tagger module is given to rule base engine. The GNP detection module detects the gender, number and person of the noun in English sentence. The tense, structure, form and type of English sentence is determined by using rules in the tense and sentence detection module. The noun and object detection module gives noun for Sanskrit of the equivalent English noun using ANN method. The adaptation rules are used to generate the word form. The root dhaatu detection module gives verb for Sanskrit of the equivalent English verb using ANN method. We apply adaptation rules to generate the required dhaatu form. The San_Tr_Rule Detection module gives the number of modules that is used in the Sanskrit translation. In this, we make input data and corresponding output data. The input data has structure, form and type of English sentence in the decimal coded form. We have stored fifty adverbs for Sanskrit of the equivalent English adverb in a database file.

In the ANN based model, we use feed forward ANN for the selection of equivalent Sanskrit word such as noun (subject or object) and verb of English sentence. In feed forward ANN, the information moves in only one direction: forward; from the input nodes, through the hidden nodes (if any) and to the output nodes. The use of feed forward ANN overcomes the output vector limitation (Thrun, S.B., 1991). Our motivation behind use of feed forward ANN in language processing tasks are work of Nemec, Peter (2003) and Khalilov, Maxim et al. (2008). We basically perform three steps in ANN based system such as encoding User Data Vector (UDV), input-output generation of UDV and decoding of UDV. The name of our data sets have called UDV here, which is used in feed forward ANN for the selection of equivalent Sanskrit word such as noun (subject or object) and verb of English sentence. English alphabet consists of twenty-six characters which can be represented by five bit binary (2^5 =32, it ranges from 00000 to 11111). First, we write alphabet (a-z) into five bit binary in which alphabet "a" as 00001, to avoid the problem of divide by zero and alphabet "z" as 11010. For the training into ANN system, we make the alphabet to decimal coded form which is obtained by dividing each to thirty-two. In the Encoding of UDV, we provide the data values to the noun (subject or object) and verb. In input-output generation of UDV, we prepare UDV of noun (subject or object) and verb. After preparing the UDV, we train the UDV through feed forward ANN and then test the UDV. We get the output of Sanskrit word in the UDV form. The output given by ANN model is in decimal coded form. From the verb table, each values of a data set is compared with the values of English alphabet, one by one and the values with minimum difference is taken with its corresponding alphabet from English alphabet. We have generated verb form and word form using rules.

4   Implementation and Results

Our EST system has been implemented on windows platform using Java. The ANN model is implemented using MATLAB 7.1 neural Networks tool. We use feed forward ANN that gives matching of equivalent Sanskrit word of English word which handles noun and verb. We have a data set of 250 input-output pair for verb. The input, hidden and output values for verb is taken 5, 38 and 6. For the noun, we have 250 input-output pair in which the input, hidden and output values are taken 5, 15 and 7. The result from our EST system for participle type of English sentence is shown in figure 1.

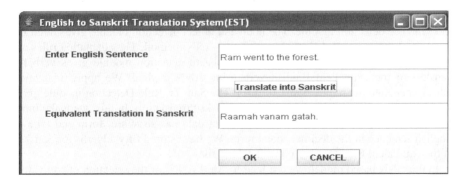

Fig. 1. Participle Type of English Sentence with their Sanskrit translation

First, English sentence is split up into tokens and tokens are matched with Sanskrit word using ANN. We check the gender, number and person (GNP) of the noun, adjective and preposition. According to suffixes of the words, we generate corresponding word form. Then, verb form is generated that depends upon the number and person of the noun. We arrange subject, adverb adjective, preposition and verb, in order to obtain the desired translation into Sanskrit language.

5 Conclusions and Future Scope

Our paper describes the handling of participles in English to Sanskrit MT that uses rule based model and the dictionary matching of equivalent Sanskrit word of English word which handles noun and verb by ANN model. The rule based model enhances the adaptation process. For the further scope of research in this direction, this integrated model would be utilized to other pair of languages for MT approach and the integration of RBMT and ANN may be utilized for EBMT.

References

1. Egenes, T.: Introduction to Sanskrit, Part one & two, 1st edn. Motilal Banarasidas Publishers Pvt. Ltd. (2000)
2. Huet, G.: Towards Computational Processing of Sanskrit. In: Recent Advances in Natural Language Processing, Proceedings of the International Conference ICON, Mysore, India, pp. 1–10 (2003)
3. Jha, G.N., et al.: Towards a Computational analysis system for Sanskrit. In: Proceedings of First National Symposium on Modeling and Shallow parsing of Indian Languages at Indian Institute of Technology Bombay, pp. 25–34 (2006)
4. Khalilov, M., et al.: Neural Network Language Models for Translation with Limited Data. In: Proceedings of 20th IEEE International Conference on Tools with Artificial, pp. 445–451 (2008)
5. Mishra, V., Mishra, R.B.: English to Sanskrit Machine Translation System: A Hybrid Model. In: Proceedings of the International Joint Conference on Information and Communication Technology, IIMT, Bhubaneswar, India, January 9th-10th, pp. 174–180 (2010a)
6. Mishra, V., Mishra, R.B.: ANN and Rule based model for English to Sanskrit Machine Translation. INFOCOMP Journal of Computer Science 9(1), 80–89 (2010b)
7. Nautiyal, C.: Vrihad Anuvaad Chandrika, 4th edn. Motilal Banarasidas Publishers Pvt. Ltd. (1997)
8. Nemec, P.: Application of Artificial Neural Networks in Morphological Tagging of Czech, pp. 1-8 (2003)
9. Ramanujan, P.: Computer Processing of Sanskrit. In: Proceedings of CALP-2, I.I.T., Kanpur, India, pp. 1–10 (2000)
10. Thrun, S.B.: The MONK's Problems: a performance comparison of different learning algorithms. Technical Report CMU-CS-91-197, Carnegie Mellon University (1991)

Web-Drawn Corpus for Indian Languages:
A Case of Hindi

Narayan Choudhary

Jawaharlal Nehru University, New Delhi
choudharynarayan@gmail.com

Abstract. Text in Hindi on the web has come of age since the advent of Unicode standards in Indic languages. The Hindi content has been growing by leaps and bounds and is now easily accessible on the web at large. For linguists and Natural Language Processing practitioners this could serve as a great corpus to conduct studies. This paper describes how good a manually collected corpus from the web could be. I start with my observations on finding the Hindi text and creating a representative corpus out of it. I compare this corpus with another standard corpus crafted manually and draw conclusions as to what needs to be done with such a web corpus to make it more useful for studies in linguistics.

Keywords: Web Corpus, Hindi Corpora, Hindi corpora for linguistic analysis.

1 Introduction

Since the advent of Unicode in Hindi and its use by the common people, the online content in Hindi has increased many folds. There was time when creating a corpus of Hindi text was an uphill task in itself. Thanks to the web, collecting text in Hindi is now easier and a sizeable text corpus of Hindi can be built in a matter of days [1][3][7]. However, this corpus then needs to be worked upon to be cleaned of inherent noise coming in from a public source such as the internet. The present paper is based on the author's personal experience of dealing with creating such a corpus and cleaning it for the purpose of extracting a lexicon and context information out of it and then comparing it with another corpus, created manually and offline from authentic sources such as books and magazines, to find out how good a web corpus can be in linguistic studies of languages like Hindi which still comparatively has scarce presence on the web.

Besides the Hindi newspapers and portals, a great lot of content is also generated through blog posts and comments made by the users. This content is most of the time full of noises. This noise reflects in different forms and at different levels. For any use of such a corpus for linguistic analysis, it has to be representative. Considering the different criterion as discussed in [4][5], an evaluation of such a web corpus is yet to be done.

2 Encoding of Hindi Content

Hindi on the web is encoded in several ways. Before the advent of Unicode in Hindi, it was the era of legacy fonts and image files. There are hundreds of such fonts

C. Singh et al. (Eds.): ICISIL 2011, CCIS 139, pp. 218–223, 2011.

available for the Devanagari script used in Hindi and several other Indo-Aryan languages. All the early contents published in Hindi or other Indian languages were created using these fonts. Most of this content has vanished from the web by now and the Unicode version has now taken over for good.

Even though Unicode is there, the problem of encoding remains at the user's end. Many people who would prefer to write in their own language and script do not know how to do that because the keyboard is Roman by default. Even when the Indian keyboards[1] have been introduced, it is mostly limited to some professionals. Transliteration mechanisms such as iTrans[2] standards are rarely followed and remain out of the knowledge of common users. In such a scenario, the default keyboard for common users remain English and if a user must need to write things on a digital platform, such as personal computers (PCs) or mobile sets, s/he has the default option of writing in Roman only.

2.1 "Hinglish"

For long since the advent of electronic communication for the common people, Hindi and other non-English languages were encoded in Roman itself. So much so that Hindi encoded in Roman and not Devanagari, the official script of the language, has earned a character of its own and has contributed much to what is now being termed as "Hinglish" [2].

The web has a lot of Hindi content written in Roman. This material has come naturally from individual users and reflects more spontaneous data in Hindi. It could be of much use for linguistic research in the coming days and warrants attention of both the linguists and other disciplinarians.

This paper however focuses more on the officially recognized Hindi, i.e. Hindi encoded in Devanagari. The observations made in other sections are for this Hindi. However, if someone wants these 'Hinglish' texts to be included into such a Hindi corpus, these would need to get back-transliterated into Devanagari encoded Hindi which is a separate task in itself. Over the years, Hinglish has attained some characteristics of its own, very distinct from Hindi, and many people have already started calling it a language in the making[3] and there are corporate initiatives[4] that have been

---

[1] A case of InSCRIPT Hindi layout can be pointed out here. This is the standard recommended by the DoE, GOI. However, it requires that one learns typing in Hindi in this particular way which is not a small for users who most of the time shifts to the comparatively easier way of transliteration instead.

[2] ITRANS (version 5.31) http://www.aczoom.com/itrans/

[3] Use of snippets of Hindi sentences has been quite old for writers in Indian English. Examples include the likes of Rushdie, Shobha De etc and a now a whole book (Mahal, 2006). However, now the reverse turn has started becoming trendy. Mainstream news papers like the Nav Bharat Times (Delhi edition) regularly use English words, that too in Roman, in the midst of news stories and headlines.

[4] There are companies who have been offering handsets with a dictionary of 'Hinglish' to create messages over the mobile quickly. This comes as a surprise as Hindi with its Devanagari script is lagging behind while Hinglish (with the grammatical skeleton of Hindi and heavy lexical borrowings from English along with snippets of English phrases and sentences) is being promoted by the likes of mobile companies such as Motorola (viz. its model *Motovuya W180*) to attract the young users.

promoting it. So, including Hinglish in a corpus of Hindi can be a separate subject of study.

2.2 Transliterated Texts

Because the users cannot type Hindi or other Indian languages using the English keyboard, the demand for transliteration has been huge. This demand has been met by several online and offline tools. Offline input method editors like Baraha[5], Microsoft's Indic IME[6], Google's Transliteration IME[7] for Hindi etc. have provided the much needed support for encoding Hindi on PCs. However, few such methods are available on mobile phones and these platforms either miss such a tool or have a tool that is rarely used by the users.

Online transliteration mechanisms have also been contributing its bit on the Hindi content creation over the web. Tools such as Google's Transliterate[8] and Quillpad[9] have been in popular use. These tools are made to provide an easy interface to the users to write in their own language and script. However, it is not always easy to manipulate these tools and users are left with little creativity of their own. For example, if one uses the Google's transliterate system and types words that are not in its dictionary (from where it suggests the words as it is not an encoding system), there will be no way to input those words and the user often resorts to tricks to get the desired character (e.g. writing a colon ':' instead of visarga '◌ः'). This results in unavoidable wrong encodings on the user's part and wrong rendering on the part of the tool being used. Here, consistency of text encoding becomes a concern.

2.3 Machine-Translated Text

With the advent of free machine translation (MT) tools such as Google Translate, a new trend on the web started. There are websites offering content in several languages most of which come translated through MT tools.

This kind of content though can bring preliminary understanding of a foreign page, it is barely worth going into a corpus of any kind (except for the test of the translation tool). As is evident, the machine translation in and out of Indian languages has to go a long way. The mechanically translated text is often misleading and if this gets included into a corpus, it will only cause depreciation in quality in the corpus.

3 Collection Method

Depending on the requirement of the corpus under creation, Hindi text can be collected from the web by some automatic tools or manually. Manual collection of Hindi text over the web is extremely labor intensive and time consuming. However, manual collection is of great value for a language like Hindi if one wants a quality corpus with the required specification such as meeting a particular domain representation.

---

[5] Baraha – Free Indian Language Software. http://www.baraha.com
[6] Microsoft Indic Language Input Tool: Hindi. http://specials.msn.co.in/ilit/Hindi.aspx
[7] Google Transliteration Input Method. http://www.google.com/ime/transliteration/index.html
[8] Google Transliteration. www.google.com/transliterate
[9] Quillpad Editor – www.quillpad.in

For the purposes of this paper, the author collected a corpus manually for the obvious reasons of avoiding any unwanted chunk of text coming from the web. Manual collection also ensured that only those documents from the web were taken that can be said to be representative Hindi (and not snippets inserted in between). Manual collection also ensured that the text selected from the web represented standard Hindi and did not deviate much from what is recommended in encoding it (i.e. had less spelling errors and largely followed the standards by the Central Hindi Directorate [8]).

4 Text Variety

The variety of Hindi text on the web is yet to diversify. The most available variety of Hindi on the web is in the form of news reports, thanks to the online edition of the several news portals. The second most type of available content on the web in Hindi is in the form of blog posts, forums and comments on the blog posts or on articles related to news. All the other domains of text on the web are very scarce on the web. For example, if one looks for text in the health domain, there are only a few portals that caters to this in Hindi and that also is too few and limited. Similar is the case with the texts in the domains of laws and legalities, business and almost all of the academic disciplines. One can say this scarceness is in fact a reflection of the poverty of such literature in the language itself. But to a greater extent that would not hold true because those types of text are available off the web.

5 Text Quality

The quality in terms of lesser noise (spelling errors, foreign words/characters etc) of the text collected online depends on the source from where the text has been taken. As a rule, the text quality over the web is going to be of lower quality than that of those that are taken from other authentic sources like that of publishers' copies or e-books. The quality then goes down as one moves to collect text from the blogs and comments sections of the web pages because these are mostly unedited texts and might have problems like spelling errors etc.

An Experiment to Test the Quality of the Text on the Web

The author had manually collected text from the web and created a representative corpus to extract a basic lexicon of Hindi. The corpus had the following domain breakup to reflect the representation of different types of text. Table 1 summarizes the domain representation of this online corpus.

This corpus had an overall size of 4.2 megabytes and the total number of words it had was 415372. Total unique number of words in this corpus was 37088, including numbers, both in Arabic and Nagari notation and foreign words which included English words written in Roman and a few words of other languages in their own script. After removing these 'noise' words and a few notation issues e.g. wrong placement of *nuktas*, done easily through a simple sort method, we embarked upon selecting the valid words out of it. Over a period of three months of editing and proof reading, we came up with a total number of valid words at 24154 without numerical words and 26222 with numerical words. This could have been done if there were a spell checker available that could validate these words. But, unfortunately, we could not find any such tool for Hindi that can be counted.

Table 1. Domain-wise break-up of the online corpus

Domain	Size (in KBs)	Domain	Size (in KBs)
Legal	216	Science & Technology	419
Trade & Business	229	Culture, Art & History	430
Music & Film	300	Geography	517
Sports	323	Literature	594
Politics & Economics	406	Others: fashion, food, hobby etc.	870

We compared this with another corpus that was created manually by selecting text from various sources barring the internet. This was a corpus created by the Hindi team of Indian Languages Corpora Initiative (ILCI) [6] team. Out of the two domains of health and tourism covered comprehensively, I have taken the domain of health to compare with corpus created through content from the web. This ILCI corpus created manually for the purpose of parallel corpora in 12 Indian languages is supposed to have 0% noise because all the text taken in this corpus is from authenticated source and is further validated by the language editors and linguists.

Table 2. A comparison of the corpus collected from the web and a manual corpus

	Example		% of Total Words		% of Unique Words	
	C1	C 2	C1	C2	C1	C2
Total Number of Words	415372	419420	-	-	-	-
Total Number of Unique Words	37088	21466	8.92	5.11	-	-
Valid Words (Without numerals)	24154	20576	5.81	4.90	65.12	95.85
Valid Words (With Numerals)	26222	20968	6.31	4.99	70.70	97.68
Valid Numerals	2068	392	0.49	0.09	5.57	1.82
Words in English (Roman)[10]	1380	496	0.33	0.11	3.72	2.31

As shown in Table 2 above, we find that unique number of words extracted from the web corpus (C1) is much greater in number than that of the manual corpus which is our standard here (C2). However, this unfortunately, does not mean that the corpus is rich in terms of diversity. In fact it only increases the job by warranting the validation of each of the words if one is supposed to extract a word list out of it. If one is to put this web corpus to some other use like that of doing some sort of

---

[10] The English words are not verified whether they are correct or not. It includes any words that constitutes of Roman character. The English words in the Health domain data of the ILCI are basically words that come in the original text and it is a common practice to include the English words in brackets after their Hindi equivalents. Inclusion of English words (in Roman script) seems to be inevitable even in a high standard supervised corpus like that of ILCI.

linguistic analysis or making it as a source for parallel corpora creation, such a kind of corpus will undergo heavy editing as the error rate with regard to words is about 26% which is pretty high for an ideal corpus.

6 Conclusions

In this paper I have tried to show that though Hindi now has a fair presence on the web, the content taken inadvertently from the web needs heavy editing before the language in such a corpus can be said to be representative.

Acknowledgments. Thanks a lot to the ILCI Hindi Team at JNU who prepared the comprehensive Hindi corpora in two domains and to Dr. Girish Nath Jha for allowing me to use the corpora even before it has been formally released.

References

1. Kilgarriff, A., Reddy, S., Pomikálek, J., Avinesh, P.V.S.: A Corpus Factory for Many Languages. In: Proceedings of Asialex, Bangkok (2009)
2. Mahal, B.K.: The Queens English: How to Speak Pukka. Collins (2006)
3. Biemann, C., Heyer, G., Quasthoff, U., Matthias, R.: The Leipzig Corpora Collection: Monolingual Corpora of Standard Size. In: Proceedings of Corpus Linguistics Birmingham, UK (2007)
4. Biber, D.: Representativeness in Corpus Design. Literary and Linguistic Computing, 8(4) (1993)
5. Leech, G.: New resources or just better old ones? The Holy Grail of Representativeness. In: Mair, C., Meyer, C.F. (eds.) Corpus Linguistics and the Web, Rodopi, Amsterdam, New York (2007)
6. Jha, G.N.: The TDIL Program and the Indian Language Corpora Initiative (ILCI). In: Calzolari, N., et al. (eds.) Proceedings of the Seventh Conference on International Language Resources and Evaluation (LREC 2010). European Language Resources Association (ELRA). (2010)
7. Baroni, M., Bernardini, S.: BootCaT: bootstrapping corpora and terms from the web. In: Proceedings of the 4th International Conference on Language Resources and Evaluation (LREC-2004), Lisbon (2004)
8. Taneja, P., et al. (eds.): Devanagari Lipi Tatha Hindi Vartani ka Manakikaran. Central Hindi Directorate, New Delhi (2006)

Handwritten Hindi Character Recognition Using Curvelet Transform

Gyanendra K. Verma, Shitala Prasad, and Piyush Kumar

Indian Institute of Information Technology, Allahabad, India
Allahabad, India - 211012
{gyanendra,ihc2009011,ihc2009017}@iiita.ac.in

Abstract. In this paper, we proposed a new approach for Hindi character recognition using digital curvelet transform. Curvelet transform well approximate the curved singularities of images therefore very useful for feature extraction to character images. A Devanagari script contains more than 49 characters (13 vowels and 33 consonants) and all the characters are rich in curve information. The input image is segmented first then curvelet features are obtained by calculating statistics of thick and thin images by applying curvelet transform. The system is trained with K-Nearest Neighbor classifier. The experiments are evaluated with in-house dataset containing 200 images of character set (each image contains all Hindi characters). The results obtained are very promising with more than 90% recognition accuracy.

Keywords: Hindi Character Recognition, Curvelet Transforms, K-Nearest Neighbor (K-NN).

1 Introduction

Natural language is the prime mode of communication for Human beings. Current technologies have more emphasis on the better human-computer interaction. It is convenient for human being to work with their native language. India is multi-lingual country comprises many languages like Hindi, Bengali, Gujarati, Malayalam, Tamil, Telugu, Urdu and so on. Hindi character recognition has been draw attention to the researchers in last few decades due to its various application potentials. There are twenty two languages in India and eleven scripts are used to write these languages. Hindi is the most popular language in India. This paper proposed a novel approach for hand-written Hindi character recognition using curvelet transform. Nowadays Curvelet transform is used due to its multiple scale analysis property in various pattern recognition problems i.e. Face recognition, finger print recognition, signature recognition etc. The K-NN is a method for classifying objects based on closest training examples in the feature space, has been used here to classify Hindi characters.

Various character recognition approaches has been employed in recent years such as Neural Network [1], Hidden Markov Model and Fuzzy Model Based Recognition of Handwritten Hindi Characters [2]. S.Abirami and Dr. D.Manjula [3] proposed two categories namely global and local approach for script identification.

C. Singh et al. (Eds.): ICISIL 2011, CCIS 139, pp. 224–227, 2011.

2 Devanagari Script Characteristics

Many Indian languages such as Hindi, Nepali, Sanskrit and Marathi are based on Devanagari script. Devanagari script contains 13 vowels and 33 consonants make it different from other scripts. The complexity of Devanagari scipt is due to many isolated dots which are vowel modifiers such as "*anushwar*", "*visharga*" and "*chandra bindu*". Devanagari script has following characteristics.

- A very scientific script. Read and written from left to right
- Developed in medieval India from ancient brahmi script.
- One symbol having one sound (syllable) and vise-versa. Only slight and minor exceptions

3 Proposed Approach

We implement the system using curvelet features obtained form curvelet coefficients of character images. The character recognition process comprises the following steps. Step1: Pre-processing of the character image. This step includes normalization, noise removal and gray scale conversion.

Step2: the hand written character image is segmented into single character by segmentation process. Step3: the thinning and thickening of the characters is done in this

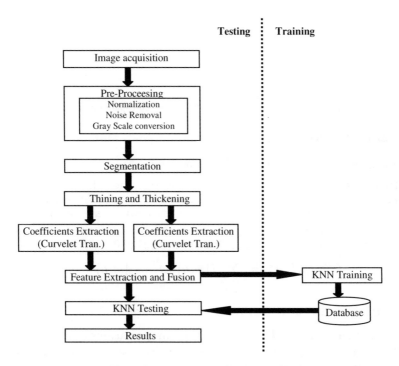

Fig. 1. Proposed system block diagram

step. Step4: digital curvelet transform is applied over thick and thin image in order to obtain curvelet coefficients. Step5: fusion of the curvelet coefficients and feature extraction by applying general statistics. Step6: train the features using KNN classifier. Step7: obtain result based on similarity matching between features of input character image and set of reference image. The block diagram of the proposed system is illustrated in figure1.

3.1   The Curvelet Transform

Curvelet transform is extension of the wavelet concept. The discrete curvelet transform can be applied using two algorithms namely Unequispaced FFT transform and Wrapping transform. In unequispaced Fast Fourier transform, the curvelet coefficients are formed by irregularly sampling the Fourier coefficients of the image. In wrapping algorithm, curvelet coefficients are formed by using a series of translations and a wraparound technique. The performance of wrapping based algorithm is fast in computation and more robust as compare to USFFT however both algorithms give the same output [4]. Figure 2 shows curvelets in frequency as well as spatial domain. The shaded area is one of the wedges, formed by dividing the frequency plane into different partitions and the spatial plane is divided in respect to θ (angular division). The angular division divides each subband image into different angles.

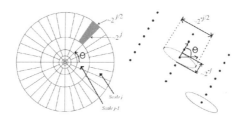

Fig. 2. Curvelets in frequency domain (left) and spatial domain (Right)

3.2   Feature Extraction Algorithm

Algorithm for finding the curvelet features are as follows:

STEP1: Take the Fourier transform of the character image
STEP2: Divide the frequency plain into polar wedges
STEP3: Find the curvelet coefficients at a particular scale (j) and angle (θ) by taking the inverse FFT of each wedge at scale j and oriented at angle θ.

$$C^D(j,l,k_1,k_2) = \sum_{\substack{0 \leq m < M \\ 0 \leq n < N}} f[m,n] \; \varphi^D_{j,l,k_1,k_2}[m,n] \tag{1}$$

Here C^D is digital curvelet transform. j is scale, l is orientation and (k_1, k_2) are location parameters.
STEP4: apply the common statistics at each subband to obtain the curvelet features.

4 Experimental Results and Discussion

We have generated In-House Hindi character database for experiments. The Hindi characters were written by hundred individual authors twice in their own hand writing. The written characters are scanned and saved in jpeg format then segmented to extract single character from scanned image. We applied fast discrete curvelet transform on each images to extract the curvelet coefficients. 4th and 6th level decomposition performed using Curvelab-2.1.2 [5]. All the experiments were performed on MATLAB 7.6.0 (R2008a).

Given an image, both the image and the curvelet are transformed into Fourier domain, and then the convolution of the curvelet with the image in spatial domain becomes the product in FD. Finally the curvelet coefficients are obtained by applying inverse Fourier transform on the spectral product. Once the curvelet coefficients have been obtained, the mean and standard deviation are computed to obtain curvelet features. The similarity matching among query feature and set of reference features are done using Euclidean distance through K-NN classifier. The classification accuracy is shown in table 1. Our results are better than other shape based technique [6] for handwritten character recognition.

Table 1. Classification accuracy

No. of characters		10	20	30	40
Recognition	6 level curvelet	84	84	88	**90.2**
Accuracy (%)	4 level curvelet	84	81	82.6	86.5

5 Conclusion and Future Work

A Curvelet transform based Hindi character recognition approach is proposed in this paper. The discrete curvelet transform is used to extract the curvelet coefficients for feature extraction. KNN classifier has been used for pattern classification in this study. The experiments result shows more than 90% accuracy.

References

1. Wang, X., Huang, T.-l., Liu, X.-y.: Handwritten Character Recognition Based on BP Neural Network. In: 3rd International Conference on Genetic and Evolutionary Computing, pp. 520–524. IEEE, Los Alamitos (2009)
2. Hanmandlu, M., Ramana Murthy, O.V., Madasu, V.K.: Fuzzy Model Based Recognition of Handwritten Hindi Characters. In: 9th Biennial Conf. on Digital Image Computing Tech. and Appli., Glenelg, Australia, pp. 454–461 (2007)
3. Abirami, S., Manjula, D.: A Survey of Script Identification techniques for Multi-Script Document Images. International Journal of Recent Trends in Engineering 1(2), 246–249 (2009)
4. Curvelet Litrature, http://en.wikipedia.org/wiki/Curvelet
5. Curvelet Software, http://www.curvelet.org/software.html
6. Prachi, M., Rege Priti, P.: Shape Feature and Fuzzy Logic Based Offline Devnagari Handwritten Optical Character Recognition. J. Pattern Recognition Research 4, 52–68 (2009)

Challenges in Developing a TTS for Sanskrit

Diwakar Mishra[1], Girish Nath Jha[2], and Kalika Bali[3]

[1,2] Special Centre for Sanskrit Studies
Jawaharlal Nehru University, New Delhi
[3] Microsoft Research Lab India, Bangalore
{diwakarmishra,girishjha,kalikabali}@gmail.com

Abstract. In this paper the authors present ongoing research on Sanskrit Text-to-Speech (TTS) system called 'Samvachak' at Special Centre for Sanskrit Studies, JNU. No TTS for Sanskrit has been developed so far. After reviewing the related research work, the paper focuses on the development of different modules of TTS System and possible challenges. The research for the TTS can be divided into two categories – TTS independent linguistic study, TTS related Research and Development (R&D). The TTS development is based on the Festival Speech Synthesis Engine.

Keywords: TTS, Speech Synthesis, Festival, normalization, word recognition, sentence recognition, phonotactics, POS, annotation, speech database.

1 Introduction

Sanskrit's heritage status requires that we develop digital libraries which can be 'read' by the machine. So far, no speech synthesizer system has been developed for Sanskrit. Among the obvious user groups for a Sanskrit TTS, we can count the visually challenged, children and elderly and spiritual with difficulty in reading. One can convert books into sound files to be accessed through music devices. It can be used for e-learning of pronunciations. Besides, any research for TTS – phonotactics, annotated speech database etc., can be put to use in further research in Sanskrit.

2 TTS for Other Indian Languages

A Text-to-Speech system for Hindi was developed at HP Labs India. It was developed on the Festival framework under Local Language Speech Technology Initiative project. [1][12] C-DAC Kolkata is developing two synthesis systems – *Bangla Vani* for Bengali and *Nepali Boli* for Nepali language. For the *Nepali Boli*, they are using ESNOLA based indigenous technology. [4] IIIT-Hyderabad and Sanskrit Academy, Osmania University are developing a prototype Sanskrit TTS using Festvox.

Dhvani, an ambitious umbrella project for Indian languages TTS started from IISc Bangalore. [6] It currently supports 10 Indian languages (Bengali, Gujarati, Hindi, Kannada, Malayalam, Marathi, Oriya, Punjabi, Tamil and Telugu) and Pushto. [7]

C. Singh et al. (Eds.): ICISIL 2011, CCIS 139, pp. 228–231, 2011.
© Springer-Verlag Berlin Heidelberg 2011

Government of India has sponsored a consortia project on Text to Speech for Indian Languages (TTS-IL) with IIT Chennai as the leader. The languages covered in this project are – Hindi, Bengali, Marathi, Tamil, Telugu and Malayalam. [8]

3 Requirements for the Sanskrit TTS

The proposed Sanskrit TTS will be developed in the Festival environment. [2][3] It is a free multilingual speech synthesis workbench, developed by the University of Edinburgh and CMU that runs on multiple-platforms. It offers a TTS system as well as an open architecture for research in speech synthesis. Though other engines for TTS are available but none of them provides a research framework like Festival.

Developing a TTS for a new language requires language specific modules. There has been fewer research activity directly oriented to Sanskrit speech technology. This research in fact depends upon various researches, some of which are richly worked and some very less. These can be classified into two categories-

a) **Linguistic research** which requires text pre-processing, normalization, phonotactics, word/sentence recognition, study of prosody patterns, morphology analysis, and light processing - POS tagging and sandhi analysis.

b) **TTS specific R&D** which will require the following – (i) phone listing and Grapheme-to-Phoneme (G2P) rules (phonology and phonotactics); (ii) prosodic phrase recognition; (iii) text preparation for recording; (iv) recording the sound and to annotate the sound database.

3.1 Text Processing, Normalization and Word/Sentence Recognition

Text processing includes the task of recognizing the readable text - cleaning of non-linguistic information, language recognition/script recognition. The input for the proposed TTS will be UTF-8 encoded. For wide usability of the system a script converter to Devanagari Unicode will also be required. Text normalization is to convert all non-words like numbers, dates, acronyms, quotes, abbreviations etc. into word form. In the case of number in Sanskrit, for example, 2116 can be pronounced in two ways - '*ṣoḍaśottaraikaśatottaradvisahasram*' and '*dvisahasraikaśataṣoḍaśa*'.

Word recognition, a basic requirement for the TTS, cannot rely upon space as the tokenizer in Sanskrit. Words are conjugated with sandhi, and in many places, where there is no sandhi, the words are written continuously, like – *ahamapi, tatkimiva*. Even for reading conjugated words in sandhi condition, word recognition is must for putting appropriate pauses or utterance boundary. In human practice, the sound is lowered down on meaningful boundaries, for example, '*rūparasagandhasparśa.....*' the listing of 24 *guṇas* of Nyāya philosophy. [6] This will obviously need sandhi and morph analyses. Prosodic phrase recognition also depends on this word recognition.

Sentence demarcation is also important for a TTS. Sandhi and morph and *kāraka* analysis have an additional role in sentence recognition if it cannot be recognized with the punctuations as word order is not a dependable phenomenon. Many of these tasks are certainly not new challenges as they have been done by various other applications done for Sanskrit. [9][10][11]

3.2 Phonotactics and Prosody

For TTS, a list of phones and phonotactic combinations – all possible or permissible phoneme sequences – is required. Generally, in Sanskrit, no two vowels occur continuously in a word but there are exceptions, for example, cases of *prakṛtibhāva* (P. 6.1.115-128) and other sandhi, as, *sa eva, amū atra*. This also includes syllable structure. Sanskrit permits a maximum of five consonants in a cluster, as in *kārtsnyena,* where the first two consonants are the coda of the first syllable and remaining three are the onset of the second.

There are very few people in India who have Sanskrit as mother tongue. Sanskrit is not a naturally spoken language in major society. Therefore finding a Sanskrit speaker who is authentic for the prosody of speech is difficult. Sanskrit *śāstra*s discuss a lot about the prosody of poetry but not much on the prosody of prose utterance.

Vedic Sanskrit has the tradition of supra-segmental features of vowels (vowels correspond to syllables), high, low and circumflex. These are different from the stress and tone. In the course of time, these supra-segmental features lost their phonemic status and their use in the language became less prevalent. In the grammar of *laukika* Sanskrit, these are defined with a very short description (P 1.2.29-31). A thorough investigation in the prosodic models of Sanskrit would involve study of *prātiśākhya* tradition, fieldwork, and observation from speech of Sanskrit speakers.

3.3 Grapheme-to-Phoneme (G2P) Rules

The script of Sanskrit corresponds to its pronunciation more than other languages, but its character sequence is not parallel to the phoneme sequence. Prototype Sanskrit TTS at IIIT Hyderabad uses G2P rules developed for Hindi TTS. That is not useful for Sanskrit TTS as pronunciation habit and syllable structure of Sanskrit are different from Hindi. Phoneme splitter used in sandhi analysis at JNU (of which an example is given below) will be useful only after some modification. After that ASCII characters need to be assigned to each of the phonemes because the proposed platform – Festival – accepts only ASCII characters. The example of phoneme splitter is following.

Input: युधिष्ठिरस्तु त्वरयति

Output: य् उ ध् इ ष् ठ् इ र् अ स् त् उ त् व् अ र् अ य् अ त् इ

3.4 Text Preparation, Sound Recording and Annotation

The speech database should have coverage of all linguistic entities at all levels – all the phones, all phonotactic combinations, all tonal and intonation patterns, all kinds of sentences, etc. So, before recording the sound, a text has to be prepared which covers all of these. Imaginary nonsense words and sentences are not good for this purpose. It is ideal if the database contains 2-5 tokens of each sound combination, phrase and sentence model. Most of this text will consist of Pañcatantra and include some text of Kādambarī and Daśakumāracaritam to cover the rare phoneme sequences. This prepared text will be useful for developing a different TTS for Sanskrit.

For recording, first task is to find a speaker who speaks Sanskrit naturally and also his/her voice is fit for TTS. For naturality and consistency in sound, the person to record voice should be trained and all recording should be done in studio environment with similar speaker conditions. Sanskrit being a less spoken language, a natural and professional speaker is relatively difficult to find.

The entire database has to be annotated on three levels- phonetic, word, and prosody. Utmost care has to be taken while annotating speech on phonetic level. The phone boundaries should be exact or almost exact, because if it gets the colour of the neighbouring sounds, then on adding other sounds, it may sound corrupted.

4 Conclusion

The authors of the paper are developing a Sanskrit TTS in Festival environment. Developing a TTS in itself is a challenging task. We have pointed out some Sanskrit specific challenges for the TTS. Most important of them are –finding a natural and professional speaker, modelling prosody and linguistic analysis for word and sentence recognition. Also some basic research for the language like study of accents in linguistic environment, intonation study, and phonotactic study is not easily available.

References

1. Bali, K., Ramakrishnan, A.G., Talukdar, P.P., Krishna, N.S.: Tools for the Development of a Hindi Speech Synthesis System. In: Proc. 5th ISCA Speech Synthesis Workshop, pp. 109–114. CMU (2004)
2. Taylor, P.A., Black, A., Caley, R.: The architecture of the festival speech synthesis system. In: The Third ESCA Workshop in Speech Synthesis, pp. 147–151. Jenolan Caves (1998)
3. Richmond, K., Strom, V., Clark, R., Yamagishi, J., Fitt, S.: Festival multisyn voices. In: Proc. Blizzard Challenge Workshop (in Proc. SSW6), Bonn (2007)
4. Annual Report of C-DAC (2008-2009)
5. Thottingal, S.: Dhvani: Indian Language Text to Speech System. In: FOSS.IN, National Science Symposium Centre, IISc, Bangalore (2007)
6. Vangiya, S.: Tarkasaṅgraha of Annambhaṭṭa, with Nyāyabodhinı of Govardhana Misra, Chaukhambha Sanskrit Sansthan, Varanasi (1976)
7. Dhvani – Sourceforge website, http://dhvani.sourceforge.net
8. C-DAC Mumbai website, http://www.cdacmumbai.in
9. Sanskrit Heritage site of Gerard Huet, http://sanskrit.inria.fr
10. Department of Sanskrit website, University of Hyderabad, http://sanskrit.uohyd.ernet.in
11. Computational Linguistics R&D website, JNU, http://sanskrit.jnu.ac.in
12. The Local Language Speech Technology Initiative website, http://www.llsti.org

A Hybrid Learning Algorithm for Handwriting Recognition

Binu P. Chacko and P. Babu Anto

Department of Information Technology
Kannur University, India
{binupchacko,bantop}@gmail.com

Abstract. Generally, gradient based learning algorithms have showed reasonable performance in the training of multi layer feed forward neural networks; but they are still relatively slow in learning. In this context, a hybrid learning algorithm is used by combining differential evolution (DE) algorithm and Moore-Penrose (MP) generalized inverse to classify handwritten Malayalam characters. DE is used to select the input weights and biases in a single layer feed forward neural network (SLFN), and the output weights are analytically determined with MP inverse. A new set of features known as division point distance from centroid (DPDC) is used to generate patterns for the classifier. The system could provide overall recognition accuracy 83.98% by spending only 184 seconds for training.

Keywords: Differential evolution, MP generalized inverse, Division point feature, Neural network.

1 Introduction

This research work on hybrid learning is a combination of differential evolution (DE) algorithm and Moore-Penrose (MP) generalized inverse. DE is one of the most powerful stochastic real parameter optimization algorithms in current use. It operates through similar computational steps as employed by a standard evolutionary algorithm (EA). However, unlike traditional EAs, DE employs difference of the parameter vectors to explore the objective function landscape. It is applicable to multi objective, constrained, large scale, and unconstrained optimization problems. On non separable objective functions, the gross performance of DE in terms of accuracy, convergence speed, and robustness make it attractive for various applications where finding an appropriate solution in reasonable amount of computational time is much weighted. The space complexity of DE is low when compared to some other real parameter optimizers like covariance matrix adaptation evolution strategies (CMA-ES). Compared to other EAs, DE is simple and straight forward to implement. The number of control parameters in DE is also very few (CR, F, and NP only) [3].

This paper deals with recognition of handwritten Malayalam characters using a hybrid learning algorithm. In the literature, a variety of network architectures have been tried for hybrid handwriting recognition, including multilayer perceptrons, time delay

C. Singh et al. (Eds.): ICISIL 2011, CCIS 139, pp. 232–235, 2011.
© Springer-Verlag Berlin Heidelberg 2011

neural networks, and recurrent neural networks. A more successful of these approaches has been to combine neural networks with hidden markov models [1]. Hybrid learning approach involving MP inverse method tends to obtain good generalization performance with dramatically increased learning speed. This classifier uses a feature vector known as division point distance from centroid (DPDC), which is generated by dividing the character image into sub images so that each sub image contains approximately same amount of foreground pixels. Section 2 describes this feature extraction method. The hybrid learning algorithm is explained in section 3. Experimental results are presented in section 4, and the paper concluded in last section.

2 Division Point Distance Feature

The idea of recursive subdivision of character image is used to find the DPDC feature. This process is applied on a size normalized ($N \times N$) and binarized image. The feature extraction method is based on different levels of granularity. At the first iteration step, the binary image is divided into four sub images using a vertical and a horizontal line. A vertical line is initially drawn that minimizes the absolute difference of the number of foreground pixels in the two sub images to its left and to its right. Subsequently, a horizontal line is drawn which minimizes the absolute difference of the number of foreground pixels in the two sub images above and below. The pixel at the intersection of these lines is referred to as the division point (DP). At further iteration steps, each sub image obtained at the previous step is further divided into four sub images using the same procedure [5]. Once the process is finished, the distance of each division point in the last level is measured from the centroid of the image. This will constitute the DPDC feature vector to be used in the classification stage.

3 Hybrid Learning Algorithm

In this section, a hybrid learning approach is introduced by combining DE and MP generalized inverse. Initially, the population is randomly generated. Each individual in the population is composed of a set of input weights and hidden biases:

$$\theta = [w_{11}, w_{12}, ..., w_{1N}, w_{21}, w_{22}, ..., w_{2N}, ..., w_{n1}, w_{n2}, ..., w_{nN}, b_1, b_2, ..., b_N]$$

where N and n are the number of hidden nodes and input parameters respectively [4]. All w_{ij} and b_j are randomly initialized within the range [-1, 1].

Next, the output weights of each set of input weights and biases are analytically determined using MP generalized inverse. Consider a set of M distinct samples ($\mathbf{x}_i$, $\mathbf{y}_i$) with $\mathbf{x}_i \in \mathbb{R}^{d_1}$ and $\mathbf{y}_i \in \mathbb{R}^{d_2}$. Then, an SLFN with N hidden neurons is modeled as:

$$\sum_{i=1}^{N} \beta_i f(\mathbf{w}_i \mathbf{x}_j + b_i), \qquad 1 \leq j \leq M \qquad (1)$$

with f being the activation function, $\mathbf{w}_i$, the input weights, b_i the biases, and β_i, the output weights. If SLFN perfectly approximates the data, the errors between estimated outputs $\hat{\mathbf{y}}_i$ and actual outputs $\mathbf{y}_i$ are zero and the relation is

$$\sum_{i=1}^{N} \beta_i f(\mathbf{w}_i \mathbf{x}_j + b_i) = \mathbf{y}_i \qquad (2)$$

which writes compactly as $\mathbf{H}\beta = \mathbf{Y}$,

$$\beta = \mathbf{H}^{\dagger}\mathbf{Y} \qquad (3)$$

where

$$\mathbf{H} = \begin{pmatrix} f(w_1 x_1 + b_1) & \cdots & f(w_N x_1 + b_N) \\ \vdots & \vdots & \vdots \\ f(w_1 x_M + b_1) & \cdots & f(w_N x_M + b_N) \end{pmatrix} \qquad (4)$$

and $\beta = (\beta_1^T, \ldots, \beta_N^T)^T$, $\mathbf{Y} = (y_1^T, \ldots, y_M^T)^T$ and $\mathbf{H}^{\dagger}$ is the MP generalized inverse of matrix $\mathbf{H}$. Equation (3) is the minimum norm least square (LS) solution to the linear system $\mathbf{H}\beta = \mathbf{Y}$. Minimum training error can be reached with this solution [2].

Finally, calculate the fitness (validation error) of all individuals in the population and apply mutation, crossover and selection of DE. During selection, the mutated vectors are compared with the original ones, and the vectors with better fitness values are retained to the next generation. Generally, neural networks show good performance with smaller weights. To incorporate this, one more criteria is added to the selection – the norm of output weights, $\|\beta\|$. If the difference of the fitness between different individuals is small, the one resulting in smaller $\|\beta\|$ is selected [4].

4 Experiment

In order to attain the reasonable recognition accuracy (RA), several experiments have been conducted and found the suitable size of the character image, number of hidden neurons and the mutation method. The offline character used in our experiments consists of grayscale images scanned from handwritten forms with a scanning resolution of 300 dpi and a gray-scale bit depth of eight. The preprocessing begins with image size normalization. It is found that an image of size 128 x 128 pixels can give good feature vector to the classifier. These normalized images are further undergone a process known as thresholding. DPDC features are then extracted from these size normalized binary images. The process is gone up to level 3, and generated 16 division points. The distance of each point from the centroid is measured and formed the feature vector. DPDC in other levels shown less accuracy due to lack of discriminative features (less than level 3) or ran out of memory as the number of hidden neurons increase (beyond level 3).

A lot of research work has been undertaken to improve the performance of DE by turning its control parameters. A good initial choice of scalar number F is 0.5, and the effective range is usually between 0.4 and 1. No single mutation method is turned out to be the best for all problems. The parameter CR (crossover rate) controls how many parameters in expectation are changed in a population member. With this view, controls parameters are tuned and selected as F=1, CR=0.8 and the mutation scheme DE/rand/1 with exponential crossover. In addition, classifier is trained with different number of neurons in the hidden layer, and the result thus obtained is shown in Table 1. We have used 13200 characters of 44 different classes for this experimentation. This character set is divided into training and test set in the ratio 3 to 2.

Table 1. Recognition accuracy of handwritten Malayalam characters

Learning algorithm	Hidden neurons	Training time (sec)	RA (%)
Hybrid learning	500	79	80.12
	600	119	81.88
	700	156	83.24
	750	184	83.98
BP	10	2403	73.75

The experimental results show that the highest RA of 83.98% is obtained when the network is trained with 750 neurons in the hidden layer. Beyond this limit, the algorithm ran out of memory. The result is also compared with that of the network trained with back propagation (BP) algorithm. The latter case could produce only 73.75% accuracy by spending more time in training. We have also obtained more RA (82.35%) than an ANFIS based work reported in 44 class handwritten Malayalam character recognition [6].

5 Conclusion

The work on handwritten character recognition has achieved a great success in reducing training time of the classifier. This is mainly due to the hybrid approached adapted in the classification process. Even though the algorithm was fast, it is necessary to take further steps to improve the RA.

References

1. Graves, A., Liwicki, M., Fernández, S., Bertolami, R., Bunke, H., Schmidhuber, J.: A Novel Connectionist System for Unconstrained Handwriting Recognition. IEEE Transactions on Pattern Analysis and Machine Intelligence 31(5), 855–868 (2009)
2. Miche, Y., Sorjamma, A., Bas, P., Simula, O.: OP-ELM: Optimally Pruned Extreme Learning Machine. IEEE Transactions on Neural Networks 21(1), 158–162 (2010)
3. Das, S., Suganthan, P.N.: Differential Evolution: A Survey of the State-of-the-Art. IEEE Transactions on Evolutionary Computation, doi:10.1109/TEVC.2010.2059031
4. Zhu, Q.-Y., Qin, A.K., Suganthan, P.N., Huang, G.-B.: Evolutionary Extreme Learning Machine. Pattern Recognition 38, 1759–1763 (2005)
5. Vamvakas, G., Gatos, B., Perantonis, S.J.: Handwritten character recognition through two stage foreground sub sampling. Pattern Recognition 43, 2807–2816 (2010)
6. Lajish, V.L.: Adaptive Neuro-Fuzzy Inference Based Pattern Recognition Studies on Handwritten Character Images. Ph.D thesis, University of Calicut, India (2007)

Hindi to Punjabi Machine Translation System

Vishal Goyal and Gurpreet Singh Lehal

Department of Computer Science, Punjabi University, Patiala
{vishal.pup,gslehal}@gmail.com

Abstract. Hindi-Punjabi being closely related language pair, Hybrid Machine Translation approach has been used for developing Hindi to Punjabi Machine Translation System. Non-availability of lexical resources, spelling variations in the source language text, source text ambiguous words, named entity recognition and collocations are the major challenges faced while developing this syetm. The key activities involved during translation process are preprocessing, translation engine and post processing. Lookup algorithms, pattern matching algorithms etc formed the basis for solving these issues. The system accuracy has been evaluated using intelligibility test, accuracy test and BLEU score. The hybrid syatem is found to perform better than the constituent systems.

Keywords: Machine Translation, Computational Linguistics, Natural Language Processing, Hindi, Punjabi. Translate Hindi to Punjabi, Closely related languages.

1 System Architecture

Hindi-Punjabi Language pair being closely related language pair[1] and on the basis of related works[2,7,8,9], Hybrid (direct and rule based) approach has been used for developing Machine translation System for this language pair. The System Architecture of this machine translation includes Pre Processing Phase, Translation Engine and Post Processing Phase. **Pre Processing Phase** is a collection of operations that are applied on input data to make it processable by the translation engine. Here, various activities incorporated include text normalization [3], replacing collocations and replacing proper nouns. Tokenizers (also known as lexical analyzers or word segmenters) segment a stream of characters into meaningful units called tokens. The tokenizer takes the text generated by pre processing phase as input. This module, using space, a punctuation mark, as delimiter, extracts tokens (word) one by one from the text and gives it to translation engine for analysis till the complete input text is read and processed.

Translation Engine Phase of the system involves various sub phases that includes Identifying titles, Identifying surnames, word-to-word translation using lexicon lookup, Word sense disambiguation and handling out-of-vocabulary words. All the modules have equal importance in improving the accuracy of the system. Identifying Titles locates proper nouns where titles are present as their previous word like श्री (*shrī*), श्रीमान (*shrīmān*), श्रीमती (*shrīmtī*) etc. There is one special character '॰' in Devanagari script to mark the symbols like डा॰, प्रो॰. If tokenizer found this symbol

C. Singh et al. (Eds.): ICISIL 2011, CCIS 139, pp. 236–241, 2011.
© Springer-Verlag Berlin Heidelberg 2011

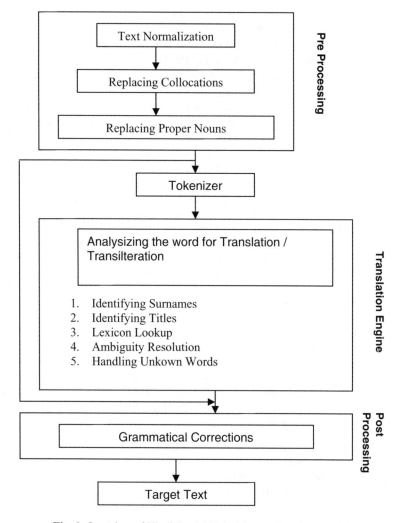

Fig. 1. Overview of Hindi-Punjabi Machine Translation System

during reading the text, the word containing it, will be marked as title and thus will be transliterated not translated. The title database consists of 14 entries. This database can be extended at any time to allow new titles to be added. Identifying Surnames locates proper names having surnames. If found, the word previous to the surname is transliterated. Thus, this module plays an important role in translation. The surnames database consists of 654 entries. Word-to-Word translation using lexicon look up phase searches Hindi to Punjabi lexicon for translating Hindi word to Punjabi directly. The HPDictionary database consists of 54,127 entries. If not found in lexicon, it is sent to Word Sense Disambiguation module using N-gram Approach[4]. If the word is not an ambiguous word, it is considered to be Out-of-Vocabulary word. In linguistics, a suffix (also sometimes called a *postfix* or *ending*) is an affix which is placed after the stem of a word. Common examples are case endings, which indicate the grammatical case of nouns or adjectives, and verb endings. Hindi is a (relatively) free word-order and highly inflectional language. Because of same origin, both languages have very similar structure and grammar. The difference is only in words and in pronunciation e.g. in Hindi it is लड़का and in Punjabi the word for boy is ਮੁੰਡਾ and even sometimes that is also not there like ਘਰ *(ghar)* and ਘਰ *(ghar)*. The inflection forms of both these words in Hindi and Punjabi are also similar. In this activity, inflectional analysis without using morphology has been performed for all those tokens that are not processed in the previous activities of pre -processing and translation engine phases. Thus, for performing inflectional analysis, rule based approach has been followed. When the token is passed to this sub phase for inflectional analysis, If any pattern of the regular expression (inflection rule) matches with this token, that rule is applied on the token and its equivalent translation in Punjabi is generated based on the matched rule(s). There is also a check on the generated word for its correctness. We are using correct Punjabi words database for testing the correctness of the generated word. This generated Punjabi word is matched with some entry in punjabiUnigrams database. The database punjabiUnigrams is a collection of about 2,00,000 Punjabi words from large Punjabi corpus analysis. Punjabi corpus has been collected from various resources like online Punjabi newspapers, blogs, articles etc. If there is a match, the generated Punjabi word is considered a valid Punjabi word. If there is no match, this input token is forwarded to the transliteration activity.

The advantage of using punjabiUnigrams database is that ingenuine Punjabi words will not become the part of translation. If the wrong words are generated by inflectional analysis module, it will not be passed to translation rather it will be treated as out-of vocabulary and will be transliterated. It has been analyzed that when this module was tested on the Hindi corpus of about 50,000 words, approx. 10,000 distinct words passed through this phase. And out of these 10,000 words, approx. 7,000 words were correctly generated and even accepted by Punjabi unigrams database. But rest was either generated wrong and was simply transliterated [5].

Post-Processing phase includes Grammar Corrections. In spite of the great similarity between Hindi and Punjabi languages, there are still a number of important grammatical divergences: gender and number divergences which affect agreement. The grammar is incorrect or the relation of words in their reference to other words, or their dependence according to the sense is incorrect and needs to be adjusted. In other words, it can be said that it is a system of correction for ill-formed sentences. The output generated by the translation engine phase becomes the input for post-processing phase. This phase will correct the grammatical errors based on the rules implemented in the form of regular expressions.

For example: In a typical Punjabi sentence, within verb phrase, all the verbs must agree in gender and number.

    Incorrect: ਨਿਰਮਲਾ ਦੀ ਅਵਾਜ ਸੁਣਦੇ ਹੀ ਭੱਜਦੀ ਹਨ।
            (nirmalā dī avāj suṇdē hī bhajjdī han.)
    Correct: ਨਿਰਮਲਾ ਦੀ ਅਵਾਜ ਸੁਣਦੇ ਹੀ ਭੱਜਦੀਆਂ ਹਨ।
            (nirmalā dī avāj suṇdē hī bhajjdīāṃ han.)

We have formulated 28 regular expressions for correcting such grammatical errors. Following table shows the distribution of regular expressions on the basis of error categories discussed above:

Table 1. Grammatical Error Category wise Regular Expression Distribution

S.No.	Grammatical Error Category	Regular Expression Count
1.	Within Verb Phrase agreement	12
2.	Noun's Oblique Form before Postpositions	02
3.	Subject Verb Agreement	13
4.	Verb Object noun phrase agreement if there is ਚਾਹੀਦਾ (*cāhīdā*) in verb phrase	01

The analysis was done on a document consisting of 35500 words. It was found that 6.197% of the output text has been corrected grammatically using these regular expressions. Following table shows the contributions of various regular expression categories in correcting the grammatical errors:

Table 2. % Contribution of Regular Expressions on the basis of Grammatical Error Categories

S.No.	Grammatical Error Category	Regular Expression Count
1.	Within Verb Phrase agreement	38.67%
2.	Noun's Oblique Form before Postpositions	3.20%
3.	Subject Verb Agreement	35.63%
4.	Verb Object noun phrase agreement if there is ਚਾਹੀਦਾ (*cāhīdā*) in verb phrase	9.84%

2 Evaluation and Results

The evaluation [6] of the system shows that the accuracy percentage for the system is found out to be 87.60%. Further investigations reveal that out of 13.40%:

- 80.6 % sentences achieve a match between 50 to 99%
- 17.2 % of remaining sentences were marked with less than 50% match against the correct sentences.
- Only 2.2 % sentences are those which are found unfaithful.

A match of lower 50% does not mean that the sentences are not usable. Intelligibility percentage of the system comes out to be 94%. After some post editing, they can fit properly in the translated text. As there is no Hindi –Parallel Corpus was available, thus for testing the system automatically, we generated Hindi-Parallel Corpus of about 10K Sentences. The BLEU score comes out to be 0.7801.

3   Comparison with Other Existing Systems

The accuracy score is comparable with other similar systems:

Table 3. Comparative analysis of %age accuracy

MT SYSTEM	Accuracy	Test Used
RUSLAN	40% correct 40% with minor errors. 20% with major error.	Intelligibility Test
CESILKO (Czech-to-Slovak)	90%	Intelligibility Test
Czech-to-Polish	71.4%	Accuracy Test
Czech-to-Lithuanian	69%	Accuracy Test
Punjabi-to-Hindi	92%	Intelligibility Test
Hindi-to-Punjabi (http://sampark.iiit.ac.in)	70% 50%	Intelligibility Test Accuracy Test
Hindi-to-Punjabi(Our System)	*94% 90.84%*	Intelligibility Test Accuracy Test

4   Conclusion

In this paper, a hybrid translation approach for translating the text from Hindi to Punjabi has been presented. The architecture used has shown extremely good results, Hence, This MT systems between closely related language pairs may follow this approach.

References

1. Goyal, V., Lehal, G.S.: Comparative Study of Hindi and Punjabi Language Scripts, Napalese Linguistics. Journal of the Linguistics Society of Nepal 23, 67–82 (2008)
2. Goyal, V., Lehal, G.S.: Advances in Machine Translation Systems. Language In India 9, 138–150 (2009)

3. Goyal, V., Lehal, G.S.: Automatic Spelling Standardization for Hindi Text. In: 1st International Conference on Computer & Communication Technology, Moti Lal Nehru National Institute of technology, Allhabad, Sepetember 17-19, pp. 764–767. IEEE Computer Society Press, California (2010)
4. Goyal, V., Lehal, G.S.: N-Grams Based Word Sense Disambiguation: A Case Study of Hindi to Punjabi Machine Translation System. International Journal of Translation (2011) (accepted, To be published in January-June 2011 Issue)
5. Goyal, V., Lehal, G.S.: A Machine Transliteration System for Machine Translation System: An Application on Hindi-Punjabi Language Pair. Atti Della Fondazione Giorgio Ronchi (Italy), vol. LXIV(1), pp. 27–35 (2009)
6. Goyal, V., Lehal, G.S.: Evaluation of Hindi to Punjabi Machine Translation System. International Journal of Computer Science 4(1), 36–39 (2009)
7. Dave, S., Parikh, J., Bhattacharyya, P.: Interlingua-based English- Hindi Machine Translation and Language Divergence. Journal of Machine Translation 16(4), 251–304 (2001)
8. Chatterji, S., Roy, D., Sarkar, S., Basu, A.: A Hybrid Approach for Bengali to Hindi Machine Translation. In: 7th International Conference on Natural Language Processing, pp. 83–91 (2009)
9. Jain, R., Sinha, R.M.K., Jain, A.: ANUBHARTI. Using Hybrid Example-Based Approach for Machine Translation. In: Symposium on Translation Support Systems (SYSTRAN 2001), Kanpur, February 15-17, pp. 123–130 (2001)

Cascading Style Sheet Styling Issues in Punjabi Language

Swati Mittal[1], R.K. Sharma[2], and Parteek Bhatia[1]

[1] Department of Computer Science and Engineering, Thapar University, Patiala, India
[2] School of Mathematics and Computer Applications, Thapar University, Patiala, India

Abstract. This paper describes the styling issues in Punjabi Websites using Cascading Style Sheets (CSS) in various web browsers. Seven different styling issues for Indian Languages have been identified by other researchers. It has been noted that most Punjabi websites make use of only underline and hyperlinks for styling the web content. To test all the styling issues, we developed our own testing website for Punjabi with the use of CSS. We have checked all the styling issues in six different browsers. The results of comparative study in different browsers are presented in this paper.

Keywords: Cascading Style Sheets, Styling Issues.

1 Introduction

The websites make use of Cascading Style Sheets (CSS) for styling the web content. This paper presents a comparative study of various CSS styling issues in displaying the Punjabi Language text in different browsers. A style sheet simply holds a collection of rules that we define to enable us to manipulate our web pages [4], [5]. It is worth mentioning that CSS styling issues are designed according to English Language. It does not support number of issues which are applicable in Indian Languages. We have worked on these issues regarding Punjabi Language which may be applicable in all other Indian Languages as well. This paper contains three sections. In Section 2, we have presented the analysis of these issues when explored with different websites using different browsers. Section 3 concludes our findings and presents the future work to be done.

2 Analysis of Different Styling Issues

We have tested ten different Punjabi websites to analyze different styling issues. It has been found that most Punjabi websites do not use first letter styling, over lining, line through of characters and horizontal spacing between characters features. So, in order to test all these issues we have designed testing website for Punjabi Language using CSS. We have tested these websites along with testing website in six browsers, namely, Google Chrome, Mozilla Firefox, Netscape Navigator, Safari, Internet Explorer and Opera [4], [5]. These issues have been discussed below.

C. Singh et al. (Eds.): ICISIL 2011, CCIS 139, pp. 242–245, 2011.
© Springer-Verlag Berlin Heidelberg 2011

2.1 Styling of First Letter

When some styling feature is applied to the starting character, then it should be applied to either single character, conjunct character or a syllable [3], [5]. Table 1 depicts first letter styling in Punjabi Language for example browsers. It is worth mentioning that only Mozilla Firefox browser displayed the correct output and applied the styling to conjunct characters. All other browsers applied the styling to single character only.

2.2 Underlining of the Characters

It has been found that during underlining of characters in a Punjabi website, the matras like ੁ Aunkar and ੂ Dulanukar become unreadable due to its overlap with underline. It creates problem in reading the information correctly. All the browsers had this problem in reading matras. Table 1 shows the underlining issue.

2.3 Link Displayed While Mouse Over

In Punjabi websites, the information is not clearly readable when we hyperlink the text because some modifiers (matras) are cut and the line overlaps with matras as shown in Table 1.

2.4 Over Lining of the Characters in Different Browsers

It has been found that no Punjabi Website uses this feature. As such, we have tested this issue on our testing website. When we use the CSS text decoration for over lining, the line overlaps the matras, like ਿ sia(h)ri, ੀ bia(h)ri, ੈ la(n), ੌ Dulai(n), ੋ HoRa and ੰ kAnauRa in Punjabi Language. Table 1 shows the results of various browsers. It has been noted that, in Internet Explorer the line is distorted and overlaps with each character individually.

2.5 Line-through of the Characters

It has been found that most Punjabi websites do not use line-through feature. We again analyzed this feature on our test website and the results for this are shown in Table 1. It has been noted that in Mozilla, the line is not exactly in the center and for Internet Explorer the line is thicker than in other browser, thus reducing the readability of letters.

2.6 Horizontal Spacing

The Horizontal spacing is used to give the space between each character [1], [5]. In case of Punjabi language the space has not be given in every character but after some portion of the character sequence. Table 1 shows horizontal spacing for our test website in different browsers. It has been noted that for most browsers the vowel gets separated from the letter. The horizontal spacing was better in Internet Explorer and the best in case of Mozilla Firefox.

2.7 Title Bar Display for Punjabi Letters

If a website has a title in Punjabi it should be displayed in Punjabi. But it has been found that, few browsers do not support this feature as indicated in Table 1. The Punjabi letters are not recognized by the browsers.

The results of our analysis are given in Table 1. Here, "YES" indicates that the corresponding issue has a display problem with corresponding browser and "NO" indicates that the issue is not having any display problem with the mentioned browser.

Table 1. Comparative study: Styling Issue Vs. Web Browser

Issues	Google Chrome	Mozilla Firefox	Netscape Navigator	Safari	Internet Explorer 7.0 or above	Opera
Styling First letter	YES	NO	YES	YES	YES	YES
Underlining of Character	YES	YES	YES	YES	YES	YES
Link while mouse over	YES	NO	NO	YES	NO	YES
Over lining of Characters	NO	NO	NO	NO	YES	NO
Line through Characters	NO	YES	NO	NO	YES	NO
Horizontal spacing between characters	YES	NO	YES	YES	YES	YES
Title bar display for Punjabi letters	YES	NO	NO	NO	YES	NO

The problem of styling issues for Punjabi language with respect to percentage of browsers is presented in Figure 1. The issues are ordered in accordane with the percentage of browsers that has problems in display. We have also analyzed the usefulness of different browsers vis-à-vis the issues discussed in this paper. The findings are presented in Figure 2. One can note from this figure that Mozilla Firefox is probably the browser that we should use for browsing Punjabi websites. In our study, we have noted that Internet Explorer has the least performance, when we consider the issues discussed in this paper, for browsing Punjabi websites.

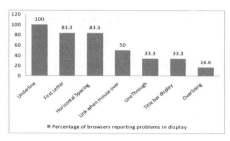
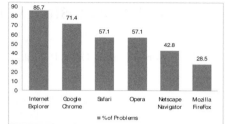

Fig. 1. Bar Graph for Styling Issues problems v/s the percentage of browsers having that problem

Fig. 2. Bar Graph to indicate the percentage of styling issue problems in each browser

3 Conclusion and Future Work

It has been noted that, most of the Punjabi websites do not make use of CSS for styling. Most sites only use underlining, link while mouse over and Title bar display characteristics. To identify more styling issues we have designed a test website in Punjabi using CSS. It has been found that, the problem of underlining of characters is present in all browsers. First letter and horizontal spacing problem is present in 83% of browsers. Over lining problem is only present in Internet Explorer. It has also been found that Mozilla Firefox has resolved a number of issues for browsing of Punjabi websites. There is a need to modify the CSS standards to accommodate Indian Languages as they are designed only for English. This has been recommended by us to W3C India. Because not all browsers comply identically with CSS code, a coding technique [6] known as a CSS filter can be used to show or hide parts of the CSS to different browsers, either by exploiting CSS-handling quirks or bugs in the browser, or by taking advantage of lack of support for parts of the CSS specifications.

References

1. Etemad, E.J.: Robust Vertical Text Layout, Technical report. In: 27th Internationalization and Unicode Conference, Berlin, Germany, pp. 15–16 (April 2005)
2. Keller, M., Nussbaumer, M.: Cascading Style Sheets: A Novel Approach Towards Productive Styling with Today's Standards. In: WWW 2009, Madrid, Spain, pp. 1161–1162 (April 2009)
3. The World Wide Web Consortium India, http://www.w3cindia.in
4. CSS Template - The World Wide Web Consortium -Indian Office, http://www.w3cindia.in
5. CSS Draft- The World Wide Web Consortium -Indian Office, http://www.w3cindia.in/cssdocument.html
6. CDAC Pune, http://iplugin.cdac.in/CSS-indic-problems/CSS-indic-problems.html

Translation of Hindi *se* to Tamil in a MT System

Sobha Lalitha Devi, P. Pralayankar, V. Kavitha, and S. Menaka

AU-KBC Research Centre
MIT Campus of Anna University
Chrompet, Chennai
sobha@au-kbc.org

Abstract. The paper attempts to describe how a word like *se* of Hindi can be a challenging task to a Machine Translation (MT) system. In most of the literature of Hindi and Urdu, a noun marked with *se* is assigned instrumental or ablative case. So it is called instrumental and ablative case marker. But a close look at its distribution shows that apart from instrumental and ablative case function, it denotes other functions also and in each of these types, it is translated differently in Tamil.

Keywords: Case, Hindi, Tamil, Machine Translation, Rule Based Approach.

1 Introduction

The Indian languages Hindi and Tamil follow postpositional method to assign a case system. Sinha et al [5] discussed about case transfer and its divergence from English to Hindi and vice-versa. Pralayankar et al [3] discussed about the case transfer pattern from Hindi to Tamil MT system. Sobha et al [6] have worked on the nominal transfer from Tamil to Hindi. In this paper we consider *se*, one of the case markers in Hindi and how it transfers into Tamil in a MT system following a rule based approach.

2 Case Marking Pattern in Hindi and Tamil

Table (1) shows a comparative list of Hindi and Tamil case markers. *se* denotes instrumental/ablative case in Hindi but Tamil has *–aal* and *koNtu* for instrumental and *–ilirunthu* and *-itamirunthu* for ablative.

Table 1. Hindi-Tamil case markers [1,2,4]

	Nom.	Erg.	Dative	Acc.	Ins.	Abl.	Loc.	Genitive
Hindi	Ø	*ne*	*ke liye*	*ko*	*se*	*se*	*meN,par*	*kaa,ke,kii*
Tamil	Ø	Ø	*-kku*	*-ai*	*-aal, koNtu*	*-ilirunthu ,itamirunthu*	*-il, itam*	*-in, athu,*

3 Distribution of *se* in Hindi

In most of the literature of Hindi, we find that the noun marked with case marker *se* is assigned instrumental or ablative case but a closer look at its distribution shows that it is more versatile than any other postpositions of Hindi. This can be seen as below:

C. Singh et al. (Eds.): ICISIL 2011, CCIS 139, pp. 246–249, 2011.
© Springer-Verlag Berlin Heidelberg 2011

Instrumental *se*. The inanimate noun gets instrumental case *se* in simple and causative constructions (Ex.1). In (1) the grammatical function assigned to this instrumental case is of adjunct. This will be translated to instrumental *–aal* in Tamil.

1. *vah*   *kalam   se*   *likhegaa*   (H)
    avan   *peenaavaal*   *ezuthuvaan*   (T)
    'He will write with pen.'

Ablative *se*. In Ex. (2), *se* is used with noun as source. In Tamil, the ablative marker *-ilirunthu* is used with inanimate noun (Ex. 2a) and *-itamirunthu* is used with animate nouns. Ex.2b is exception to this generalization where the noun is animate, but still takes *–ilirunthu*. This happens only in case with *se-tak* (from-to) construction.

2. a. *kitaab/pitaajii   se*   *paDho*   (H)
       puththakaththilirunthu/appaavitamirunthu pati   (T)
       'Read/learn from the book/father.'

   b. *aanand se   ravi   tak   sab ko   bulaao*   (H)
       aanantilirunthu ravi   varai   ellaaraiyum   kuuppiut   (T)
       'Call everyone from Anand to Ravi.'

Agent *se*. The animate noun marked with case marker *se* is assigned agent case and occupies subject position in the verb's argument structure in the passive voice (Ex. 3).

3. *ritaa   se   gaanaa   nahiN   gayaa   jaataa*   (H)
    riittaav.aal   paattu   paata   iyalavillai   (T)
    'Rita is not able to sing song.'

Agentive form of animate noun with case marker *se* is also found in the argument structure of causative sentences in Hindi as in (4a-b).

4. a. *mai-ne   raaju   se   peDa   kata-aa-yii*   (H)
       naan   raajuvai   maraththai   vettaceytheen (T)
       'I made Raju cut a tree.'

   b. *maaN ne   shikSaka se   raaju ko paatha   paDa-waa-yaa*   (H)
       ammaa aaciriyaraik koNtu raajuuviRku paatam kaRpikka ceythaaL   (T)
       'Mother caused the teacher to teach lesson to Raju.'

If SUB2 *shikSaka* in Ex. (4b) is replaced with *chamaccha* (spoon), it will get instrumental role rather than the agent role taken by *shikSaka se* because 'spoon' cannot be the agent, performing the action on its own. The mother is the actual performer of the action. The spoon is used as an instrument to perform the action.

Temporal *se*. If *se* follows a temporal noun denoting specific point of time, it is translated to the ablative marker *-ilirunthu* in Tamil. But if the temporal noun denotes a period of time, it is translated to adverbial marker *–aaka* in Ex. (5).

5. *rimmi   do dinoN/mangalvaar   se   bimmar hai*   (H)
    rimmikku iraNtu naatkaLaaka/cevvaayilirunthu   utalnilai cariyillai (T)
    'Rimmi is sick since two days/Tuesday.'

Mutual *se*. *se* is also used to mark animate nouns at object position, which are experiencers of the action involved (Ex. 6). Such verbs denote action performed mutually between two animate nouns; subject and object and so are called 'mutual'.

6. *manohar   ne   nikki   se   baat   kii*   (H)
    manokar   nikkiyitam   pecinaan   (T)
    'Manohar talked to Nikki.'

Movement *se*. When path or vehicle or spatial nouns are used with a motion verb to denote movement (Ex. 7), they are marked with case marker *se*. Such *se* is translated to locative *–il* in Tamil.

7.	*ritaa*	*car*	*se*	*kochi*	*jaaegii*	(H)
	riitaa	*kaaril*		*kocci*	*celvaaL*	(T)
	'Ritaa will go to Kochi by car.'					

Comparative *se*. *se* when used for comparison between two nouns (Ex. 8) is translated to accusative marker *–ai* followed by postposition *vita* in Tamil.

8.	*rimmi*	*se*	*riyaa*	*adhika sundar hai*	(H)
	rimmiyai	*vita*	*riyaa*	*mikavum azakaanavaL*	(T)
	'Riyaa is more beautiful than Rimmi.'				

When infinitive verb uses *se* in Hindi (Ex. 9), it is translated to dative case in Tamil.

9.	*richaa*	*kahane se*	*Darati hai.*	(H)
	riccaa	*colvathaRku*	*payappatukiRaaL*	(T)
	'Richa is afraid of saying.'			

Adverbial *se*. When conceptual nouns take case marker *se* (Ex. 10), it is translated to adverbial marker *–aaka*.

10.	*riyaa*	*jor se*	*haNsane*	*lagii*	(H)
	riyaa	*caththamaaka*	*ciriththaaL*		(T)
	'Riya started laughing loudly.'				

Clausal *se*. If *se* comes after *jab* and *tab*, which is considered as a clause marker, this is clausal *se* (Ex.11). Here both the *se* are translated to one ablative *-ilirunthu* and other structural transfer takes place [7].

11.	*jab se*	*mai chennai aayaa huN tab se*	*tumko*	*jaantaa huN*	(H)
	naan	*cennai vanthathilirunthu unnai aRiveen*			(T)
	'Ever since I came to Chennai, I know you.'				

4 Rules for Disambiguation of *se*

The rules for the disambiguation of discussed various types of se are as follows:

1. Instrumental/Agent *se*: a. NP *se* …V -> NP *aal*… V, where NP is an inanimate direct object, V is in/ditransitive Verb (Ex. 1)
b. NP *se* …Vp -> NP *aal* …Vp, where NP is animate noun in agentive role, located at subject position, and Vp is Passive verb (Ex 3)
c. NP *se*…VC-> NP*ai* …VC, where NP is animate noun , VC is causative verb (Ex 4)
2. Temporal *se*: NP *se* -> NP *aaka*, where NP is temporal noun (Ex 5)
3. Mutual *se*: NPs NP1 *se* -> NPs NP1*il*, where NPs and NP1 are subject and object noun phrases respectively (Ex. 6)
4. Movement *se*: NP *se*…V -> NP *il*... V, where V is finite verb of motion, NP is an inanimate noun (Ex. 7)
5. Comparative *se*: a. NP1 *se* NP2 INT/ADJ -> NP1 *ai* *vita* NP2 INT/ADJ, where NP1 and NP2 are noun phrases, INT is Intensifier, ADJ is adjective (Ex. 8)
b. VP1 *se* VP2 -> *VP1 ai vita* VP2, where VP1 and VP2 are nonfinite verb (Ex.9)
6. Adverbial *se*: ADV *se* -> N *aaka*,   where ADV is adverb, N is noun (Ex. 10)

7. Clausal *se*: (Ex 11)   *jabse… tab se -> -iliruntu,*
8. Default *se*: NP *se* ->NP1 *-iliruntu* (or) NP2 *–itamirunthu,* where   NP   is   noun
phrase, NP1 and NP2 are inanimate and animate noun phrases respectively.

5 Results and Discussion

After implementation of the above rules we noticed a significant improvement regarding the transfer of *se* in a MT system, where earlier [3] it was considered that *se* is either instrumental or ablative in Hindi. But later we got many types of *se*, which forced us to do further study and implement the new rule for this. But still in certain cases like *maine Daaka se patra bhejaa hai* (I have sent a letter through post), the system is transferring *se* into locative *–il* because the verb in the sentence is motion verb but in real this should be transferred to *–aal* and postposition *mulam*.

6 Conclusion

The paper presents a micro study of *se* and rules for how to disambiguate different types of *se* for a MT system following rule based approach. We are testing the system with large number of sentences and getting more types of distribution like, locative *se* (*vah divaal se lagakar khaDaa hai*), but its distribution is very less.

References

1. Ganesan, S.N.: A Contrastive Grammar of Hindi and Tamil, University of Madras, 260-291, Chennai, India (1975)
2. Lehmann, T.: A Grammar of Modern Tamil, Pondicherry Instritute of Linguistics and Culture, Pondicherry, pp. 23–47 (1989)
3. Pralayankar, P., Kavitha, V., Sobha, L.: Case Transfer pattern from Hindi to Tamil MT. PIMT Journal of Research 2(1), 26–31 (2008)
4. Pope, R.G.U.: A Handbook of the Tamil Language, 7th edn., New Delhi, pp. 29–32. First published Oxford (1904); Asian Educational Services, Chennai (2006)
5. Sinha, R.M.K., Thakur, A.: Translation Divergence in English-Hindi MT, EAMT, Budapest, Hungary, May 30-31, pp. 245–254 (2005)
6. Lalitha Devi, S., Kavitha, V., Pralayankar, P., Menaka, S., Bakiyavathi, T., Vijay Sundar Ram, R.: Nominal Transfer from Tamil to Hindi. In: To presented in International Conference on Asian Language Processing (IALP), Harbin, China, December 28-30 (2010)
7. Lalitha Devi, S., Vijay Sundar Ram, R., Pralayankar, P., Bakiyavathi, T.: Syntactic Structure Transfer in a Tamil to Hindi MT System - A Hybrid Approach. In: Gelbukh, A. (ed.) CICLing 2010. LNCS, vol. 6008, pp. 438–450. Springer, Heidelberg (2010)

Preprocessing Phase of Punjabi Language Text Summarization

Vishal Gupta[1] and Gurpreet Singh Lehal[2]

[1] Assistnt Professor, Computer Science & Engineering,
University Institute of Engineering & Technology,
Panjab University Chandigarh, India
vishal@pu.ac.in
[2] Professor, Department of Computer Science,
Punjabi University Patiala, Punjab, India
gslehal@yahoo.com

Abstract. Punjabi Text Summarization is the process of condensing the source Punjabi text into a shorter version, preserving its information content and overall meaning. It comprises two phases: 1) Pre Processing 2) Processing. Pre Processing is structured representation of the Punjabi text. This paper concentrates on Pre processing phase of Punjabi Text summarization. Various sub phases of pre processing are: Punjabi words boundary identification, Punjabi language stop words elimination, Punjabi language noun stemming, finding Common English Punjabi noun words, finding Punjabi language proper nouns, Punjabi sentence boundary identification, and identification of Punjabi language Cue phrase in a sentence.

Keywords: Punjabi text summarization, Pre Processing, Punjabi Noun stemmer.

1 Introduction to Text Summarization

Text Summarization[1][2] is the process of selecting important sentences, paragraphs etc. from the original document and concatenating them into shorter form. Abstractive Text Summarization is understanding the original text and retelling it in fewer words. Extractive summary deals with selection of important sentences from the original text. The importance of sentences is decided based on statistical and linguistic features of sentences. Text Summarization Process can be divided into two phases: 1) Pre Processing phase [2] is structured representation of the original text. Various features influencing the relevance of sentences are calculated. 2)In Processing [3][4][12] phase, final score of each sentence is determined using feature-weight equation. Top ranked sentences are selected for final summary. This paper concentrates on Pre processing phase, which has been implemented in VB.NET at front end and MS Access at back end using Unicode characters [5].

2 Pre Processing Phase of Punjabi Text Summarization

2.1 Punjabi Language Stop Word Elimination

Punjabi language Stop words are frequently occurring words in Punjabi text. We have to eliminate these words from original text, otherwise, sentences containing them can

C. Singh et al. (Eds.): ICISIL 2011, CCIS 139, pp. 250–253, 2011.

get influence unnecessarily. We have made a list of Punjabi language stop words by creating a frequency list from a Punjabi corpus. Analysis of Punjabi corpus taken from popular Punjabi newspapers has been done. This corpus contains around 11.29 million words and 2.03 lakh unique words. We manually analyzed unique words and identified 615 stop words. In corpus of 11.29 million words, the frequency count of stop words is 5.267 million, which covers 46.64% of the corpus. Some commonly occurring stop words are ਦੀ dī, ਤੋਂ tōṃ, ਕਿ ki ,ਅਤੇ atē, ਹੈ hai, ਨੇ nē etc.

2.2 Punjabi Language Noun Stemming

The purpose of stemming [6][7] is to obtain the stem or radix of those words which are not found in dictionary. If stemmed word is present in dictionary, then that is a genuine word, otherwise it may be proper name or some invalid word. In Punjabi language noun stemming[9][10][13], an attempt is made to obtain stem or radix of a Punjabi word and then stem or radix is checked against Punjabi dictionary [8], if word is found in the dictionary, then the word's part of speech is checked to see if the stemmed word is noun. An in depth analysis of corpus was made and various possible noun suffixes were identified like ੀਆਂ īāṃ, ਿਆਂ iāṃ, ੁਆਂ ūāṃ, ਾਂ āṃ, ੀਏ īē etc. and the various rules for noun stemming have been generated. Some rules of Punjabi noun stemmer are ਫੁੱਲਾਂ phullāṃ → ਫੁੱਲ phull with suffix ਾਂ āṃ, ਲੜਕੀਆਂ laṛkīāṃ → ਲੜਕੀ laṛkī with suffix ੀਆਂ īāṃ, ਮੁੰਡੇ muṇḍē → ਮੁੰਡਾ muṇḍā with suffix ੇ ē etc.

An In depth analysis of output is done over 50 Punjabi documents. The efficiency of Punjabi language noun stemmer is 82.6%. The accuracy percentage of correct words detected under various rules of stemmer are: ੀਆਂ īāṃ rule1 86.81%, ਿਆਂ iāṃ rule2 95.91%, ੁਆਂ ūāṃ rule3 94.44%,ਾਂ āṃ rule4 92.55%, ੇ ē rule5 57.43%, ੀਂ īṃ rule6 100%, ੋਂ ōṃ rule7 100% and ਵਾਂ vāṃ rule8 79.16%. Errors are due to rules violation or dictionary errors or due to syntax mistakes. Dictionary errors are those errors in which, after noun stemming, stem word is not present in noun dictionary, but actually it is noun. Syntax errors are those errors, in which input Punjabi word is having some syntax mistake, but actually that word falls under any of stemming rules. Overall error % age, due to rules voilation is 9.78%, due to dictionary mistakes is 5.97% and due to spelling mistakes is 1.63%.

Graph1 depicts the percentage usage of the stemming

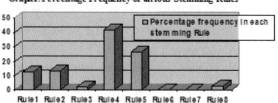

Graph1.Percentage Frequency of various Stemming Rules

2.3 Finding Common English-Punjabi Noun Words from Punjabi Corpus

Some English words are now commonly being used in Punjabi. Consider a sentence such as ਟੈਕਨਾਲੇਜੀ ਦੇ ਯੁੱਗ ਵਿਚ ਮੋਬਾਈਲ *Technology de yug vich mobile*. It contains ਟੈਕਨਾਲੇਜੀ *Technology* and ਮੋਬਾਈਲ *mobile* as English-Punjabi nouns. These should obviously not be coming in Punjabi dictionary. These are helpful in deciding sentence importance. After analysis of Punjabi corpus, 18245 common English-Punjabi noun words have been identified. The percentage of Common English-Punjabi noun words in the Punjabi Corpus is about 6.44 %. Some of Common English Punjabi noun words are ਟੀਮ team, ਬੋਰਡ board, ਪ੍ਰੈੱਸ press etc.

2.4 Finding Punjabi Language Proper Nouns from Punjabi Corpus

Proper nouns are the names of person, place and concept etc. not occurring in diction-ary. Proper Nouns play important role in deciding a sentence's importance. From the Punjabi corpus, 17598 words have identified as proper nouns. The percentage of these proper noun words in the Punjabi corpus is about 13.84 %. Some of Punjabi language proper nouns are ਅਕਾਲੀ akālī, ਅਜੀਤ ajīt, ਭਾਜਪਾ bhājapā etc.

2.5 Identification of Cue Phrase in a Sentence

Cue Phrases [11] are certain keywords like In Conclusion, Summary and Finally etc. These are very much helpful in deciding sentence importance. Those sentences which are beginning with cue phrases or which contain these cue phrases are generally more important than others Some of commonly used cue phrases are ਅੰਤ ਵਿੱਚ/ ਅੰਤ ਵਿਚ ant vicc/ant vic, ਕਿਉਂਕੀ Kiukī, ਸਿੱਟਾ siṭṭā, ਨਤੀਜਾ / ਨਤੀਜੇ natījā/natījē etc.

3 Pre Processing Algorithm for Punjabi Text Summarization

Pre Processing phase algorithm proceeds by segmenting the source Punjabi text into sentences and words. Set the scores of each sentence as 0. For each word of every sentence follow following steps:

- Step1: If current Punjabi word is stop word then delete all the occurrences of it from current sentence.
- Step2:If Punjabi word is noun then increment the score of that sentence by 1.
- Step3: Else If current Punjabi word is common English-Punjabi noun like ਹਾਊਸ house then increment the score of current sentence by 1
- Step4: Else If current Punjabi word is proper noun like ਜਲੰਧਰ jalandhar then increment the score of current sentence by 1.
- Step5: Else Apply Punjabi Noun Stemmer for current word and go to step 2.

Sample input sentence is ਤਿੰਨਾਂ ਸ਼ਰਤਾਂ ਤੇ ਜੋ ਪੂਰਾ ਉਤਰਦਾ ਹੈ ਉਸ ਨੂੰ ਹੀ ਵੋਟ ਦਿੱਤਾ ਜਾਣਾ ਚਾਹੀਦਾ ਹੈ। tinnāṃ shartāṃ 'tē jō pūrā utradā hai us nūṃ hī vōṭ dittā jāṇā cāhīdā hai.

Sample output sentence is ਤਿੰਨ ਸ਼ਰਤ ਉਤਰਦਾ ਵੋਟ ਚਾਹੀਦਾ tinn sharat utradā vōṭ cāhīdā with Sentence Score is 2 as it contains two noun words.

4 Conclusions

In this paper, we have discussed the various pre-processing operations for a Punjabi Text Summarization System. Most of the lexical resources used in pre-processing such as Punjabi stemmer, Punjabi proper name list, English-Punjabi noun list etc. had to be developed from scratch as no work had been done in that direction. For developing these resources an indepth analyis of Punjabi corpus, Punjabi dictionary and Punjabi morph had to be carried out using manual and automatic tools.

This the first time some of these resources have been developed for Punjabi and they can be beneficial for developing other NLP applications in Punjabi.

References

1. Berry, M.W.: Survey of Text Mining Clustering, Classification and Retrieval. Springer Verlag, LLC, New York (2004)
2. Kyoomarsi, F., Khosravi, H., Eslami, E., Dehkordy, P.K.: Optimizing Text Summarization Based on Fuzzy Logic. In: Proceedings of Seventh IEEE/ACIS International Conference on Computer and Information Science, pp. 347–352. IEEE, University of Shahid Bahonar Kerman, UK (2008)
3. Fattah, M.A., Ren, F.: Automatic Text Summarization. Proceedings of World Academy of Science Engineering and Technology 27, 192–195 (2008)
4. Kaikhah, K.: Automatic Text Summarization with Neural Networks. In: Proceedings of Second International Conference on Intelligent Systems, pp. 40–44. IEEE, Texas (2004)
5. Unicode Characters Chart,
   http://www.tamasoft.co.jp/en/
   general-info/unicode-decimal.html
6. Zahurul Islam, M., Nizam Uddin, M., Khan, M.: A light weight stemmer for Bengali and its Use in spelling Checker. In: Proceedings of 1st International Conference on Digital Comm. and Computer Applications (DCCA 2007), Irbid, Jordan, pp. 19–23 (2007)
7. Kumar, P., Kashyap, S., Mittal, A., Gupta, S.: A Hindi question answering system for E-learning documents. In: Proceedings of International Conference on Intelligent Sensing and Information Processing, Banglore, India, pp. 80–85 (2005)
8. Singh, G., Gill, M.S., Joshi, S.S.: Punjabi to English Bilingual Dictionary. Punjabi University Patiala, India (1999)
9. Gill, M.S., Lehal, G.S., Joshi, S.S.: Part of Speech Tagging for Grammar Checking of Punjab. The Linguistic Journal 4(1), 6–21 (2009)
10. Punjabi Morph. Analyzer,
    http://www.advancedcentrepunjabi.org/punjabi_mor_ana.asp
11. The Corpus of Cue Phrases,
    http://www.cs.otago.ac.nz/staffpriv/alik/papers/apps.ps
12. Neto, J., et al.: Document Clustering and Text Summarization. In: Proc. of 4th Int. Conf. Practical Applications of Knowledge Discovery and Data Mining, London, pp. 41–55 (2000)
13. Ramanathan, A., Rao, D.: A Lightweight Stemmer for Hindi. In: Workshop on Computational Linguistics for South-Asian Languages, EACL (2003)

Comparative Analysis of Tools Available for Developing Statistical Approach Based Machine Translation System

Ajit Kumar[1] and Vishal Goyal[2]

[1] Multani Mal Modi College, Patiala
ajit8671@gmail.com
[2] Department of Computer Science, Punjabi University, Patiala
vishal.pup@gmail.com

Abstract. Statistical Machine Translation model take the view that every sentence in the target language is a translation of the source language sentence with some probability. The best translation, of course, is the sentence that has the highest probability. A large sample of human translated text (parallel corpus) is examined by the SMT algorithms for automatic learning of translation parameters. SMT has undergone tremendous development in last two decades. A large number of tools has been developed for SMT and put to work on different language pairs with fair accuracy. This paper will give brief introduction to Statistical Machine Translation; tools available for developing Statistical Machine Translation systems based on Statistical approach and their comparative study. This paper will help researcher in finding the information about SMT tools at one place.

Keywords: Statistical Machine Translation, SMT Tools, MOSES, EGYPT, PHARAOH, Whittle, Thot, Systran, GIZA, GIZA++, SLMT, YASMET, MALLET, MARIE, SRILM, IRSTLM, CMU-SLM, Joshua, ReWrite Decoder, MOOD, RAMSES, CARMEL, CAIRO, OpenNLP.

1 Introduction to SMT

Statistical approach to Machine Translation was first introduced by Warren Weaver in 1949 taking ideas from information theory. In recent year IBM's Thomas J. Watson Research Center contributed significantly to SMT [W1].

Basic idea behind SMT is that if we want to translate a sentence f in source language F to a sentence e in the target language E, then it is translated according to the probability distribution P(e|f). One approach to find P(e|f) is to apply Baye's Theorem, that is

$$P(e|f) \propto P(f|e)*P(e) \qquad (1)$$

Where P(e) and P(f|e) are termed as Language model and Translation Model respectively. [P1] More detail about Statistical Machine translation can be found at http://www.statmt.org/

C. Singh et al. (Eds.): ICISIL 2011, CCIS 139, pp. 254–260, 2011.
© Springer-Verlag Berlin Heidelberg 2011

2 Steps in Statistical Machine Translation

Basic steps involved in the development of Statistical Machine Translation System are (1) development, cleaning and alignment of parallel corpus in the required language pair. (2) Find the probability P(e) using a Language modeling toolkit like SRILM, SLMT, MALLET, YASMET, IRSTLM (3) Find the probability P(fle) using Translation modeling toolkit like GIZA, GIZA++, cairo and cairoize. (4) Use decoder to convert source language sentence to target language using Language Model and Translation Model. The resultant translation can be evaluated using evaluation tools like BLEU.

3 Overview of Available Tools for Developing Statistical Machine Translation System

3.1 Complete Toolkits

These toolkits contain multiple tools used in different phases of translation process. Moses, EGYPT, PHARAOH, Thot, MOOD and OpenNLP tools are common toolkits.

Moses[P3][W2] was developed in C++ for Linux and Intel MAC by Hieu Hoang and Philipp Koehn at the University of Edinburgh. It is licensed under the LGPL (Lesser General Public License). Tools used in this toolkit are GIZA++, SRILM and IRSTLM.The tool has been successfully used for translating English to {French, Spanish, German, Russian} [W3] Czech to English [W3], English to Bangla [P2] etc.

EGYPT [W4] was developed by the Statistical Machine Translation team during the summer workshop in 1999 at the Center for Language and Speech Processing at Johns-Hopkins University (CLSP/JHU). The kit includes the following tools: **Whittle** (A tool for preparing and splitting bilingual corpora into training and testing sets),

GIZA (Training program that learns statistical translation models from bilingual corpora.), **cairo** (Word alignment visualization tool) and **cairoize** (A tool for generating alignments files). The toolkit is tested for English-French [W4] translation.

PHARAOH [P4] was developed by Philipp Koehn as part of his PhD thesis at the University Of Southern California and Information Sciences Institute. It is written for Linux. It requires some additional tools like Parallel corpus, **SRILM** (to train Language Model), **Carmel** (Finite State Toolkit for the generation of word lattices and n-best list). The use of product for Non Commercial use is allowed under agreement with University of Southern California. A training system is required to generate translation model that is not available in the toolkit and need to written.

Thot toolkit for SMT [W6][P5] was developed in C++ by Daniel Ortiz at the Pattern Recognition and Human Language Technology (PRHLT) research group of the Universidad Politécnica de Valencia (UPV) and the Intelligent Systems and Data Mining (SIMD) research group of the Universidad de Castilla-La-Mancha (UCLM). The main purpose of the toolkit is to provide an easy, effective, and useful way to train phrase-based statistical translation models to be used as part of a statistical machine translation system, or for other different NLP related tasks.

MOOD [P6] stands for Modular Object-Oriented Decoder. It is implemented in C++ programming language and is licensed under the GNU General Public License (GPL). Two major goals of the design of this toolkit are: offering open source, state of- the-art decoders and providing architecture to easily build these decoders.

OpenNLP [W5]: It is a collection of natural language processing tools. It is developed by Organizational Center for Open Source Projects related to Natural Language Processing. OpenNLP hosts a variety of java-based NLP tools which perform sentence detection, tokenization, pos-tagging, chunking and parsing, named-entity detection, and co-references. The objective of OpenNLP project is to bring NLP research community on to a common platform. The OpenNLP machine learning package is Maxent. It is Java based package for training and using maximum entropy models.

3.2 Language Modeling Tools

Statistical Language Modeling Toolkit (SLMT) [W7]: SLMT is a suite of UNIX based software tools to facilitate the construction and testing of statistical language models. This toolkit provides support for n-grams of arbitrary size. In addition the toolkit is used to count word n-grams, vocabulary n-grams and id n-grams. It is written with the objective to increase the speed of operation.

SRILM [W8][P7]: SRI Language Modeling Toolkit is mainly used in speech recognition, statistical tagging, segmentation, and machine translation. It has been under development in the 'SRI Speech Technology and Research Laboratory' since 1995. SRILM is freely available for noncommercial purposes. It is used as part of Moses and Pharaoh for language modeling.

IRST LM Toolkit [W9]: It is a tool for the estimation, representation, and computation of statistical language models. It has been integrated with Moses, and is compatible with language models created with SRILM.

YASMET [W10]: YASMET stands for Yet Another Small MaxEnt Toolkit. This is a tiny toolkit for performing training of maximum entropy models. YASMET is free software under the 'GNU Public License' and is distributed without any warranty. *Complete information about this tool is not available.*

MALLET [W11][P8]: MALLET stands for Machine Learning for Language Toolkit. It is a Java-based package based on statistical approach for natural language processing, document classification, clustering, topic modeling, information extraction, and other machine learning applications.

CMU-SLM [W12] [P9]: The CMU Statistical Modeling toolkit was written by Roni Rosenfeld at Carnegie Mellon University and released in 1994 in order to facilitate the construction and testing of bigram and trigram language models.

3.3 Translation Modeling Tools

GIZA++ [W13]: This tool is the extension of the program GIZA. GIZA++ has a lot of features such as support for Model 4, Model 5 and Alignment models depending on word classes. It implements the HMM alignment model: (Baum-Welch training,

Forward-Backward algorithm, empty word, dependency on word classes, transfer to fertility models). This tool is used as part of Moses and Egypt for the training of translation model.

CARMEL: [W14] Carmel is a finite-state transducer package written in C++ by Jonathan Graehl at USC/ISI. The tool is used to develop translation model for PHARAOH. The supporting software requires for its working are GNU Make (at least version 3.8), or Microsoft Visual C++ and .NET. The product can be used for research purpose under license agreement with University of Southern California.

3.4 Decoders

ReWrite Decoder [W15][P11]: ReWrite decoder use the Translation Model Trained using GIZA++. It takes input in XML format, plain text is not supported. While performing swapping It takes care of phrase boundaries.

MARIE [P10]: It is N-gram-based Statistical Machine Translation decoder [P8]. It has been developed at the TALP Research Center of the Universitat Politècnica de Catalunya (UPC) by Josep M. Crego as part of his PhD thesis, with the aid of Adrià de Gispert and under the advice of Professor José B. Mariño. It was specially design to deal with tuples (bilingual translation units) and a translation model trained as a typical N-gram language model (N-gram-based SMT). In addition, MARIE can use phrases (bilingual translation units) and behave as a typical phrase-based decoder (phrase-based SMT). In order to perform better translations, the MARIE decoder can make use of a target language model, a reordering model, a word penalty and any additional translation models.

Joshua [W16]: It is an open source decoder for statistical translation models based on synchronous context free grammars. Joshua Decoder was released on June 12 2009 by Chris Callison Burch. Joshua uses the synchronous context free grammar (SCFG) formalism in its approach to statistical machine translation.

RAMSES [P6]: RAMSES is a clone of PHARAOH developed as part of MOOD. The main goal behind the development of RAMSES was to develop a decoder similar to PHARAOH and make it freely available to the research community.

Phramer [W17]: Is an Open-Source Statistical Phrase-Based Machine Translation Decoder. Version 1.1 was released on July 5, 2009. It is available at http://www.phramer.org

3.5 Evaluation Tools

BLEU [P12][P13][P14]: **BLEU (Bilingual Evaluation Understudy)** is an algorithm for evaluating the quality of text which has been machine-translated from one natural language to another. Quality is considered to be the correspondence between a machine's output and that of a human: "the closer a machine translation is to a professional human translation, the better it is" [P2]. BLEU was one of the first metrics to achieve a high correlation with human judgments of quality [P13,P14] and remains one of the most popular evaluation tool.

4 Comparison of Some of the Available Toolkits

Tool Kit	Language	OS	License	Source Code Availability	Downloadable from
Moses	C++	LINUX, Intel Mac	OSL under LGPL	Yes	http://www.statmt.org/moses/
EGYPT	C, C++, java, Perl	LINUX	OSS	Yes	http://www.clsp.jhu.edu/ws99/ projects/mt/toolkit/
SLMT	C	UNIX	OSS	Yes	http://svr-www.eng.cam.ac.uk/~prc14/ toolkit.html
MALLET	Java	Any	OSS	Yes	http://mallet.cs.umass.edu/index .php
IRST LM	C++	LINUX	OSS	Yes	http://sourceforge.net/projects/ irstlm/
YASMET	C	UNIX	GNU GPL	Yes	http://www.fjoch.com/YASME T.html
SRILM	C++	LINUX	OSS	Yes	http://www.speech.sri.com/proj ects/srilm/
OpenNLP	Java	Any	OSS	Yes	http://maxent.sourceforge.net/
PHAROH	C++	LINUX	Under License Agreement	No	http://www.isi.edu/publications/ licensed-sw/pharaoh/
Thot	C++	LINUX	OSS	Yes	http://www.info-ab.uclm.es/simd/software/thot
MOOD	C++	LINUX Windows	Under License Agreement	Yes	http://smtmood.sourceforge.net.

OSS: Open Source Software.   **LGPL:** Lesser General Public License.
GNU GPL: GNU General Public License.

5 Conclusion

Most of the tools studied in this paper are developed using C++ language for UNIX and LINUX operating systems. Source code of all the tools except PHARAOH and MOOD is available for use and modification under Open Source Software License agreement. The use of MOOD and PHARAOH is under License agreement with developing organizations. It has been found that Moses is the most widely used SMT tool. MALLET and OpenNLP are Java based tools and can be used on any platform under OSS. It is also found that a number of tools are available for developing Language Model (GIZA, GIZA++, cairo and cairoize), Translations Model (SRILM, SLMT, MALLET, YASMET, IRSTLM). These models are used by specific decoder

(PHARAOH, ReWrite Decoder, MARIE, Phramer, Ramses, Joshua) for language translation. We are not able to search any system in Indian languages using these tools. Only one system has been developed using Moses for English to Bangla Translation by Md. Zahurul Islam in August 2009. A lot of work needs to be done on Statistical Machine Translation in Indian languages.

References

[P1] Brown, P.F., Della Pietra, S.: The Mathematics of Statistical Machine Translation: Parameter Estimation. Association of Computational Linguistics 19(2), 263–311 (1993)

[P2] Zahurul Islam, M.: English to Bangla Phrase-Based Statistical Machine Translation, Master Thesis, Department of Computational linguistics at Saarland University. Internet source: lct-master.org (2009)

[P3] Koehn, P.: Moses :Statistical Machine Translation System User Manual and Code Guide, University of Edinburgh (2010),
      http://www.statmt.org/moses/manual/manual.pdf

[P4] Koehn, P.: PHARAOH a Beam Search Decoder for Phrase-Based Statistical Machine Translation Models – User Manual and Description for Version 1.2, USA Information Science Institute (2004), http://www.isi.edu/licensed-sw/pharaoh/

[P5] Ortiz-Martinez, D., Garcia-Varea, I., Casacuberta, F.: Thot: a toolkit to train phrase-based models for statistical machine translation. In: Proc. of the Tenth Machine Translation Summit, Phuket, Thailand, pp. 141–148 (2005)

[P6] Patry, A., Gotti, F., Langlais, P.: Mood at work: Ramses versus Pharaoh. In: Proceedings of the Workshop on Statistical Machine Translation, New York City, pp. 126–129. Association for Computational Linguistics (2006)

[P7] Stolcke, A.: Srilm —An Extensible Language Modeling Toolkit, Speech Technology and Research Laboratory SRI International, Menlo Park, CA, U.S.A (2002),
      http://www-speech.sri.com/papers/icslp2002-srilm.ps.gz

[P8] McCallum, K.A.: MALLET: A Machine Learning for Language Toolkit. (2002),
      http://mallet.cs.umass.edu

[P9] Clarkson, P.R., Rosenfeld, R.: Statistical Language Modeling Using the CMU-Cambridge Toolkit. In: Proceedings ESCA Eurospeech, pp. 2707–2710 (1997)

[P10] Crego, J.M., Mariño, J. B., de Gispert, A.: An Ngram-based Statistical Machine Translation Decoder. Association for Computational Linguistics, 527-549 (2006)

[P11] Crego, J.M., Mariño, J. B., de Gispert, A.: An Ngram-based Statistical Machine Translation Decoder. In: 9th European Conference on Speech Communication and Technology, Lisbon, Portugal, pp. 3185–3188 (2005)

[P12] Papineni, K., Roukos, S., Ward, T., Zhu, W.J.: BLEU: a method for automatic evaluation of machine translation. In: 40th Annual meeting of the Association for Computational Linguistics, pp. 311–318 (2002)

[P13] Callison-Burch, C., Osborne, M., Koehn, P.: Re-evaluating the Role of BLEU in Machine Translation Research. In: 11th Conference of the European Chapter of the Association for Computational Linguistics, pp. 249–256 (2006)

[P14] Doddington, G.: Automatic evaluation of machine translation quality using n-gram co-occurrence statistics. In: Proceedings of the Human Language Technology Conference (HLT), San Diego, CA, pp. 128–132 (2002)

[P15] Knight, K. : A Statistical MT Tutorial Workbook, JHU summer workshop, Internet (1990), http://cseweb.ucsd.edu/~dkauchak/mt-tutorial/

[P16] Josef, F.: Readme-File Of Yasmet 1.0 Yet Another Small MaxEnt Toolkit: YASMET (2001), http://www.fjoch.com/YASMET.html
[P17] Lopez, A.: Statistical machine translation. ACM Comput. Surv. (2008), http://doi.acm.org/10.1145/1380584.1380586

Web References

[W1] http://en.wikipedia.org/wiki/Statistical_machine_translation
[W2] http://www.statmt.org/moses/manual/ manual.pdf
[W3] http://www.statmt.org/moses/?n=Public.Demos
[W4] http://www.clsp.jhu.edu/ws99/projects/mt/
[W5] http://opennlp.sourceforge.net/
[W6] http://thot.sourceforge.net
[W7] http://svr-www.eng.cam.ac.uk/~prc14/ toolkit.html
[W8] www.speech.sri.com/projects/srilm/
[W9] http://hlt.fbk.eu/en/irstlm
[W10] http://www.fjoch.com/YASMET.html
[W11] http://mallet.cs.umass.edu/index.php
[W12] http://svr-www.eng.cam.ac.uk/~prc14/toolkit.html
[W13] http://www.fjoch.com/GIZA++.html
[W14] http://www.isi.edu/licensed-sw/carmel/
[W15] http://www.isi.edu/licensed-sw/rewrite-decoder/
[W16] http://cs.jhu.edu/~ccb/joshua/
[W17] http://www.phramer.org

Discriminative Techniques for Hindi Speech Recognition System

Rajesh Kumar Aggarwal and Mayank Dave

Department of Computer Engineering,
National Institute of Technology, Kurukshetra,
Haryana, India
rka15969@gmail.com, mdave67@gmail.com

Abstract. For the last two decades, research in the field of automatic speech recognition (ASR) has been intensively carried out worldwide, motivated by the advances in signal processing techniques, pattern recognition algorithms, computational resources and storage capability. Most state-of–the–art speech recognition systems are based on the principles of statistical pattern recognition. In such systems, the speech signal is captured and preprocessed at front-end for feature extraction and evaluated at back-end using continuous density hidden Markov model (CDHMM). Maximum likelihood estimation (MLE) and several discriminative training methods have been used to train the ASR, based on European languages such as English. This paper reviews the existing discriminative techniques like maximum mutual information estimation (MMIE), minimum classification error (MCE), and minimum phone error (MPE), and presents a comparative study in the context of Hindi language ASR. The system is speaker independent and works with medium size vocabulary in typical field conditions.

Keywords: ASR, HMM, discriminative techniques, MFCC.

1 Introduction

Speech is the most natural and comfortable means of communication for humans. To communicate with computer machines man still requires interfaces like keyboard, mouse, screen and printer, operated by sophisticated languages and software. A simple solution of this problem would be possible if machine could simulate the human production and understanding of speech. To realize this, speech technology has come into existence over the last four decades. The key components of this technology are automatic speech recognition (ASR), text to speech conversion (TTS) and speaker and language identification. Among them ASR is the most difficult task having a variety of applications such as interactive voice response system and applications for physically challenged persons [1].

The two main components, normally used in ASR, are signal processing component at front end and pattern matching component at back end. We have used Mel frequency cepstral coefficient (MFCC) [2] and continuous density hidden

C. Singh et al. (Eds.): ICISIL 2011, CCIS 139, pp. 261–266, 2011.
© Springer-Verlag Berlin Heidelberg 2011

Markov model at front-end and back-end respectively [3]. Markov models are generally trained with the help of tagged databases using the maximum likelihood estimation (MLE) [4] or discriminative training techniques [5].

For European languages such as English, standard databases like TIMIT and ATIS are available [6] but the major hurdle in speech research for Hindi or any other Indian language is the deficiency in resources like speech and text corpora. We have prepared our own database with the help of EMILLE text corpus [7].

This paper presents a comparative study of the conventional MLE technique and the recently proposed discriminative techniques [8] such as maximum mutual information estimation (MMIE), minimum classification error (MCE), and minimum phone error (MPE) for the design and development of Hindi speech recognition system. Rest of the paper is organized as follows: section 2 presents the architecture and functioning of proposed ASR. Statistical discriminative techniques are covered in section 3. Section 4 shows the experimental results with brief analysis. Finally conclusions are drawn in section 5.

2   Working of ASR

ASR operates in five phases: feature extraction, acoustic modeling (i.e. HMM), pronunciation modeling (i.e. lexicon), language modeling and decoding as shown in Fig. 1. As a first step, the speech signal is parameterized by the feature extraction block. Features are normally derived on a frame by frame basis using the filter bank approach. Using a rate of roughly 100 frames/sec, Hamming speech windows of 20-30 milliseconds are processed to extract cepstral features, which are augmented with their first and second order derivatives to form observation vectors [9].

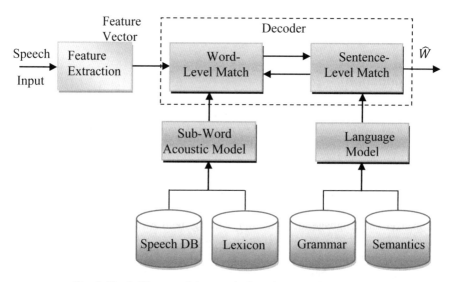

Fig. 1. Block Diagram of Automatic Speech Recognition System

These feature vectors are decoded into linguistic units like word, syllable, and phones [10]. For large vocabulary task, it is impractical to create a separate acoustic model for every possible word since it requires too much training data to measure the variability in every possible context. A word model is formed by concatenating the models for the constituent subword sounds in the word, as defined in a word lexicon or dictionary. Similarly sentences are built by concatenating word models.

Speech recognition is a pattern classification problem which requires training and testing of the system. At the time of training, acoustic and language models are generated which are used as knowledge sources during decoding [11]. The main role of acoustic model is the mapping from each sub word units to acoustic observations. In language model rules are introduced to follow the linguistic restrictions present in the language and to allow redemption of possible invalid phoneme sequences [3].

3 Discriminative Techniques in Statistical Framework

3.1 Analysis of Conventional Statistical Methods

Among the various acoustic models, HMM is so far the most widely used technique due to its efficient algorithm for training and recognition. It is a statistical model for an ordered sequence of symbols, acting as a stochastic finite state machine which is assumed to be built up from a finite set of possible states [12]. Traditionally, the parameters of Markov models are estimated from speech corpora using the maximum likelihood estimation (MLE) [4], which maximizes the joint likelihood over utterances, namely, acoustic features extracted from speech signals and their transcriptions (words or phonemes) in the training data. MLE is based on Expectation-Maximization (EM) algorithm [13] and is very appealing in practice due to its simplicity and efficiency, especially in case of large vocabulary where it is required to manage thousands of hours of acoustic data. Mathematically it can be expressed as:

$$\mathcal{F}_{ml}(\lambda) = \frac{1}{R} \sum_{r=1}^{R} log \left(p\left(\mathbf{Y}^{(r)} \middle| \mathbf{w}_{ref}^{(r)}; \lambda \right) \right) . \tag{1}$$

Where $\mathbf{Y}^{(r)}$ is the rth training utterance with transcription $\mathbf{w}_{ref}^{(r)}$ and λ refers to the model parameters. However, ML-based learning relies on the assumptions [4]:

- The correct functional form of the joint probability between the data and the class categories is known.
- There are sufficient amount of representative training data.

Which are not often realistic in practice and limits the performance of ASR.

3.2 Discriminative Techniques

Noting the weakness of MLE, many researchers have proposed discriminative techniques as an alternate solution. These are maximum mutual information estimation (MMIE), minimum classification error (MCE), and minimum phone error/ minimum word error (MPE/MWE). The aim of MMI criterion is to maximize the mutual information between the word sequence, and the information extracted by a recognizer with parameters λ from the associated observation sequence, $\mathbf{Y}$ [14]. The conventional MCE has been based on the generalized probabilistic descent (or gradient descent) method, in which we define the objective function for optimization that is closely related to the empirical classified errors [15]. The MPE criterion is a smoothed approximation to the phone transcription accuracy measured on the output of a word recognition system given the training data. The objective function in MPE, which is to be maximized, is:

$$\mathcal{F}_{MPE}(\lambda) = \sum_{r=1}^{R} \sum_S P_{\lambda}^k(S|O_r)A(S, S_r) \quad .$$ (2)

Where λ represents the HMM parameters; $P_{\lambda}^k(S|O_r)$ is defined as the scaled posterior probability of the sentence S being the correct one (given the model) and formulated by:

$$P_{\lambda}^k(S|O_r) = \frac{P_{\lambda}(O_r|S)^k P(S)^k}{\sum_u P_{\lambda}(O_r|u)^k P(u)^k}$$ (3)

Where K is the scaling factor typically less than one, O_r is the speech data for r^{th} training sentence; and $A(S, S_r)$ is the raw phone transcription accuracy of the sentence S given the reference S_r, which equals the number of reference phones minus the number of errors [16].

4   Experimental Results

The input speech was sampled at 12 kHz, and then processed at 10 ms frame rate (i.e.120 samples/frame) with a Hamming window of 25 milliseconds to obtain the 39 MFCC acoustic features. For speech signal parameterization, 24 filters were used to get 12 basic features. Thirteen static features (12 basic and one energy) were augmented with 13 delta and 13 delta delta features, thus forming a 39 dimensional standard MFCC feature vector. For word model 7-states HMM per word and for phone model 3-states HMM per phone, along with dummy initial and final nodes were used. The experiments were performed on a set of speech data consisting of five hundred words of Hindi language recorded by 10 male and 10 female speakers using an open source tool Sphinx4 [17] with Fedora 9. For recording a sound treated room was used and microphone was kept at a distance of about 10 cm from the lips of the speaker. Each time model was trained using various utterances of each word. Testing of randomly chosen hundred words spoken by different speakers is made and recognition rate (i.e. accuracy) is calculated, where

Recognition rate = Successfully detected words / Number of words in test set.

Using the frame synchronous CDHMM the following results were analyzed:

• Variation in the recognition rate with different discriminative techniques.
• Variation in the recognition rate with different modeling units.

4.1   Experiment with Discriminative Techniques

Experiments were performed for MLE-based and discriminative based HMMs with another parameter speaker dependent (SD), independent (SI) and MLLR based speaker adaptation (SA)[18] as given in Table 1. Two to five percent more accuracy was achieved by the advanced models proposed in literature in comparison to the standard HMM.

Table 1. Accuracy with Estimation Techniques

Estimation Techniques	SD	SI	SA
Conventional MLE	93%	86.9%	88.2%
Discriminative MMI	93.8%	88%	89.3%
Discriminative MCE	94.5%	89.2%	90.2%
Discriminative MPE	95%	90.4%	91.3%

4.2 Experiment with Modeling Units

Whole word model and context independent phoneme model of standard HMM with linear left-right topology was used to compute the score against a sequence of features for their phonetic transcription. In phoneme based HMM, total 48 Hindi language based phone models were used [19]. For small vocabulary the word model is enough, but as the vocabulary size increases phone model is required to achieve optimum results as shown in Fig. 2.

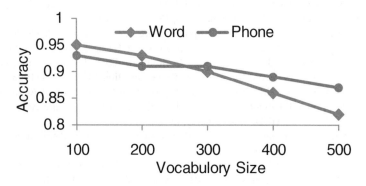

Fig. 2. Accuracy with Modeling Units

5 Conclusion

In a developing country like India, there lies vast potential and immense possibility to use speech effectively as a medium of communication between Man and Machine, to enable the common man to reap the benefits of information and communication technologies. Motivated by this, we have compared different types of statistical techniques in the context of medium size (vocabulary) Hindi speech recognition system to achieve the best results. With the help of our results we observed a significant improvement in the performance of Hindi ASR by using discriminative MPE technique with speaker adaptation. Further the results showed that the word recognition accuracy of whole word models decreases more rapidly than that of sub word models with large vocabularies.   We conclude that the discriminative approaches presented in this paper outperform the traditional model for domain specific Hindi speech recognition system in typical field conditions and can be useful in real time environment.

References

1. O'Shaughnessy, D.: Interacting with Computers by Voice-Automatic Speech Recognitions and Synthesis. Proceedings of the IEEE 91(9), 1272–1305 (2003)
2. Davis, S., Mermelstein, P.: Comparison of Parametric Representations for Monosyllabic Word Recognition in Continuously Spoken Sentences. IEEE Transactions on Acoustics, Speech and Signal Processing 28(4), 357–366 (1980)
3. Jelinek, F.: Statistical Methods for Speech Recognition. MIT Press, Cambridge (1997)
4. Juang, B.H.: Maximum-Likelihood Estimation for Mixture Multivariate Stochastic Observations of Markov Chains. AT & T Technical Journal 64(6), 1235–1249 (1985)
5. He, X., Deng, L.: Discriminative Learning for Speech Recognition: Theory and Practice. Morgan & Claypool Publishers (2008)
6. Becchati, C., Ricotti, K.: Speech Recognition Theory and C++ Implementation. John Wiley & Sons, Chichester (2004)
7. ELRA catalogue, The EMILLE/CIIL Corpus, catalogue reference: ELRA-W0037, http://catalog.elra.info/ product_info.php?products_id=696&keywords=mic
8. Jiang, H.: Discriminative Training of HMMs for Automatic Speech Recognition: A Survey. Computer Speech and Language 24, 589–608 (2010)
9. Huang, X., Acero, A., Hon, H.W.: Spoken Language Processing: A Guide to Theory Algorithm and System Development. Prentice Hall-PTR, Englewood Cliffs (2001)
10. Lee, C.H., Gauvain, J.L., Pieraccini, R., Rabiner, L.R.: Large Vocabulary Speech Recognition Using Subword Units. Speech Communication 13, 263–279 (1993)
11. Aggarwal, R.K., Dave, M.: An Empirical Approach for Optimization of Acoustic Models in Hindi Speech Recognition Systems. In: 8th International Conference on Natural Language Processing (ICON), IIT Kharagpur (2010) (accepted)
12. Rabiner, L.R.: A Tutorial on Hidden Markov Models and Selected Applications in Speech Recognition. Proc. of the IEEE 77(2), 257–286 (1989)
13. Dempster, A.P., Laird, N.M., Rubin, D.B.: Maximum Likelihood from Incomplete Data via the EM Algorithms. Journal of the Royal Statistical Society 39, 1–38 (1977)
14. Gales, M., Young, S.: The Application of Hidden Markov Models in Speech Recognition. Foundations and Trends in Signal Processing 1(3), 195–304 (2007)
15. He, X., Deng, L.: A New Look at Discriminative Training for HMM. Pattern Recognition Letters 28, 1285–1294 (2007)
16. Khe, C., Gales, M.J.F.: Minimum Phone Error Training of Precision Matrix Models. IEEE Transaction on Acoustic Speech and Signal Processing (2006)
17. SPHINX: An open source at CMU: http://cmusphinx.sourceforge.net/html/cmusphinx.php
18. Leggetter, C.J., Woodland, P.: Speaker Adaptation Using Maximum Likelihood Linear Regression. Computer Speech and Language 9(2), 171–185 (1995)
19. Kumar, M., Verma, A., Rajput, N.: A Large Vocabulary Speech Recognition System for Hindi. Journal of IBM Research 48, 703–715 (2004)

An Experiment on Resolving Pronominal Anaphora in Hindi: Using Heuristics

Kiran Pala and Rafiya Begum

International Institue of Information Technology,
Hyderabad, India
{kiranap,rafiyabegum}@gmail.com

Abstract. India is a multilingual, linguistically dense and diverse country with rich resources of information. In this paper we describe the heuristics in a pre-processor layer to an existing anaphora resolution approaches in Hindi language and tested with an experiment. The experiment was conducted on pronouns and presented the results as observations.

Keywords: Anaphora resolution, Syntactico-semantic Knowledge, Natural Language Processing, pre-processor, hybrid approach, heuristics.

1 Introduction

In recent years the Natural Language Processing (NLP) applications have increasingly demonstrated the importance of anaphora or coreference resolutions. In fact, that the successful identification of anaphoric or coreferential links is vital to a number of applications such as machine translation, automatic abstracting, dialogue systems, question answering system and information extraction [3],[6].

In Indian languages, research in the development of computational solutions to automatic anaphora resolution (AR) is not much as in English and in other European languages. Hobb's algorithm is built-up on syntactic information. It is considered to be computationally an economical algorithm. Though Hirst has reported limitations of this algorithm, the suitability of the application of Hobb's algorithm has explored for pronominal anaphoric reference for Hindi language texts [2]. For Indian languages, various experiments has been presented on Hindi language using rule based approach, a comparative system between Malayalam and Hindi, corpus based studies using centering theory and s-list algorithm. In fact we have seen Hobb's Naive algorithm being used for Turkish, similarly a machine learning based algorithm being suggested for Tamil [3], [5], [6], [7], [9]. However, have seen considerable advances in the field of AR, still a number of outstanding issues that either remain need further attention and, as a consequence, represent major challenges to the further investigation is how far the performance of AR algorithms can go and what are the limitations of knowledge-of poor methods. In particular, more research should be carried out into the factors influencing the performance of existing algorithms. Another significant problem for automatic AR systems is that the accuracy of the processing is still too low and as a result the performance of such systems remains far from ideal.

C. Singh et al. (Eds.): ICISIL 2011, CCIS 139, pp. 267–270, 2011.
© Springer-Verlag Berlin Heidelberg 2011

We aim to come up with some heuristics which may be used to quickly decypher anaphora. Functionally, we took Hindi treebank corpus as inputs (i.e. manually annotated data for verify the experiment), apply on it created heuristics, and through these heuristics identify maximum number of anaphora. Those anaphoras that miss this heuristic filter may further be identified through conventional process means which are currently available for knowledge poor and then knowledge rich methods. This is an attempt to present a framework of anaphora resolution that is based on combination of heuristics and knowledge poor and rich methods to be applied for Hindi only. The type of anaphora to be incorporated is pronominal anaphora. The refined aim then is to investigate natural language processing techniques that are applicable to pronominal AR in Hindi.

2 Experiment and Methodology

Our approach will be a hybrid– it will consider the previous approaches but be beyond them, we intend to involve heuristics. The plan is to pre-process the data with heuristics before feeding them to an existing AR approaches, like in Fig. 1.

2.1 Architecture of Framework

We hypothesize that by early pre-processing we are able to not only identify more anaphora in a discourse, but also be able to increase the efficiency of an overall proposed AR system Fig. 1. Therefore, we propose in this paper that we only conduct the experiment and establishing the possibility of a heuristics in a pre-processor layer. It would be up-to the follow-up - incorporate a complementary layers, and to fine tune the overall hybrid framework to increase resolution efficiencies.

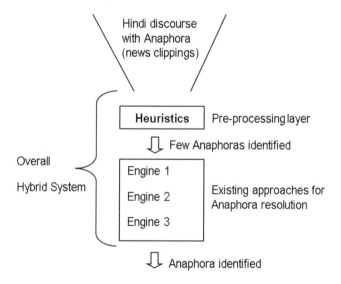

Fig. 1. Hybrid approach framework for Indian language AR system

2.2 Data for Experiment

Corpus. We have used the corpus which was collected from various reliable web resources specifically on health and tourism domains. This data have been cleaned and parsed through Hindi shallow parser [4]. The parsed data will be in Shakti Standard Format (SSF) [1], [4]. This data have been verified by language editors. For testing and evaluation, we used this parsed Hindi corpus.

2.3 Methodology and Procedure

We have chosen to use a heuristics in a pre-processor layer of AR system. We wanted to experiment with how such a system would be conceived, devised and developed. For this sake we choose to start with toy-sentences, come up with heuristics based on syntactico-semantic knowledge of human-beings, and tested with one sample heuristic with pronoun *"apne"* on standard test data, which is manually verified corpus. Choosing only one heuristic was that we wanted to see the entire process from beginning to end and resolve the system level issues first before choosing and implementing multiple heuristics. We did develop 10 heuristics but in this paper presented the results of only one [8].

We have implemented an algorithm to test heuristics. In the pre-processor layer– the given input data has parsed through parser and it has extracted features from output of parser which is in SSF. Resulted data process through given heuristics and verified the conditions if get satisfies sent to output layer, else iterate. Finally, compare the results against those that manually identified anaphora results on test data. The results of comparison, analysis was presented in observations and conclusions section.

3 Observations and Conclusions

We have implemented a heuristic on *"apne"* pronoun in given test data i.e. 10 stories for this experiment. Test data contains total 12 number of *"apne"* pronoun according to definition. Manually identified testing anaphors as baseline are 10. After successful implementation, the heuristic has identified 11 anaphors from the same test data i.e. 92% recall. The outputs have been verified manually and removed the errors, finally the precision has reported as 71%. This reported result properly tracked at first iteration of the heuristic on *"apne"* pronoun.

In this process overall 29% of errors were occurred due limitations are not working with a cataphora, not working with antecedent chaining and not going beyond current sentence for an *"apne"* pronoun resolution. This experiment was simultaneously challenging, interesting and insightful.

We faced many challenges with preparation of heuristics, execution of algorithm, collection and cleaning of Hindi text data, handling of old and new versions, etc. Our learning has been that toy grammar and sentences are one thing, and real-life sentences are another. Toy resources are good for conceptual understanding in human beings, but not for designing "industrial grade" algorithms or solutions.

We feel that the ultimate desired solution is currently only being asymptotically reached by today's popular strategies. Though lot of work has already been done on anaphora resolution (both for English and other languages), there is still lot of new work to be done for Indian languages. It is indeed an ocean for the interested linguists and researchers.

4 Future Work

We emphasize the need for the extending this work for further research into the refinement of other heuristics and integration with S-list based approach according to the proposed framework. And work with more data towards value adds. Finally, build a web interface for the proposed framework as a complete AR system to Indian languages in multilingual context.

Acknowledgments. We would like to acknowledge the valuable suggestions made by Sriram V, Samar Hussain from LTRC, IIIT, Hyderabad. We are also thankful to the reviewers for their constructive comments which we have tried to take into account as far as possible. We would like to thank the participants for their help and for kindly providing the time and data for analysis.

References

1. Akshar, B., Chaitanya, V., Sangal, R.: Natural Language Processing: A Paninian Perspective, pp. 65–106. Prentice-Hall of India, New Delhi (1995)
2. Bhargav, U.: Pronoun Resolution for Hindi', Master dissertation submitted to International Institute of Information Technology, Hyderabad, Andhra Pradesh (2009)
3. Davison, A.: Lexical anaphors in Hindi/Urdu. In: Wali, K., Subbarao, K.V., Lust, B., Gair, J. (eds.) Lexical Anaphors and Pronouns in Some South Asian Languages: a Principled Typology, pp. 397–470. Mouton de Gruyter, Berlin (2000)
4. Hindi Shallow Parser, http://ltrc.iiit.ac.in/analyzer/
5. Mitkov, R.: Outstanding issues in anaphora resolution. In: Gelbukh, A. (ed.) CICLing 2001. LNCS, vol. 2004, pp. 110–125. Springer, Heidelberg (2001)
6. Mitkov, R.: Robust pronoun resolution with limited knowledge. In: Proceedings of the 36th Annual Meeting of the Association for Computational Linguistics and 17th International Conference on Computational Linguistics, vol. 2, pp. 869–875. Association for Computational Linguistics (1998)
7. Murthy, K.N., Sobha, L., Muthukumari, B.: Pronominal resolution in tamil using machine learning. In: Proceedings of the First International Workshop on Anaphora Resolution (WAR-I), pp. 39–50 (2007)
8. Pala, K., Sai, G.: A Novel approach for Hindi Anaphora Resolution, Course report submitted to Language Technology Research Center, International Institute of Information Technology, Hyderabad, Andhra Pradesh (2009)
9. Prasad, R., Strube, M.: Constraints on The Generation of Referring Expressions, With Special Reference To Hindi, A PhD dissertation University of Pennsylvania (2003)

A Novel GA Based OCR Enhancement and Segmentation Methodology for Marathi Language in Bimodal Framework

Amarjot Singh, Ketan Bacchuwar, and Akash Choubey

Dept. of Electrical Engineering, NIT Warangal,
A.P., India
amarjotsingh@ieee.org, bacchuwarketan@gmail.com,
ee.akash@gmail.com

Abstract. Automated learning systems used to extract information from images play a major role in document analysis. Optical character recognition or OCR has been widely used to automatically segment and index the documents from a wide space. Most of the methods used for OCR recognition and extraction like HMM's, Neural etc, mentioned in literature have errors which require human operators to be rectified and fail to extract images with blur as well as illumination variance. This paper explains proposes an enhancement supported threshold based pre-processing methodology for word spotting in Marathi printed bimodal images using image segmentation. The methodology makes use of an enhanced image obtained by histogram equalization followed by followed by age segmentation using a specific threshold. The threshold can be obtained using genetic algorithms. GA based segmentation technique is codified as an optimization problem used efficiently to search maxima and minima from the histogram of the image to obtain the threshold for segmentation. The system described is capable of extracting normal as well as blurred images and images for different lighting conditions. The same inputs are tested for a standard GA based methodology and the results are compared with the proposed method. The paper further elaborates the limitations of the method.

Keywords: OCR, Genetic Algorithm, Bimodal, Blur, Illumination.

1 Introduction

Optical character recognition (OCR) is a mature field, born in 1950's initially practiced with electronic and electro-mechanical methods aimed towards machine printing while limited to fixed-format applications. A typical OCR system is widely used for character segmentation, segment reconstruction, character recognition, and word and phrase construction without human intervention or human correction. The basic step before all the application mentioned above is image enhancement and segmentations, hence the paper focuses on the same.

According to literature survey, a number of methods have been proposed for image enhancement and segmentation of OCR but the in ability of these systems to work together for different experimental conditions is a major drawback in document

C. Singh et al. (Eds.): ICISIL 2011, CCIS 139, pp. 271–277, 2011.

analysis. Besides the poor quality of images, illumination changing conditions also makes OCR recognition a daunting task. The error incurred is corrected using a human support system which is a time consuming as well as a costly proposal. This paper proposes an OCR system which makes use of histogram equalization to extract images blur as well as illumination varying conditions simultaneously. The histogram used by the mentioned algorithm is bimodal in nature hence it can be divided into two classes. Genetic algorithm is further used to select the threshold from the histogram for extracting the object from the background. The capabilities of the system are tested on images with blur, noise and change in illumination.

The paper is divided into five sections. The next section elaborates the related work to OCR enhancement and recognition developed over the past few years. Third section explains the standard and proposed methodology used in the paper to enhance and extract the characters followed by the results obtained from the simulations in the fourth section. A brief summary of the paper is presented in the last section of the paper.

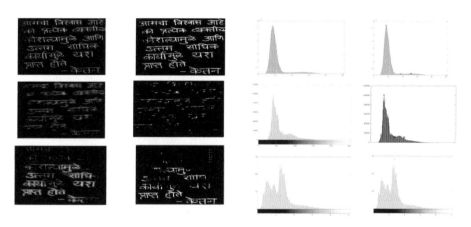

Fig. 1. (a) The input RGB image further converted into gray scale, the results along with the histogram obtained from standard algorithm (b The input blur image, the corresponding results along with the respective histograms obtained from standard algorithm (c) The input illumination variant image, the results along with the respective histograms obtained from standard algorithm

2   Related Work

In case of bimodal images, the histogram has a deep and sharp valley between two peaks representing objects and back ground respectively which can be used to select the threshold representing the bottom of this valley. Khankasikam et. al [1], [6] proposed the valley sharpening techniques which restricts the histogram to the pixels with large absolute values of derivatives where as S. Watanable et. al. [2], [7] proposed the difference histogram method, which selects threshold at the gray level with the maximal amount of difference. These techniques utilize the information concerning neighboring pixels or edges in the original picture to modify the histogram so as to make it useful for thresholding. Another class of methods deal directly with the grey level histogram by parametric techniques. The histogram is approximated in the least square sense by a sum of Gaussian distributions, and statistical decision procedures are applied [8]. However, such methods are tedious and involve high computational power. In Di Gesu [3], [8] the idea of using both intensities and spatial information has been considered to take into account local information used in human perception. A number of new strategies and methodologies have been proposed over the last couple of years to detect the global as well as local solutions in a nonlinear multimodal function optimization [4], [5], [9]. Multiple peaks can be maintained in multimodal optimization problem with the help of crowding. Crowding method is extremely efficient in detecting in detecting the two peaks on a bimodal histogram. Further, GA can be applied to discover the valley bottom between these peaks which can be used as the threshold for extracting the information from the background.

3 Algorithm

The section discusses the standard GA based algorithm being used for information extraction. In the second section, a novel histogram based methodology which can be efficiently used to extract information from bimodal images is presented. The algorithms are explained in detail below.

3.1 Standard GA Based Algorithm

The algorithm is a histogram based approach effectively used to extract useful information from bimodal images. The histogram of the digital image is a plot or graph of the frequency of occurrence of each gray level in the image across gray scale values. Genetic algorithm is applied on the histogram of the bimodal image to extract the useful information from the background. A random population of size N is initialized where the element acquire the value between 0 to 255. The crossover and mutation operations are carried out on the randomly chosen two parents. Appropriate value of crossover probability (P_c) and mutation probability (P_m) is fixed. The winner of each tournament (the one with the best fitness) is selected. After computing the fitness value of the off-springs, Tournament selection strategy is used to allow off-springs to compete with the parents. It involves running a competition among two individuals

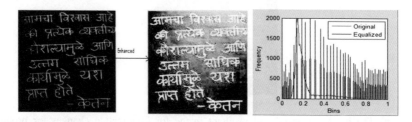

Fig. 2. (a) The input image and the corresponding enhanced image (b) The equalized histogram of the resultant enhanced image

chosen at random from the population. The fittest between both is selected. This is the method used in Genetic algorithm for selection of individual from the population. Here two parents and two-off springs compete to give two best individuals as result. The resulting selected elements are located in their respective classes. The methodology used is termed as Crowding Method. This method basically replaces the older elements in the population by the fittest elements in the resulting generation which helps to reduce replacement error. The repetition performed for all the elements results into convergence. This converged value is the gray value corresponding to the minima between two peaks. Then this gray value is used as threshold value and the image is segmented.

Fig. 3. (a) The input, enhanced and resultant image for illumination changes for proposed algorithm (b) The input, enhanced and resultant image for blur for proposed algorithm

3.1 Proposed Algorithm

The paper proposes a new method which gives better results under blur as well as illumination varying conditions as compared to the standard GA based method. The method is divided into two steps. In the first step, the histogram of the image is equalized which results into the redistribution of the intensities in the histogram as shown in fig. 2 (a). The normalized histogram for the reference input is shown in fig. 2 (b), which shows the redistribution of pixel intensities. It results into an increase in the intensities of the pixels which are low in grayscale while performs a decrement for high intensity valued pixels. The intensities with lower intensity value are upgraded to higher values and vice versa. These changes in the intensity values results into the enhancement of the bimodal image. With the enhancement in the histogram, the intensity values increase as well as decrease for every pixel due to redistribution of the histogram. This leads to an enhancement in the gray values of the class which is has gray scale values near to the upper band (129 to 255). This operation helps the pixels belonging to upper band to move towards upper band. Further genetic algorithm is applied on the equalized histogram in order to extract the threshold. The threshold selected in this case will be higher as compared to the previous methodology due to the increase in the pixel intensities because of which a higher threshold will be required to extract the lower class pixels from the higher class pixels. On applying the threshold, the pixels values corresponding to the upper class can be easily separated from the pixels of the lower band as the histogram equalization already moved the pixels which were lower in value but belonged to the upper band towards the higher value. The method is highly efficient and can be effectively used to extract the information from the image with different conditions. The technique used doesn't oblige any valley sharpening techniques.

4 Results

The results obtained from the simulation enable us to explain the capabilities of the methodology applied to bimodal images. The robustness of the method is also further tested on blurred as well as images with varying illumination on an Intel Core 2 Duo 2.20 GHz machine. The section compares the results on image with blur and illumination variance for the standard method with the proposed method. In the first phase the experimentation is performed for the standard method. Later the simulations are performed for the proposed method and the results are compared.

A RGB image of Marathi language of image size 256 by 256 as shown in Fig. 1(a) is given as input to the standard GA based method system. Selection of a global threshold now reduces to determine a suitable gray value in between the peaks of the histogram. The two peaks in the histogram are obtained using crowding. The parameters of the algorithm are selected as (i) number of population elements " N " is 20 (ii) crossover probability P_c =0.9 (iii) mutation probability P_m decreases trailing an exponential rate with starting value 0.05. Fig. 1(a) shows the maxima's or detected peaks along with the minima's obtained using genetic algorithm. Two peaks or maxima's computed using genetic algorithm shown in fig. 1(a), form the basis of image segmentation. The peaks are obtained at gray values 48 and 129 while the valley is obtained at 92. Using the gray value 117 as threshold value segmentation of the original image is carried out and the resultant image is shown in Fig. 1 (a). The segmentation of the object from the background is clearly visible using the threshold corresponding to the valley as shown in fig. 1 (a). The modes are clearly visible, separated by a long valley which can be further used to segment the image.

Further, the capabilities of the standard GA based system is tested on images with blur as shown in fig. 1(b). The histogram of the blur image is plotted as shown in fig. 1 (b). The peaks are obtained at gray values 47 and 88 while gray value 69 is used as threshold value for segmentation of the original blurred image from the background as shown in Fig. 1 (b). The same system is also tested on images with different illumination and lighting conditions as shown in fig 1(c). The histogram of the image was plotted as shown in fig. 1 (c). The peaks are obtained at gray values 67 and 129 while gray value 128 is used as threshold value for segmentation of the original image with illumination from the background as shown in Fig. 1 (c). The method is unable to extract the information from the background in case of both cases as shown in fig. 1 (a) and fig. 1(b) respectively.

The section further explains the simulations obtained for proposed method on the same inputs as for the standard GA method. The image given input to the system is enhanced by histogram equalization as show in fig 2(a). The equalized histogram is shown in fig 2(b). It is observed that the intensity of the white boundary bounding pixels are increased along with the black pixels. The image becomes much brighter as compared to the original image hence the pixels which are suppose to be in the upper band move to a higher intensity. In the second and final step, image segmentation is performed using the GA based approach described above in the paper. The initial population, the crossover as well as the mutation probabilities assigned for the current methodology use the same values as specified in the previous method. The threshold value is obtained using genetic algorithm. The threshold value for the enhanced image is 210, which is much higher as compared to the threshold obtained in the case of normal image, as the overall intensity values have been enhanced in the histogram. The results obtained from the proposed methodology are much better as compared to the results obtained from the previous methodology mentioned above. The previous methodology mentioned in the paper failed to extract the useful information from the blurred images and images affected by illumination variations while the proposed method effectively extracts the information from the background efficiently as shown in fig. 3(a) and fig. 3(b) respectively.

5   Conclusion

The section presents a brief summary and a comparison of the algorithms discussed in this paper. Image segmentation for bimodal documents is the primary focus of this paper. The problem triggers down to determine the threshold using histogram of the given image. Both the methodologies take support of genetic algorithm in order to decide the threshold while the preprocessing step is different for both the techniques. The proposed algorithm applies GA on an equalized histogram where as the standard method uses unchanged histogram of the image for the threshold determination. It is observed that the standard GA based algorithm mentioned in the paper is incapable of extracting the images with blur and illumination variations while the proposed algorithm efficiently separates the background from the useful data. The proposed method is extremely efficient to extract all kinds of bimodal images including blur and illumination but fails for images having histograms with multi-modal features. Currently, attempts are made to address two class images with noises and images requiring multiple thresholds. Attempts have been made to overcome this problem.

References

1. Weszka, J.S., Nagel, R.N., Rosenfeld, A.: A Threshold selection technique. IEEE Trans. Computer C-23, 1322–1326 (1974)
2. Watanable, S., CYBEST Group.: An automated apparatus for cancer processing. Comp. Graph. Image processing 3, 350–358 (1974)
3. Di Gesh, V.: A Clustering Approach to Texture Classification. In: Jain, A.K. (ed.) Real Time Object and Environment Measurement and Classification. NATO AS1 Series F, vol. 42, Springer, Heidelberg (1988)
4. Fu, K.S., Mui, J.K.: A Survey on Image segmentation. Pattern Recognition 13, 3–16 (1981)
5. Pal, N.R., Pal, S.K.: A Review on Image Segmentation Techniques. Pattern Recognition 26(9), 1277–1294 (1993)
6. Kalas, M.S.: An Artificial Neural Network for Detection of Biological Early Brain Cancer. International Journal of Computer Applications 1(6), 17–23 (2010)
7. IP, H.H.-S., Chan, S.-L.: Hypertext-Assisted Video Indexing and Content- based Retrieval. ACM 0-89791-866-5, 232–233 (1997)
8. Storn, R., Price, K.: Differential Evolution – A Simple and Efficient Heuristic for Global Optimization over Continuous Spaces. Journal of Global Optimization, 341–359 (1997)
9. Kao, Y.-T., Zahara, E.: A hybrid genetic algorithm and particle swarm optimization for multimodal functions. Applied Soft Computing 8(2), 849–857 (2008)

Panmozhi Vaayil - A Multilingual Indic Keyboard Interface for Business and Personal Use

H.R. Shiva Kumar, Abhinava Shivakumar, Akshay Rao,
S. Arun, and A.G. Ramakrishnan

MILE Lab, Dept of Electrical Engineering, Indian Institute of Science, Bangalore

Abstract. A multilingual Indic keyboard interface is an Input Method that can be used to input text in any Indic language. The input can follow the phonetic style making use of the standard QWERTY layout along with support for popular keyboard and typewriter layouts [1] of Indic languages using overlays. Indic-keyboards provides a simple and clean interface supporting multiple languages and multiple styles of input working on multiple platforms. XML based processing makes it possible to add new layouts or new languages on the fly. These features, along with the provision to change key maps in real time make this input method suitable for most, if not all text editing purposes. Since Unicode is used to represent text, the input method works with most applications. This is available for free download and free use by individuals or commercial organizations, on code.google.com under Apache 2.0 license.

Keywords: Input Method, IM, Indic, Localization, Internationalization (i18n), FOSS, Unicode, Panmozhi vaayil, Vishwa vaangmukha.

1 Introduction

Input method editors (IME) provide a way in which text can be input in a desired language other than English. Latin based languages are represented by the combination of a limited set of characters, and most languages have a one-to-one correspondence of each character to a given key on a keyboard. When it comes to East Asian languages (Chinese, Japanese, Korean, Vietnamese etc.) and Indic languages (Tamil, Hindi, Kannada, Bangla etc.), the number of key strokes to represent an akshara can be more than one, which makes using one-to-one character to key mapping impractical. To allow for users to input these characters, several input methods have been devised to create Input Method Editors.

2 Objective

The focus has been to develop an Indic multilingual input method editor, with a minimalistic interface providing options to configure and select various language layouts. Configurability is inclusive of addition of new layouts or languages. Inputs can be based on popular keyboard layouts or using a phonetic style [1]. We call it

C. Singh et al. (Eds.): ICISIL 2011, CCIS 139, pp. 278–283, 2011.
© Springer-Verlag Berlin Heidelberg 2011

Panmozhi Vaayil in Tamil and Vishwa Vaangmukha: in Sanskrit, both meaning entrance for many languages. It is known by the generic name, Indic Keyboard IM in English and is available for download from http://code.google.com/p/indic-keyboards/

3 Motivation

Some of the main reasons for developing Indic-keyboards are:

1. To ease inputting of any Indian language under any platform.
2. To facilitate increased use of Indian languages on the computer and internet.
3. To provide a free interface with an unrestricted license.
4. Support phonetic and other popular layouts in a single package.
5. Need for a unified multiplatform input method.
6. Ease of configurability and customizability.

3.1 Existing Works and What They Offer

Some of the existing, popular, easy, flexible input methods are:

- **Baraha IME** – Provides phonetic support for a fixed number of languages and designed for use on Microsoft Windows platform [2].
- **Aksharamala** – Similar to BarahaIME with support for MS Windows [3].
- **Smart Common Input Method (SCIM)** – Designed to work on Linux with phonetic style of input [4].

3.2 What Indic-Keyboards (Panmozhi Vaayil) Offers

The distinguishing features of our input method editor are:

- Phonetic as well as popular keyboard layouts.
- Dynamic module enabling the addition of new keyboard layouts by users.
- Both on Linux platform and Microsoft Windows.
- Phonetic key maps can be changed to meet user's requirements.
- No installation hassles.
- Open source.
- Available under Apache 2.0 License, which means even commercial companies can use our code to develop products, after acknowledging us.

4 Design

Figure 1 shows the architecture. The design can be broadly categorized into the following modules:

(1) User interface and the shell extension. (2) Capturing the keyboard events. (3) XML based Unicode processing. (4) Rendering the Unicode.

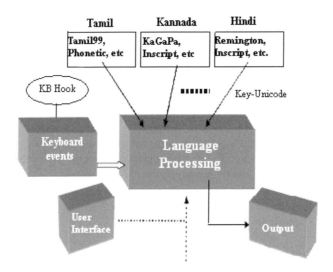

Fig. 1. System architecture of Panmozhi Vaayil showing the modules and their interaction

4.1   User Interface and Shell Extension

The User interface is a shell extension which sits in the system tray/Notification area. The main purpose of this is to allow users to interact with the input method. This mainly involves selection of the language and the particular keyboard layout. It also helps in enabling and disabling the input method, accessing help and to display the image of the keyboard layout currently selected. Apart from these, the menu also has provision for addition of new keyboard layouts.

4.2   Capturing the Keyboard Events

The input method is designed to operate globally. That is, once the input method is enabled, further key strokes will result in characters of the particular language selected being rendered system wide. This requires the capture of the key presses system wide across all processes. A keyboard hook installed in the kernel space will enable this. This module is, therefore, platform specific.

4.3   XML Based Unicode Processing

Finite Automata exists for each language and for every keyboard layout. It has been designed as XML files, where every XML file corresponds to a kBD layout. XML based processing makes it possible to add new layouts or new languages dynamically. The input key pattern is matched with the XML file to see if the pattern matched is a vowel or a consonant. For the input pattern, a sequence of Unicode(s) is returned. The structure of the XML file is as follows:

```
<pattern>
     <char>A</char>
     <unicode>0C86</unicode>
     <consonant>0</consonant>
     <uni2>0CBE</uni2>
</pattern>
```

The above XML block indicates that for the key press "A", the corresponding Unicode is 0C86. The consonant tag tells us whether it is a vowel or consonant. If it is a vowel, a second tag gives the Unicode of the associated dependent vowel (if any).

Two algorithms have been designed, one for phonetic style input and the other for keyboard layouts. Both are generic, i.e. same algorithm is used for keyboard layouts of all languages and one algorithm for phonetic input in any language. The XML key maps can be changed on the fly and the changes are reflected instantly.

4.4 Unicode Rendering

Once the key is pressed, simple grammar rules are applied to determine whether the output has to be a consonant, an independent vowel or a dependent vowel. The XML file is parsed and the corresponding Unicode is fetched. The Unicode is sent back to the process, where the keypress event took place and is rendered if any editable text area is present. The rendering of Unicode is platform specific.

5 Implementation

The following tools and languages have been used to implement the input method:

1)**Java SE** – This is used to implement the main language processing module to get easy portability. Up to 80% of the code has remained common across platforms.
2)**Eclipse SWT** – Used to implement the user interface. Eclipse SWT, which uses Java SWT, is preferred over other toolkits to get a native look and feel..
3)**XML** – Finite automata exists for every language (layout) and XML has been used to design it. Simple API for XML Parsing (SAX) has been used to parse the XML.
4)**Win32 libraries:** Windows API, is Microsoft's core set of application programming interfaces (APIs) available for MS Windows platform. Almost all Windows programs (eg. SAPI, Tablet PC SDK, Microsoft Surface) interact with the Windows API. Platform specific portions have been implemented to run on Microsoft Windows variants using the Microsoft Win32 libraries. Both keystroke capturing and Unicode rendering have been accomplished using Win32 libraries. The steps involved are:

a. Syshook.dll : Install a keyboard hook in the operating system. The hook is set up for the keyboard to listen to key presses. The Windows API used is **SetWindowsHookEx()** and the library accessed is user32.dll. (See Fig. 2)

b. opChars.dll : Responsible for putting the character on to the current active window. Sends a message to the input event queue using the Windows API **SendInput()**. The library accessed is user32.dll

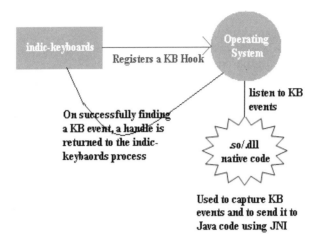

Fig. 2. JNI-Native code and keyboard hook procedure

5)Evdev - Also known as the user input device event interface/Linux USB subsystem. Used to capture keystrokes in GNU/Linux. This provides a clean interface to handle keyboard, mouse, joystick etc. in the userspace in Linux. This involves the following:

   a. Open file descriptor to the input device using the **open()** API with suitable parameters. Use **ioctl()** to identify the device.
   b. Using the open file descriptor, continuously read bytes from the input device (can be key press/release, mouse movements) using the **read()** API.

6)Xlib – Used for Unicode rendering in GNU/Linux. The steps involved are:

   a. Identify the current active window using **XGetInputFocus()**
   b. Make the window listen to all keypress events using **XSelectInput()**
   c. Using the keycodes obtained for every keypress/release event from evdev, using a mapping table to map the keycode to the keysym. Output the Unicode to the active window using **XSendEvent()** API.

7)Java Native Interface – Also known as JNI in short. The JNI enables the integration of code written in the Java programming language with code written in other languages such as C and C++. The write once, run anywhere concept arises from the fact that Java acts as an abstraction layer on top of the native implementation. All the API java provides have been natively implemented and the Java code allows the same APIs to be used across platforms.

   The native code is usually packaged as a DLL or a Shared Object. The Java method which accesses the native code is created with a keyword "native". Header files need to be created for the classes which contain these methods. At run-time, java code interacts with the native libraries using predefined interfaces. The native methods can also call Java methods. This mechanism is known as **JNI callback**.

6   Performance and Conclusion

The languages and keyboard layouts currently supported are listed in Table 1. An easy-to-use user interface has been provided to add new layouts which are Inscript

Table 1. Languages and layouts currently supported by Panmozhi Vaayil

Language	Phonetic	Layouts
Tamil	Yes	Inscript, Tamil99, Remington
Kannada	Yes	Inscript, KaGaPa
Telugu, Gujarati	Yes	Inscript
Bengali, Malayalam, Oriya, Gurumukhi	No	Inscript
Hindi, Marathi	Yes	Inscript, Remington

like. Additional phonetic or other layouts can be added based on the existing layouts by creating new XML files and following the prescribed structure. Existing layouts can be changed/customized to suit the user's needs. In phonetic layouts, a single key press to vowel mapping is used to ensure lesser key presses for the completion of the CV combination. Ex : k (க்) + Y (ஐ○) = கை instead of k + ae/ai.

The input method is multithreaded and the following runtime statistics have been obtained. Java Monitoring and Management console has been used to profile.
(1) Average Heap Memory usage : 4.0 MB (maximum : 5.0 MB). (2) CPU usage : 0.2% – 0.3% (3) Garbage Collector: Average time for one sweep – 0.05s. Average heap space freed up – 1 MB. (4) Number of threads: Peak – 15. Average live threads – 13 (2 threads are spawned by the input method).

The flexibility of adding new Indic languages on the fly, modification of the existing layouts, changing the keypress - Unicode input combination for phonetic input makes it easy to use. Thus, we have abstained from modifying any system files and relieved the user of all installation hassles. It is fast and light on system resources. The user can run it through a pen drive, CD, DVD, hard disk or any portable media. Being open source and licensed under the Apache 2.0 License, developers and users alike can modify, recompile, or rewrite the source and can also make these appendages closed source. The license also allows developers to sell the modified code. Thus, a dynamic, flexible, easy to use, multiplatform, multilingual,.clean, unrestrictive input method has been designed.

References

1. Rolfe, R.: What is an IME (Input Method Editor) and how do I use it?. Microsoft Global Development and Computing Portal (July 15, 2003)
2. Baraha - Free Indian Language, http://www.baraha.com
3. Aksharamala, http://www.aksharamala.com/
4. Smart Common Input Method SCIM, http://www.scim-im.org/

Power Spectral Density Estimation Using Yule Walker AR Method for Tamil Speech Signal

V. Radha, C. Vimala, and M. Krishnaveni

Department of Computer Science, Avinashilingam Deemed University for Women,
Coimbatore, Tamilnadu, India

Abstract. Window theory always an active topic of research in digital signal processing. It is mainly used for leakage reduction in spectral analysis. In this paper, the effect of windowing in power spectral density estimation of Tamil speech signal is analyzed. Four different window functions are implemented and their performances are evaluated based on the parameters such as sidelobe level, fall off and gain. Based on the experiments it is found that the effect of applying hamming window best suites for the Tamil speech signal. It can reduce the spectral discontinuities of the signal and this effect of hamming window is given as the potent metric for estimating the spectral power of Tamil speech signal. Here Power Spectral Density (PSD) estimation is computed by using parametric and non-parametric methods. The reduction of noise ratio in PSD is considered as the parameter and it is estimated through crest factor. Finally the paper concludes with the need of best windowing method for PSD particularly in parametric techniques. Evaluation is handled both objectively and subjectively for Tamil speech datasets.

Keywords: Power Spectral Density, Hamming, Yule Walker, Crest factor, Tamil Speech.

1 Introduction

Window choice is an important task in any digital signal processing applications. Among the different types of windows, hamming window best suites for speech signal processing. Hence its potent advantage is used for the estimation of PSD [1] of Tamil voice signal. Especially, if the signals contain a non-stationary component which is randomly distributed among the data. Since speech is non-stationary to investigate these frequency characteristics, the PSD estimation [1] is done for various Tamil spoken words. For above process, two types of PSD methods are implemented namely parametric [5] and Non-parametric in which Tamil speech signals are taken as input data.

The paper is organized as follows. Section 2 explains the subjective evaluation of windowing methods. Section 3 deals with various PSD techniques and its performance in Tamil speech signal. Section 4 explores the performance evaluation. Finally, the conclusion is summarized in section 5 with future work.

C. Singh et al. (Eds.): ICISIL 2011, CCIS 139, pp. 284–288, 2011.

2 Evaluation of Windowing Methods

Window choice [4] is crucial for separation of spectral components, where one component is much smaller than another. For this research work four different types of windows are used namely Blackman, Bartlett, Triangular and Hamming. To evaluate the effect of these windows, various characteristics are considered such as sidelobe level, environmental bandwidth etc and they are explained in section 4. The best windowing method should have minimum value for the above properties and it is clear from the figure1 hamming window gives better results than any other windows.

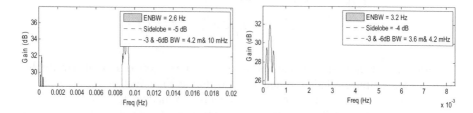

Fig. 1. (a) Performance of Blackman windows   **Fig. 1.** (b) Performance of Bartlett windows

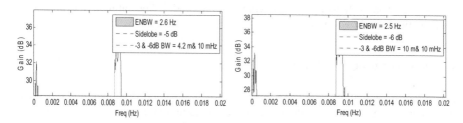

Fig. 1. (c) Performance of Triangular window   **Fig. 1.** (d) Performance of Hamming windows

3 PSD Estimation for Tamil Speech Signal

The process of estimating PSD is a very useful tool when analyzing the amplitude of a signal. This is done to avoid problems due to truncation of the signal [7]. In this paper, Welch method of nonparametric and Yule-Walker AR [3] of parametric method is implemented.

3.1 Non Parametric Welch Method

Non-parametric methods estimates the PSD directly from the signal [9] itself. One of the popular methods of this type is Welch's method [9], which consists of dividing the time series data into segments, and then averaging the PSD estimates. By default, the data is divided into four segments with 50% overlap between them. Although overlap between segments tends to introduce redundant information, this effect is diminished by the use of a hamming window, which reduces the importance or weight given to the end samples of segments (the samples that overlap).

3.2 Parametric Yule-Walker AR

Parametric methods [5] can yield higher resolution than non-parametric methods in cases where the signal length is short. Instead of estimating the PSD directly from the data, they model the data as the output of a linear system driven by white noise [6] [7] (an adaptive filter), and then attempt to estimate the parameters of that linear system. The output of such a system for white noise [7] [8] input is an autoregressive (AR) process. These methods are sometimes referred to as AR methods. All AR methods yield a PSD estimate given by

$$\hat{P}_{AR} = (f) = \frac{1}{f_s} \frac{\varepsilon_p}{\left|1 + \sum_{k=1}^{p} \hat{\alpha}_{p(k)} e - 2\pi jkf / f_s\right|^2} \tag{1}$$

The different AR methods estimate the AR parameters $a_p(k)$ slightly differ yielding different PSD estimates. This formulation leads to the Yule-Walker equations, which are solved by the Levinson-Durbin recursion. Only this method applies window [4] to data. Since parametric method [5] gives input the adaptive filter, the accurate PSD can be obtained without noise [6] [7]. It is clear form the experiments the Yule-Walker AR spectrum is smoother than Welch method especially in the start and end frame of a signal because of the simple underlying all-pole model. The subjective evaluation of these methods is shown in section 4.

4 Performance Evaluation

Performance evaluation is done for both windowing [4] and PSD methods separately. It is clear form the section 3, the performance of hamming window is best when compared with other windows. It is done based on the following characteristics.

1. *Sidelobe Level:* To minimize the effects of spectral leakage, a window function's FFT should have low amplitude sidelobes away from the centre, and the fall off to the low sidelobes should be rapid.
2. *Worst case processing loss:* It is defined as the sum of scalloping loss and processing Loss. This is a measure of the reduction of output signal to noise ratio resulting from the combination of the window function and the worst case frequency location.
3. *Equivalent noise bandwidth:* A given FFT bin includes contributions from other frequencies including accumulated broadband noise. To detect a narrow band signal in the presence of noise, the noise should be minimized. This can be done by using a narrow bandwidth window function.

Based on these characteristics it is clear that the hamming window satisfies all the above parameters. It has low amplitude sidelobes, has minimum worst case processing loss and has less Equivalent noise bandwidth.

For this implementation ten different Tamil speech signals are used. The Figure 2 presents the performance evaluation of different windows.

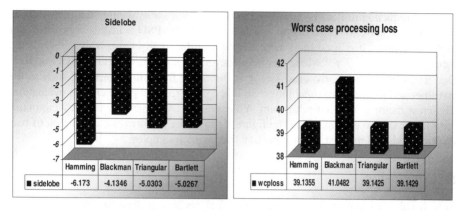

Fig. 2. (a) Comparison of windows based on side lobe   (b) Worst Case Processing Loss

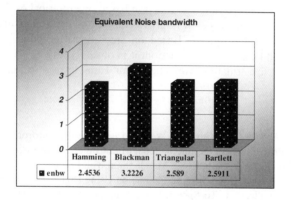

Fig. 2. (c) Equivalent Noise Bandwidth

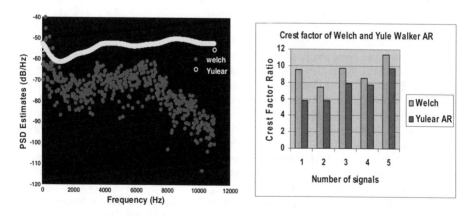

Fig. 3. (a) Comparison of Welch and Yulear Method (b) Comparison based on crest factor

Noise reduction in PSD is considered as a parameter for evaluating the performance of PSD. Pink noise [6] [7] is a signal or process with the PSD is inversely proportional to the frequency. It is used little more loosely to refer to any noise with a PSD of the form $S(f) \propto 1/f^\alpha$ where f is frequency and $0 < \alpha < 2$, with α usually close to 1. One parameter of noise, the crest factor, is important for testing purposes, such as for amplifier and loudspeaker capabilities. So it is considered as a parameter for evaluation. The crest factor is reduced with using Yule Walker (AR) method. It gives minimum crest factor than Welch method. The figure 3 shows the performance evaluation of two PSD methods and its crest factor reduction for 5 speech samples.

5   Conclusion

Speech recognition problems use spectrum analysis as a preliminary measurement to perform speech bandwidth reduction and further acoustic processing. This paper investigates the use of possible window functions that minimize some of the difficulties encountered with default rectangular function based on empirical evaluation. The best window was selected from which it improves the power spectral estimation of speech signal which is also investigated. Based on smoothness of power spectral density estimation curve and crest factor reduction, it is observed that the Yulear Walker-AR method of parametric estimation works better for Tamil speech recognition system.

References

1. Saheli, A.A., Abdali, G.A., Suratgar, A.A.: Speech Recognition from PSD using Neural Network. In: Proceedings of the International MultiConference of Engineers and Computer Scientists, IMECS 2009, Hong Kong, March 18-20, vol. I (2009)
2. Shannon, B.J., Paliwa, K.K.: Spectral Estimation Using Higher-Lag Autocorrelation Coefficients With Applications To Speech Recognition, 0-7803-9243-4/05/$20.00. IEEE, Los Alamitos (2005)
3. Hernando, J., Nadeu, C.: AR modeling of the speech autocorrelation to improve noisy speech recognition. grant TIC 92-0800-C05/04
4. Sedlacek, M., Stoudek, Z.: Design of DSP Windows Using Window Spectrum Zeros Placement. IMEKO (2009) ISBN 978-963-88410-0-1
5. Nagi Reddy, K., Narayana Reddy, S., Reddy, A.S.R.: Parametric Methods of Spectral Estimation of MST Radar Data. IUP Journal of Telecommunications II(3), 55–74 (2010)
6. Sysel, P., Smékal, Z.: Enhanced estimation of power spectral density of noise using the wavelet transform. In: IFIP International Federation for Information Processing, vol. 245, pp. 521–532 (2007), doi:10.1007/978-0-387-74159-8_52
7. Martin, R.: Noise Power Spectral Density Estimation Based on Optimal Smoothing and Minimum Statistics. IEEE Transactions on Speech And Audio Processing 9(5) (July 2001)
8. Suresh Reddy, K., Venkata Chalam, S., Jinaga, B.C.: Efficient Power Spectrum estimation using prewhitening and post coloring technique. International Journal of Recent Trends in Engineering 2(5) (November 2009)
9. Xiao, X., Chng, E.S., Li, H.: Temporal Structure normalization of Speech Feature for Robust Speech Recognition. IEEE signal processing letters 14(7) (July 2007)
10. Haraa, Y., Matsumotob, M., Miyoshia, K.: Method for estimating pitch independently from power spectrum envelope for speech and musical signal. J. Temporal Des. Arch. Environ. 9(1) (December 2009)

Challenges in NP Case-Mapping in Sanskrit Hindi Machine Translation

Kumar Nripendra Pathak and Girish Nath Jha

Special Centre for Sanskrit Studies,
Jawaharlal Nehru University, New Delhi -110067
nri.pathak@gmail.com, girishjha@gmail.com

Abstract. Sanskrit and Hindi are considered structurally close owing to genea-logical relations. However, on a closer look, Hindi appears to have diverged significantly more in terms of structure than in lexical ingenuities. Gender, number distinctions, ergative, postposition, verb group, double causative, echo are some (among many) remarkable structural innovations that Hindi has gone through over the ages. While the structure of Sanskrit vibhakti was fairly organized, the same may not be true for Hindi. The present paper is a study in mapping Sanskrit Noun Phrase (NP) case markers with Hindi for Machine Translation (MT) purposes with a view to evolve cross-linguistic model for Indian languages.

Keywords: case, case marker, *vibhakti*, divergence, transfer, contrastive study.

1 Introduction

It is a well-known fact that Machine Translation (MT) is one of the most difficult areas under computational Linguistics (CL). While translating from Source Language (SL) to Target Language (TL), a human translator may try to translate sentences as close to the meaning of source language and the structure of the target language as possible but the machines have a hard time translating these structures. A machine has to be taught the grammars of the two languages and the significant differences as well for the output to be as close to the input as possible. Though Hindi has descended from Sanskrit and has been heavily influenced by Sanskrit in many ways, yet there are significant differences between them in the structure and behavior of noun phrases. For example, if (1) *bālikā āmram khādati* is translated into Hindi then the accusative marker *am* is not realized in Hindi (2) *bālikā ām khātī hai*. In another example *am* in *vṛkṣam* (3) *vānarah vṛkṣam ārohati,* translates to *par* in Hindi. Similarly the *am* in *gṛham* (4) *sah gṛham adhitiṣṭhati* translates to *me* in Hindi translation. It is therefore important to study these comprehensively and evolve a reasonable Transfer Grammar (TG).

2 Nature of Sanskrit and Hindi

Sanskrit and Hindi belong to the Indo-Aryan family – Sanskrit belongs to and is also referred to as the 'Old Indo-Aryan (OIA)' and Hindi is a Modern Indo-Aryan (MIA)

C. Singh et al. (Eds.): ICISIL 2011, CCIS 139, pp. 289–293, 2011.
© Springer-Verlag Berlin Heidelberg 2011

language. Sanskrit has a well defined grammar thanks to Paṇini (7th BCE). According to Sanskrit grammar, a *pada* is defined as *suptiṅantaṃ padaṃ* i.e a word with inflectional suffix *subanta* or *tiṅanta* is *pada*. *Subanta* is a noun phrase and *tiṅanta* is the verb phrase. According to *ek tiṅ vākyaṃ*, a sentence may have only a verb phrase. Due to gender, number mismatches between Sanskrit and Hindi, the translation from the former into the latter needs significant structural alignments. The neuter in Sanskrit got arbitrarily assigned to masculine or feminine in Hindi. For example, *agni* is masculine in Sanskrit but feminine in Hindi. Sanskrit has three numbers including the dual while Hindi has only two. As a result, the duals in Sanskrit got assigned to plurals in Hindi. For example *dvau aśvau* is marked for dual number but in Hindi *do ghoḍe* is marked for plural. There are more structural differences between the two at the level of verb forms like compound/complex verb forms, verb groups, and also at the level of NPs in terms of ergative, postpositions among many others.

The syntactic and semantic functions of noun phrases are expressed by case-suffixes, postpositions and various derivational and inflectional processes. There are two cases in Hindi: direct and oblique. The direct case is used for nouns and is not followed by any postposition. The oblique case is used for nouns followed by a postposition. Adjectives qualifying nouns in the oblique case in the same phrase will inflect in the same manner. Case-suffixes and postpositions are used to express syntactico-semantic roles. Case suffixes are bound morphemes and are added only to the noun phrases. The NPs in Hindi become oblique when followed by postpositions. The vocative address forms may be preceded by the vocative morphemes *o/he/are*. Except for ergative, dative and passive subjects, the default case marker is null. Ergative marker *ne* is used with subject in perfective aspect with transitive verbs. When *ne* follows a noun, it is written separately and when it follows a pronoun, it is written as one word. *ko* case marker is used in a larger context in Hindi. It is generally assumed as *karma kāraka* (accusative) when verb is in *kṛdanta* (primary derived noun) and shows *anivāryatā* (necessity) of the verb. The *ko* may be used in many cases like agentive (5) *rām ko ghar jānā hai*, accusative (6) *pitā putra ko dekh rahā hai*, dative (7) *mohan rām ko pustak detā hai*, locative (8) *somvār ko paḍhāī hogī*. In imperative, there is no case marker with agent as in *āp baiṭhiye, tum pānī lāo*. etc. But in the sense of '*cāhiye*', there must be '*ko*' case marker in the imperative sentence as in (9) *bālak ko paḍhanā cāhiye*. etc. The *se* is used with instrumental and ablative case frequently but it is also used in other cases like agentive (10) *ram se ab uṭhā nahī jātā hai* which is *sāmarthya-bodhaka*. With accusative, instrumental, ablative, locative, in the sense of eating with subsidiary items (11) *sonu caṭanī se roṭī khātā hai*, with negligence (12) *tū mat paḍh, merī balā se*, with direction (13) *ayodhyā se Mithilā gayā*, with attention (14) *dhyān se suntā hai*, with time (15) *tumhen kitane samay se ḍhūṇḍha rahā hūn*. The marker *ke dvārā* is used in passive sentences (16) *rām ke dvārā rāvaṇ mārā gayā*.The marker *ke liye* is used with dative case (17) *bālak ke liye pāni lao*. The markers *kā, ke, kī, nā, ne, nī, rā re, rī* are genitive case markers (Vajpeyee, 1976).

3 Contrast between Sanskrit and Hindi Case Marking

NPs in Sanskrit have *sup* suffixes as bound morphs while in Hindi they are represented as free morphs and therefore are assigned many roles in the language leading to ambiguities. For example, in Hindi, *ø/ko/se* case markers are used in other cases as well. Similarly in Sanskrit, *kāraka* may have *karma-samjñā* in seven senses - desired, undesired, gambling, locative (*adhi upasarga* with √*sīṅ*, √*sthā* and √*ās*, *abhi* + *ni* + √*viś* , *up/anu/adhi/āṅ*+√*vas*), *akathita kāraka* (by *akathitaṃ ca*), motion etc (by Pāṇini 1.4.52), anger etc (by Pāṇini 1.4.38). In these senses, Hindi may have *ko/me/se/par/ø* case markers.

As Sanskrit is inflectional in nature, *sup pratyaya* (suffix) handles these Hindi postpositions without any ambiguity. Here, in each action, the *vibhaktis,* are assigned by specific rules from the Aṣṭādhyāyī. Therefore, one *vibhakti* may be used to show other *kāraka* (case marker) in Hindi. Examples of accusative in Sanskrit are (18) *āmraṃ khādati,* (19) *vṛkṣam ārohati,* (20) *sītāṃ pariṇināya,* (21) *grāmaṃ adhitiṣṭhati,* (22) *grāmaṃ gacchati,* (23) *krośaṃ gacchati,* (24) *āvāṃ krīḍāṃ paśyāmaḥ* etc. A *vibhakti* in Sanskrit may realize as multiple postpositions in Hindi. In above examples (18-21), *āmraṃ, vṛkṣam, sītāṃ* and *grāmaṃ* have *dvitīyā vibhakti* (accusative), but in Hindi, the meaning is *ām+ø (khātā hai), vṛkṣa par (caḍhatā hai), sītā se (vivāha karatā hai)* and *grām me(rahatā hai)* respectively. In sentences (22) and (23), *grāmaṃ* and *krośaṃ* have also the same case marker and verb *gacchati* is also the same but their translation *gāv jātā hai* and *kos bhar jātā hai* have different sentence structures and meanings. Here *bhar* is inserted with *kos* to denote the distance. In (24), the accusative has no case marker (*hamdono khel dekhte hain*).

Generally for accusative in Hindi, *ko* case marker is used. But *ko* case marker depends on semantic conditions like the object being animate and whether specificity is being intended. *ko* will not be used if the object is in-animate or non specific marking is intended. For example in (25) *grāmaṃ gacchati (gāv jātā hai),* *gāv* is in-animate and non-specific.

In Sanskrit, the 3rd case is used for subject and instrument and *se* or *ke dvārā* case marker comes with them. But in the following example (26) *guṇaiḥ ātmasadṛśīṃ kanyāṃ udvahe (guṇ me apne samān kanyā se tū vivāh kar), guṇaiḥ (tṛtīyā vibhakti)* is used in the sense of *me* case marker in Hindi translation instead of *saptamī* case *guṇeṣu.* When 4th case in Sanskrit is translated into Hindi, *ko/ke liye* case marker is assigned in Hindi. But translating the same *vibhakti* which is used in a particular sense of action or verb, it may have different case marker in Hindi. For example (27) *bālakāya phalaṃ dadāti (bālaka ko phal detā hai)* and (28) *vṛtrāy vajraṃ prāharat (vṛtra par vajra phenkā).* In both, the 4th case is used with *bālaka* and *vṛtra* in Sanskrit, but Hindi translation has *ko* and *par* postpositions respectively. Sentences with genitive case in Sanskrit take *kā/ke/kī/nā/ne/nū/rā/re/rī* case markers in Hindi sentences. But in the example (29) *naiṣa bhāro mama (yah mere liye bojh nahī hai), mama* is *ṣaṣṭhī* and its translation is *mere liye* which appears to be dative in Hindi. In another example (30) *nūtana eṣa puruṣāvatāro yasya bhagavān bhṛgunandano'pi na vīraḥ (ye koī nayā hī puruṣ kā avatār hai jisake liye bhagavān bhṛgunandan bhī vīr nahīn hain),* translation of *yasya* is *jiske liye* in Hindi, whereas literal meaning of *yasya* is *jisakā.* Here *kṛte pada* is assumed in Sanskrit.

Generally locative in Hindi has *me* and *par* case markers, but in this example (31) *vṛkṣaśākhāsvabalambante yatīnāṃ vāsāṃsi (vṛkṣ kī śākhā se muniyon ke vastr laṭake hain) śākhā* is locative and 7th case is used in Sanskrit and its translation has *se* case marker in Hindi.

These are some contrastive properties in case mapping between Sanskrit and Hindi. Some other contrastive properties are given below in the table1.

Table 1. Contrastive property of Sanskrit and Hindi language

Sanskrit	Hindi
Inflectional	Post positional
Three numbers	two numbers: singular & plural
Three genders: masculine, feminine & neuter.	Two genders: masculine and feminine.
No explicator compound verb	Explicator compound verb
Relatively free word order	Less free word order
No ergative	Ergative as special feature
No verb groups	Verb groups (*jātā hai or jā rahā hai*)
No echo	Echo
No double causatives	Have double causatives
No Subject-Verb agreement for gender	Have gender agreement at the level of verbs
Two *vibhaktis* cannot come together	Two *vibhaktis* can come together (*usme se*)
karma-pravacanīya can come after or before a word	*karma-pravacanīya* can come only after a word.
Many *vibhaktis* with *upapada*	Only genitive case with *upapada*
Conjunct *ca* at the end	Conjunct in between words
Adjectives and nouns both have same case markers	Only noun has case marker (except oblique cases)
Accusative is used in multiple *kārakas*	*ko* case marker is used only in accusative and dative

4   Conclusion

For Sanskrit scholars, who know Hindi syntax and different kinds of Hindi usages, case handling may be a matter of common sense, but for computational linguists, making a compatible MT System is a tough task because of problems discussed above. A careful study of case systems of both the languages can be useful in mapping the differences at the transfer grammar level. Tagged lexicons with semantic information like animate/inanimate and verb valency will be needed. Without this, machine will have difficulty in translating sentences like *gṛham gacchati* (*ghar jātā hai*) and *Rāmam āhvayati* (*Rām ko bulātā hai*). As we find the usage of *anugacchati* in different contexts, it is clear that different *upasargas* with the same verb-root plays an important role in case marking in Hindi translation. It can also be found in the use of *nipāta*. For correct output, machine must understand the category and the usage of the words in different senses. After such detailed studies only, transfer grammar rules (rules for structural transfer) can be written for Sanskrit Hindi Machine Translation.

References

1. Chandra, S.: Machine Recognition and Morphological Analysis of Subanta-padas, submitted for M.Phil degree at SCSS, JNU (2006)
2. Guru, K.P.: Hindi-vyakaran, panchshil prakshana, Jaypur (2010)
3. Jha, G.N.: Morphology of Sanskrit Case Affixes: A Computational analysis, Dissertation of M.Phil submitted to Jawaharlal Nehru University (1993)
4. Jha, G.N.: Proposing a computational system for Nominal Inflectional Morphology in Sanskrit. In: Proc. of national seminar on Reorganization of Sanskrit Shastras with a view to prepare their computational database (1995)
5. Kachru, Y.: Aspects of Hindi Grammar. Manohar Publications, New Delhi (1980)
6. Kale, M.R.: A Higher Sanskrit grammar. Motilal Banarasidas, Delhi (1972)
7. Kapoor, K.: Dimensions of panini Grammar: the Indian Grammatical System. D.K. Printworld (P) Ltd., New Delhi (2005)
8. Tripathi, A.R.: Hindi Bhashanushasan, Bihar Hindi Grantha Acadmy, Patna (1986)
9. Vajpayee, K.D.: Hindi abdnusan, Nagari pracrani sabha, Varanasi (1976)

Modified BLEU for Measuring Performance of a Machine-Translation Software

Kalyan Joshi and M.B. Rajarshi

Department of Statistics, University of Pune, Pune 411007, India
`kalyan.joshi@gmail.com`, `m.b.rajarshi@gmail.com`

Abstract. The BLEU score compares various n-grams of words of a MT software with an ideal or reference translation of a sentence. We suggest a Weighted BLEU (WBLEU) which is probably more suitable for translation system for English into an Indian language. Weights are obtained so that the correlation between the weighted BLEU scores and a human evaluator's scores is maximum.

Keywords: BLEU, correlation, optimal weights.

1 Introduction

It is customary to use Bilingual Evaluation Understudy (BLUE) scores (cf. Papineni, K. *et al* (2002)) for judging performance of a Machine Translation Software (MTS). A BLUE score is based on comparing various n-grams (i.e., a set of n consecutive words) in an output of MTS with those in a reference translation, assumed to be an ideal translation. Frequently, the MTS is initially being developed for a corpus which includes translations of sentences already available from various sources.

Limitations of the BLUE scores have been noted in the literature; see, for example, Koehn (2010), p. 229. The BLUE scoring system has been particularly found to have further limitations when applied to MTS translations from English into an Indian language (Ananthakumar *et al* (2007)). This is so for many reasons. Firstly, a sequence of patterns of words in a target Indian language is very much different than the sequence of the same words in English. In the simplest case, the SVO (Subject-Verb-Object) pattern in English leads to the SOV pattern into Hindi, Marathi and a number of Indian languages. Secondly, Indian languages have a high rate of inflexions, for example, changes in a noun to be made in view of prepositions. A minor mistake in such inflexions leads to an incorrect word in the BLEU system and thus at least two incorrect 2-grams. As we illustrate, a BLEU score with equal weights to n-grams, seems to be a rather harsh critique of the MTS. Ananthakumar *et al* (2007) also note possibility of a poor correlation of BLEU scores with human scores, whereas Coughlin (2003), based on a study involving multiple languages, concludes that the BLEU scores are highly correlated with human scores. Other scoring systems such as METEOR can be used, however they are quite complicated to apply, cf. Koehn (2010), p. 228.

C. Singh et al. (Eds.): ICISIL 2011, CCIS 139, pp. 294–298, 2011.

A BLEU score typically gives equal weights to the various n-grams. Though different weights for different n-grams have been mentioned in the literature, it is not clear how to choose the various weights. Here, we propose to choose differential weights for various n-grams. We suggest that the weights be chosen, so that the weighted BLEU score is, on an average, nearest to human scores of sentences. This is equivalent to choosing weights which maximize correlation between the weighted BLEU scores and HS (human scores) for a group of sentences/corpus for which reference translations are available. It needs to be pointed out that an evaluator of MT translations need not be a linguistic expert. It is enough that such a person is in a position to see whether meaning of a sentence in the source language has been conveyed adequately and whether the flavor of the target language has been reasonably maintained. In our experiment below, the set of reference translations were not made available to the human examiner.

Though this is a rather limited study, we believe that it throws some light on the mechanism of the human scoring system.

2   Modifications to a BLEU Score

A simple BLEU score can sometimes result in a heavy penalty for a simple mistake. Consider the following three examples. A full-stop is included in computing BLEU scores.

(i)   English sentence: *She wound the thread around the pencil.*

Output translation: तिने पेन्सिलभोवती दोरा गुंडाळला.   /tine pensilabhovatI doraa guMDaaLalaa.

Reference translation: तिने पेन्सिलीभोवती दोरा गुंडाळला.   /tine pensilIbhovatI doraa guMDaaLalaa.

There is an incorrect inflexion in the output translation for the words "around the pencil". The 1-gram, 2-gram and 3-gram scores are respectively given by 4/5, 2/4 and 1/3. The BLEU score given by $\ln(\text{BLEU}) = 1/3\ln(0.80 * 0.50 * 0.33)$ (where ln refers to the natural logarithm) and equals 0.51. We notice that the weights for logarithms of 1-gram, 2-gram and 3-gram are the same and equal to 1/3.

(ii)   English sentence: *Ram killed Ravan.*

Output translation: रावणाला रामाने मारले. / raavaNaalaa raamaane maaraLe.

Reference translation: रामाने रावणाला मारले. / raamaane raavaNaalaa maaraLe.

Here, the 1-gram and 2-grams are 1 and 1/3 respectively. The BLEU score is 0.5772. Actually, the meaning has been clearly conveyed, thanks to the inflexions, which clearly identifies the subject and the object of the verb "to kill".

(iii)   English sentence: *Complete this work in fifteen minutes.*

Output translation: हे काम पंधरा मिनिटात पूर्ण कर/करा. / he kaama paMdharaa miniTaata pUrNa kara/karaa.

Reference translation: हे काम पंधरा मिनिटांत पूर्ण कर/करा. / he kaama paMdharaa miniTaaMta pUrNa kara/karaa.

Here, there is a single incorrect word wherein a single letter is incorrect. This is so since the Devanaagari script for Marathi indicates plural of "minute" in Marathi as given in the reference translation. The scores are: 1-gram=6/7, 2-gram= 4/6 and 3-gram=2/5 and the BLEU is 0.6114. A human expert would probably assign a very high score for the output translation.

To overcome such problems, we suggest weights which are calculated so as to minimize the distance between the SME score and weighted BLEU score. This is an objective choice of weights. The WBLEU (Weighted BLEU) is defined by

$$\ln(WBLEU) = w(1)\ln G(1) + w(2)\ln G(2) + \ldots + w(r)\ln G(r),$$

where $G(i)$ is the i-gram score, $w(i)$ is the associated weight, $i=1,2,\ldots,r$ and r is the chosen maximum length of a gram. The sum $w(1)+w(2)+\ldots W(r)$ equals 1.The weights $w(i)$'s are obtained so that the Karl Pearson product moment correlation coefficients between the WBLEU scores and human expert's scores for N sentences in a corpus is maximum. For a BLEU score, $w(i)= 1/r$ for all i. Further, if a MT perfectly matches with the reference translation of a sentence, the BLEU coincides with WBLEU.

3   An Experiment Involving Saakava, an English-Marathi Machine Software

Saakava, a system for English into Marathi Machine translation, can be accessed from www.saakava.com. For the present study, we consider the book "My First English-Marathi Dictionary"(2005), published by the Maharashtra Textbook Bureau. It is not only a dictionary; but also illustrates various meanings of a word by giving sentences involving such words and their translations into Marathi. The sentences have been translated so as a student with Marathi mother-tongue can easily see the meaning of the words involved and also how they are to be used in English. This set of translations naturally gives us a set of reference translations of N=2224 sentences. Thus, the number is moderately large. Moreover, the book is at an introductory level. The set is thus ideal for a MTS group to pick up at the entry level.

An individual was asked to score these sentences. She has a good command over English and Marathi both though she is not a linguist. The Mean HS is 0.65 whereas the mean BLEU is 0.50. (Performance of *Saakava* at this stage may not seem to be satisfactory.) The BLEU score is quite lower than the HS.

To compute WBLEU, the optimal weights $w(i)$'s are computed as follows. We first fix a set of $w(i)$'s and compute the correlation coefficient. We then allow $w(i)$'s to vary over a grid in the simplex (a set of non-negative numbers whose total is one). The value of r is the length of an **English sentence**. For the corpse of sentences discussed above, we get following table which gives optimal weights for each gram for a given sentence length r=1,2,…,7.

It seems that the human evaluator is assigning more weights for 1-grams, which implies that one is satisfied if each word has been correctly outputted and that the appropriate sequencing is not so important. Further, there are groups of sentences, where 2-grams carry significant weights. The importance of 2-gram matching is well reflected in the new BLEU system. But, for longer sentences, there is a certain shift to the entire sentence itself.

Table 1. Optimal weights for each gram

Sentence Length(r)	1gram	2gram	3gram	4gram	5gram	6gram	7gram
2	0.77	0.23					
3	0.77	0.1	0.13				
4	0.48	0.52	0	0			
5	0.69	0.31	0	0	0		
6	0.89	0	0	0	0	0.11	
7	0	0	0	1	0	0	0

The WBLEU score is 0.59 thus closer to the HS (0.65). The three correlation coefficients (Cor) between the scores are as given below. They indicate the strength of linear relationship between the various scores. We have Cor (BLEU, HS) = 0.586, Cor (BLEU, WBLEU) = 0.925 and Cor (WBLEU, HS) =0.634.

The correlation between the WBLEU and HS (though higher than the correlation between BLEU and HS) is not very satisfactory, particularly if we wish to take such a score as a reference point. We discuss this point in the next section.

Using the weights as obtained above, we now compute the WBLEU for the three sentences given in the Section 2. For example, for the first sentence, the weights as read from the Table 1, are w(1)=0.77, w(2)= 0.1 and w(3)=0.13 and the WBELU is. 0.6811. For the second sentences, the optimal weights are 1-gram=0.77 and 2-gram= 0.23 and the WBLEU is 0.7767. The third sentence has the same weights as the first sentence and the WBLEU is 0.7570. All the WBLEU scores are higher than the BLEU and probably describe the reality in a better manner.

4 Concluding Remarks

We have demonstrated that a system of non-uniform weights can result in a better association between BLEU scores with unequal weights for various grams and scores of a human evaluator. Though more studies are needed, the present study indicates that for shorter sentences, 1-grams and 2-grams are more important. However, both the BLEU and WBLEU are much less than the HS. We discuss now why this is so.

Firstly, if there is a mistake in a single letter of a word (see examples 1 and 3 above), both the BLEU and WBELU assign a zero score for the 1-gram.This results in a further reduction in the 2-grams, 3-grams etc. A human evaluator seems to regard such an error as less serious than the two machine scoring systems. We suggest that, as we compare sentences based on n-grams, we may compare a word based on the various n-letter-grams. Thus, instead of a zero score, in the second word of the Example 1, the score can be taken to be 5/6, indicating that only one letter has been incorrectly produced by the MT. This may increase the correlation. Work in this direction is in progress.

Secondly, in some situations, there are two equivalent words in Marathi. For example, the word 'door' has two equivalent words in Marathi : 'daara' and 'daravaajaa'. If the MT uses a word different than the one in the reference translation, this results in a substantially low score for the MT. The human evaluator sees no loss in the meaning of a sentence and gives a perfect score. We need to modify the BLEU (and WBLEU) to deal with such a situation. Koehn(2010), p. 227 suggests that we need to have multiple translations and should be willing to give a high or perfect score, if a MT matches with one of the multiple translations. As is well known, even for simple sentences, two or more individuals can come up with different translations, though all convey exactly the same and can be regarded as correct. We are also in the process of developing methods to deal with a situation where there are two or more different words in the target language for a word in the source language. Problem regarding translation the word "you" and its versions ("your", "(to) you") pose similar problems in Hindi and Marathi: a MT may use "tumhI", whereas a reference translation may use "tu" (or vice versa).

References

1. Papineni, K., Roukos, S., Ward, T., Zhu, W.J.: BLEU: a method for automatic evaluation of machine translation ACL-2002. In: 40th Annual meeting of the Association for Computational Linguistics, pp. 311–318 (2002)
2. Coughlin, D.: Correlating Automated and Human Assessments of Machine Translation Quality in MT Summit IX, New Orleans, USA, pp. 23–27 (2003)
3. Ananthakrishnan, R., Bhattacharya, P., Sasikumar, M., Shah, R.M.: Some Issues in Automatic Evaluation of English-Hindi M. In: Bendre, S.M., Mani, I., Sangal, R. (eds.) More Blues for BLEU Natural Language Processing ICON (2007)
4. Koehn, P.: Statistical Machine Translation. Cambridge University Press, Cambridge (2010)
5. My First English-Marathi Dictionary, Maharashtra Raajya Paathyapustak Nirmiti and Abhyaskram Samshodhan Mandal, Pune 411 004 (2005)

A System for Online Gurmukhi Script Recognition

Manoj K. Sachan[1], Gurpreet Singh Lehal[2], and Vijender Kumar Jain[1]

[1] Sant Longowal Institute of Engineering & Technology, Longowal, India
[2] Punjabi University Patiala, India
manojsachan@gmail.com

Abstract. Handwriting recognition is the task of transforming a language represented in its spatial form of graphical marks into its symbolic representation. There are two type of handwriting recognition, offline and online. In offline handwriting recognition, the user writes on the paper which is digitized by the scanner. The output of the scanner is presented as an image to the system which recognizes the writing. In contrast, the online handwriting recognition requires that the user's writing is captured through digitizer pen and tablet before recognition. Online handwriting recognition assumes importance as it is still much more convenient to write with pen as compared to typing on the keyboard. Secondly, these days so many PDAs and handheld devices are used where it is easier to work with stylus then using keyboard. This has motivated research in online handwriting recognition in different languages of the world including Indic scripts such as Tamil, Telgu, Kannada, Devanagari and Gurmukhi. In our work, a system for recognition of Gurmukhi Script is presented. In this work, the input of the user's handwriting is taken as a sequence of packets captured through the movement of stylus or pen on the surface of the tablet. The packet consists of x,y position of the stylus, button(tip of stylus), pressure of the stylus and the time of each packet. The user's writing is preprocessed and is segmented into meaniningful shapes. The segmented shapes are processed to extract features which are Distributed Directional Feature. The feature data is fed to the recognition engine which is a Nearest Neighbor Classifier. The average recognition accuracy is 76% approximately. The block diagram of the system for Online Gurmukhi Script recognition is shown in Fig 1 below. The main strengths of this system is that it takes complete word for segmentation and recognition.

C. Singh et al. (Eds.): ICISIL 2011, CCIS 139, pp. 299–300, 2011.
© Springer-Verlag Berlin Heidelberg 2011

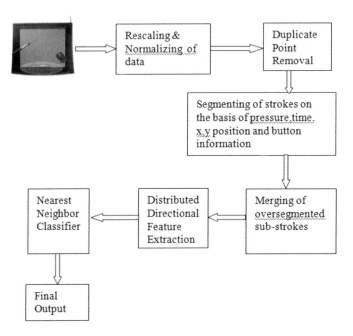

Fig. 1. Block Diagram showing recognition of Online Gurmukhi Script

Spoken Isolated Word Recognition of Punjabi Language Using Dynamic Time Warp Technique

Ravinder Kumar[1] and Mohanjit Singh[2]

[1] Department of Computer Science & Engineering,
Thapar Univeristy, Patiala – 147004, India
ravinder@thapar.edu
[2] Tech., Shaw Communications,
4950 47 St NE, Calgary, Canada
mohanjitsd@gmail.com

Abstract. This research work is to develop a speech recognition system for speaker dependent, real time, isolated words of Punjabi language. The methods used for speech recognition have since been developed and improved with increasing accuracy and efficiency leading to a better human machine interface. In this work, I have developed a speech recognition system, which has a medium size dictionary of isolated words of Punjabi language. The study involved the detailed learning of the various phases of the signal modeling process like preprocessing and feature extraction as well as the study of multimedia API (Application Programming Interface) implemented in Windows 98/95 or above. Visual C++ has been used to program sound blaster using MCI (Media Control Interface) commands. In this system the input speech can be captured with the help of microphone. I have used MCI commands and record speech. The sampling frequency is 16 kHz, sample size is 8 bits, and mono channels. The Vector Quantization and Dynamic Time Warping (DTW) have been used for the recognition system and some modifications have been proposed to noise detection, word detection algorithms. In this work, vector quantization codebook of size 256 is used. This size selection is based on the experimental results. The experiments were performed with different size of the codebook (8, 16, 32, 64, 128, and 256). In DTW, there are two modes: one is training mode and other is testing mode. In training mode the database of the features (LPC Coefficients or LPC derived coefficients) of the training data is created. In testing mode, the test pattern (features of the test token) is compared with each reference pattern using dynamic time warp alignment that simultaneously provides a distance score associated with the alignment. The distance scores for all the reference patterns are sent to a decision rule, which gives the word with least distance as recognized word. Symmetrical DTW algorithm is used in the implementation of this work. The system with small isolated word vocabulary on Punjabi language gives 94.0% accuracy. System can recognize 20 – 24 words per minute of interactive nature with recording time 3 – 2.5 seconds respectively.

Keywords: dynamic time warp (DTW), linear predictive coding (LPC), Punjabi language, vector quantization (VQ), application programming interface (API), media control interface (MCI).

C. Singh et al. (Eds.): ICISIL 2011, CCIS 139, p. 301, 2011.
© Springer-Verlag Berlin Heidelberg 2011

Text-To-Speech Synthesis System for Punjabi Language

Parminder Singh[1] and Gurpreet Singh Lehal[2]

[1] Dept. of Computer Sc. & Engg., Guru Nanak Dev Engg. College, Ludhiana (Pb.) – India
parminder2u@rediffmail.com
[2] Dept. of Computer Science, Punjabi University, Patiala (Pb.) – India
gslehal@gmail.com

Abstract. A Text-To-Speech (TTS) synthesis system has been developed for Punjabi text written in Gurmukhi script. Concatenative method has been used to develop this TTS system. Syllables have been reported as good choice of speech unit for speech databases of many languages. Since Punjabi is a syllabic language, so syllables has been selected as the basic speech unit for this TTS system, which preserves within unit co-articulation effects. The working of this Punjabi TTS system can be divided into two modules: Online Process and Offline Process. Online process is responsible for pre-processing of the input text, schwa deletion, syllabification and then searching the syllables in the speech database. Pre-processing involves the expansion of abbreviations, numeric figures and special symbols etc. Schwa deletion is an important step for the development of a high quality Text-To-Speech synthesis system. Phonetically, schwa is a very short neutral vowel sound, and like all vowels, its precise quality varies depending on the adjacent consonants. During utterance of words not every schwa following a consonant is pronounced. In order to determine the proper pronunciation of words, it is necessary to identify which schwas are to be deleted and which are to be retained. Grammar rules, inflectional rules and morphotactics of language play important role for identification of schwa those are to be deleted. A rule based schwa deletion algorithm has been developed for Punjabi having accuracy of about 98.27%. Syllabification of the words of input text is also a challenging task. Defining a syllable in a language is a complex task. There are many theories available in phonetics and phonology to define a syllable. In phonetics, syllables are defined based upon the articulation. However in phonological approach, syllables are defined by the different sequences of the phonemes. In every language, certain sequences of phonemes are recognized. In Punjabi seven types of syllables are recognized – V, VC, CV, VCC, CVC, CVCC and CCVC (where V and C represents vowel and consonant respectively), which combine in turn to produce words. A syllabification algorithm for Punjabi has been developed having accuracy of about 96.7%, which works on the output of the schwa deletion algorithm.

The Offline process of this TTS system involved the development of the Punjabi speech database. In order to minimize the size of speech database, effort has been made to select a minimal set of syllables covering almost whole Punjabi word set. To accomplish this all Punjabi syllables have been statistically analyzed on the Punjabi corpus having more than 104 million words. Interesting and very important results have been obtained from this analysis those helps to select a relatively smaller syllable set (about first ten thousand syllables

C. Singh et al. (Eds.): ICISIL 2011, CCIS 139, pp. 302–303, 2011.
© Springer-Verlag Berlin Heidelberg 2011

(0.86% of total syllables)) of most frequently occurring syllables having cumulative frequency of occurrence less than 99.81%, out of 1156740 total available syllables. The developed Punjabi speech database stores the starting and end positions of the selected syllable-sounds labeled carefully in a wave file of recorded words. As the syllable sound varies depending upon its position (starting, middle or end) in the word, so separate entries for these three positions has been made in the database for each syllable. An algorithm has been developed based on the set covering problem for selecting the minimum number of words containing above selected syllables for recording of sound file in which syllable positions are marked.

The syllables of the input text are first searched in the speech database for corresponding syllable-sound positions in recorded wave file and then these syllable sounds are concatenated. Normalisation of the synthesized Punjabi sound is done in order to remove the discontinuities at the concatenation points and hence producing smooth, natural sound. A good quality sound is being produced by this TTS system for Punjabi language.

Hand-Filled Form Processing System for Gurmukhi Script

Dharam Veer Sharma and Gurpreet Singh Lehal

Department of Computer Science, Punjabi University, Patiala
dveer72@hotmail.com, gslehal@gmail.com

Abstract. A form processing system improves efficiency of data entry and analyses in offices using state-of-the-art technology. It typically consists of several sequential tasks or functional components viz. form designing, form template registration, field isolation, bounding box removal or colour dropout, field-image extraction, segmentation, feature-extraction from the field-image, field-recognition. The major challenges for a form processing system are large quantity of forms and large variety of writing styles of different individuals.

Some of the Indian scripts have very complex structures e.g. Gurmukhi, Devnagari and Bengali etc. Use of head line, appearance of vowels, parts of vowel or half characters over headline and below the normal characters (in foot) and compound characters makes the segmentation and consequently recognition tasks very difficult.

The present system is a pioneering effort for developing a form processing system for any of the Indian languages. The system covers form template generation, form image scanning and digitization, pre-processing, feature extraction, classification and post-processing. Pre-processing covers form level skew detection, field data extraction by field frame boundary removal, field segmentation, word level skew correction, word segmentation, character level slant correction and size normalization. For feature extraction Zoning, DDD and Gabor filter have been use and for for classification, kNN and SVM have been put to use. A new method has been developed for post processing based on the shape similarity of handwritten characters.

The results of using kNN classifier for different values of k with all features combined are 72.64 percent for alphabets and 93.00 percent for digits. With SVM as classifier and all the features combined, the results improve marginally (73.63 percent for alphabets and 94.83 percent for digits). In this demo we shall demonstrate the working of the whole system.

Keywords: Hand-filled form processing system, form processing system, Gurmukhi script, OCR.

C. Singh et al. (Eds.): ICISIL 2011, CCIS 139, p. 304, 2011.
© Springer-Verlag Berlin Heidelberg 2011

Urdu to Hindi and Reverse Transliteration System

Gurpreet Singh Lehal[1], Tejinder Singh Saini[1], and V.S. Kalra[2]

[1] Punjabi University Patiala 147 002, Punjab, India
[2] Sociology, SOSS, Manchester University, UK
{gslehal,tej74i}@gmail.com, kalra@manchester.ac.uk

Abstract. In spoken form Hindi and Urdu are mutually comprehensible languages but they are written in mutually incomprehensible scripts. This research work aims to bring the Urdu and Hindi speaking people closer by developing a transliteration tool for the two languages. Even though the language is same, but still developing a high accuracy transliteration system is not a trivial job. The accuracies of statistical and rule based Urdu-Hindi and reverse transliteration systems are 97.12% and 99.46% at word level.

1 About the System

South Asia is one of those unique parts of the world where single languages are written in different scripts. This is the case for example with Urdu and Hindi spoken by more than 600 million people, but written in India (500 million) in Devnagri script (a Left to Right script) and in Pakistan (80 million), it is written in Urdu (a Right to Left script based on Arabic). This research work aims to bring the Urdu and Hindi speaking people closer by developing a transliteration tool for the two languages. Even though the language is same, but still developing a high accuracy transliteration system is not a trivial job. The main challenges, in transliteration from Urdu to Hindi and reverse are:

- Missing short vowels and diacritic marks in Urdu text
- One Urdu character can correspond to multiple Devnagri characters.
- No half characters in Urdu corresponding to Devnagri.
- Multiple Hindi words for an Urdu word.
- Word segmentation issues in Urdu including broken and merged words.
- Many combinations of Urdu words are written as a multi-word expression in Hindi.
- Many Urdu words from foreign language particularly English are frequently used, for which separate dictionaries had to be developed.
- No exact equivalent mappings in Hindi for some Urdu Characters.
- Decrease in use of *Nukta* Symbols in Hindi Text
- Difference between Pronunciation and Orthography
- Transliteration of Proper Nouns
- Special rules for handling nasalized sound characters

To meet these challenges, for the first time a hybrid statistical and rule based Urdu-Hindi and reverse transliteration system has been developed. The systems makes use of Hindi and Urdu Corpus, Hindi-Urdu and reverse dictionaries, bigram, trigram

C. Singh et al. (Eds.): ICISIL 2011, CCIS 139, pp. 305–306, 2011.

tables, Urdu/Hindi/English Morphological analysis, Urdu-Hindi character and word lookup tables and rule bases for resolving character and word ambiguity and handling Urdu word segmentation problems. The Urdu to Hindi system has been tested on more than 150 pages of text taken from newspapers and Urdu articles. A transliteration accuracy of more than 97.12% has been achieved, which to the best of our knowledge is the best accuracy achieved so far. The reverse transliteration system has been tested on about 100 pages of text and has 99.46% accuracy at word level.

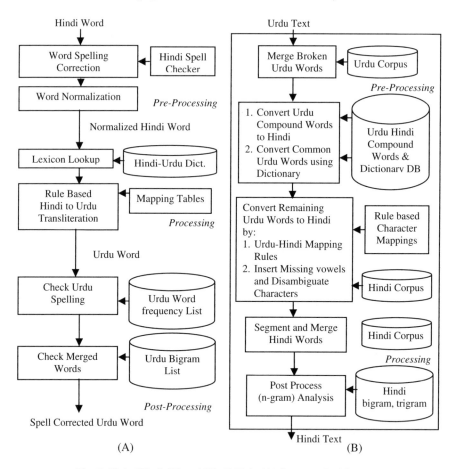

Fig. 1. Urdu-Hindi (B) and Hindi-Urdu (A) Systems Architecture

The main features of the system (http://uh.learnpunjabi.org) are:

1. Direct support for Urdu InPage files, Urdu and Hindi Unicode typing interface with Romanized, Remington, phonetic keyboards and mouse driven typing facility
2. Tool tip support for difficult or complex Urdu word meaning in Hindi text
3. Sending e-mails in Hindi or Urdu Script before or after transliteration
4. Urdu/Hindi Unicode web page transliteration support

The authors will like to acknowledge the support provided by *ISIF grants* for carrying out this research work.

iPlugin: Indian Language Web Application Development Tool

Anup Kanaskar and Vrundesh Waghmare

Centre for Development of Advanced Computing
7th Floor, NSG IT Park,
Sarja Hotel Lane, Aundh
Pune, India – 411007

iPlugin is an Indian Language web application development software tool. iPlugin allows user to type in Indian Languages in web pages over the internet. iPlugin software helps to develop unique interactive applications for users in Indian languages. iPlugin is ideal for creating interactive applications such as online chat, localize database query applications, type blogs in your language, feedback and e-mail in Indian languages, reports or any other such application which requires support of typing in Indian Languages over Web. iPlugin helps in creation of Indian language web content for front end and back end of internet and intranet portal solutions.

iPlugin software allows user to type in browser in various languages of India such as Assamese, Bengali, Bodo, Dogri, Gujarati, Hindi, Kannada, Konkani, Malayalam, Marathi, Maithali, Manipuri (Bengali), Nepali, Oriya, Punjabi (Gurumukhi/Panjabi), Santali, Sanskrit, Tamil and Telugu with download of Indian language tools on client system. All the client side components are digitally signed and hence safe for download.

iPlugin software provides INSCRIPT (DoE), EasyPhonetic (transliteration), Phonetic and Typewriter Keyboard Layouts for user to type Indian languages in web browsers such as IE, FireFox. Client user can type on OS such as Win9x, WinXP, Vista, Windows 7 irrespective of whether they are language enabled.

Design Time controls of iPlugin for ASP, .Net and JSP tag libraries reduce development efforts. Server side support for conversion and storing Indian language data in ISCII as well as in UNICODE on various platforms is possible.

Strengths of iPlugin

- iPlugin supports IE6,E7, IE8, Firefox.
- iPlugin supports 19 Indian lauguages.
- Support for Enhanced INSCRIPT : 4 layer keyboard with UNICODE 5.1 characters.
- Light weight digitally signed components safe for download.
- Phonetic Assistant (Transliteration).
- Semi Transparent On-screen Keyboard.
- Advertisement support in iPlugin Toolbar.

C. Singh et al. (Eds.): ICISIL 2011, CCIS 139, pp. 307–308, 2011.
© Springer-Verlag Berlin Heidelberg 2011

- XPI based tools for Firefox on Windows.
- Support for Indian Rupee symbol in Inscript keyboard layout.
- Support UNICODE and ISCII base storage at database.
- Design time controls in JSP, .Net for faster web application development.
- Works with internet and intranet applications.

Limitation

Components needs to be downloaded once for enabling particular Indian language.

Comparison

iPlugin	Other
Keyboard layout supported Inscript, phonetic, easy phonetic, typewriter.	Phonetic or Inscript layout is given.
Supports 19 Indian language	Supports Less number of Indian language.
Organized keyboard for old as well as advanced user.	Phonetic or Inscript keyboard support is provided.
Design time controls for web application development with major technology.	No such design time controls are provided.
Toolbar is provided for ease of use.	No toolbar is provided.

An OCR System for Printed Indic Scripts

Tushar Patnaik

CDAC Noida

The project 'Development of Robust document Analysis and Recognition for printed Indian Scripts' is a Department of Information Technology sponsored project to develop OCR for printed Indian scripts. A consortia led by IIT Delhi has completed the phase –I in OCR .The consortia members include

1. IIT Delhi
2. IISC Bangalore
3. ISI Kolkatta
4. IIIT Hyderabad
5. Central University ,Hyderabad
6. Punjabi University,Patiala
7. MS University,Baroda
8. Utkal University,Bhubaneswar
9. CDAC Noida
10. CDAC Pune

Different consortia members are responsible for different language OCRs like Punjabi University has contributed. Gurumukhi OCR ,IIIT Hyderabad for Malalayam OCR etc.CDAC Noida has done the integration of OCRs with pre processings.

OCR System is developed for printed Indian scripts, which can deliver desired performance for possible conversion of legacy, printed documents into electronically accessible format and handle seven Indian scripts (Devanagari, Gurumukhi, Malayalam, Tamil, Telugu, Kannada, Bangla).

The GUI provides various options for the user to play with the integrated system

- Basic image enhancement and editing tools (cropping, rotation, zoom in/zoom out, orientation, binarization, noise
  removal etc.).
- Running individual modules successively for obtaining final OCR output.
- End-to-End OCR.
- Workflows for different combination of pre processing routines.
- Text editing tool coupled with dictionary.

Strengths:-

1) OCR System Supports Seven Indian Scripts.
2) Character level accuracy of OCRs is above 95%
3) Block Segmentation and Layout retention Engine
4) User Interactive

C. Singh et al. (Eds.): ICISIL 2011, CCIS 139, pp. 309–310, 2011.
© Springer-Verlag Berlin Heidelberg 2011

Limitations:-

1) Works only on single column text documents.
2) OCR accuracy varies with font size,broken characters,local skew .

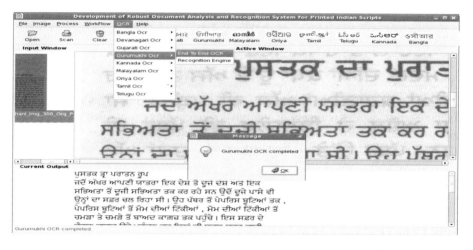

OCR Graphical User Interface

Gujarati Text – To – Speech System

Samyak Bhuta and S. Rama Mohan

The M S University of Baroda, Vadodara, Gujarat, India

The need for text-to-speech systems in various languages is obvious with the current fast paced technology development in information and communication technology. Keeping the Gujarati language, intimately used by 55 odd million people in India and abroad, abreast with technology development is not just logical but heartfelt.

The Gujarati Text-to-speech system should be an apparatus that should take arbitrary Gujarati text as input and should produce the equivalent speech sounds keeping the phonetic and prosodic concerns intact. The quality of the generated sound output should be determined by two parameters viz., intelligibility and naturalness.

With these objectives in mind, a Gujarati Text-to-Speech system has been developed at the Resource Center for Indian Language Technologies for Gujarati at The M. S. University of Baroda, Vadodara with financial support from the Ministry of Communications and Information Technology of Government of India.

The approach used in synthesizing the speech is that of concatenation of segments of pre-recorded speech sound. The system's core engine has been developed to operate with any type of concatenation unit i.e., partneme, diphone, disyllable etc. (or mix of them) to form a desired output. It is to be noted that the speech engine has been developed from the ground level instead of borrowing an existing engine, for the sake of greater flexibility that it would provide.

The engine is designed to take the speech input in International Phonetic Alphabet. This also helps in achieving the separation of concerns with respect to the language independent aspects of the system from those of the language dependent parts. In our experiments we have found that working on the disyllable level as concatenation units provides better results than working with the partneme or diphoneme as the concatenation units. This happens since most of the concatenations occur at the midpoints of the vowel sounds.

Results of the output sound from the Gujarati TTS system are encouraging, being both intelligible and natural to a good extent. The system also takes care of common abbreviations and pronouncing numbers by pre-processing. But the database of disyllables is still short of being complete. Efforts to complete the data base are continuing.

Many improvements can be brought to the system in addition to completing the data base of disyllables. These include making system more efficient from computational point of view. Another direction in which the quality of synthesized speech can be significantly improved is the introduction of prosodic inputs that could be garnered by studying the Gujarati phonology.

C. Singh et al. (Eds.): ICISIL 2011, CCIS 139, p. 311, 2011.
© Springer-Verlag Berlin Heidelberg 2011

Large Web Corpora for Indian Languages

Adam Kilgarriff and Girish Duvuru

Lexical Computing Ltd.,
Brighton, UK

Abstract. A crucial resource for language technology development is a corpus. It should, if possible, be large and varied: otherwise it may well simply fail to cover all the core phenomena of the language, so tools based on it will sometimes fail because they are encountering something which was not encountered in the development corpus. They are critical for the development of morphological analysers because they show a good sample of all the words, in all their forms, that the analyser might be expected to handle. Since the advent of the web, corpora development has become much easier: now, for many languages, of many text types, vast quantities of text are available by mouse-click (Kilgarriff and Grefenstette 2003).

'Corpora for all' is our company's mission. We want to encourage corpus use for linguists, language learners and language technologists in all sorts of contexts. To that end we have set up a 'Corpus Factory' which quickly and efficiently creates general corpora, from the web, for any widely-spoken languages (Kilgarriff et al 2010). To start with we addressed the large languages of Europe and East Asia: now we have large corpora for all of those and have moved on to the many large languages (Hindi, Bengali, Malayalam, Telugu, Kannada, Urdu, Gujarati, Tamil etc) of the subcontinent. At time of writing, we have corpora that are all multi-million-word. We believe these corpora are larger and more varied than any others available for the languages in question.

Once we have created a corpus, we load it into the Sketch Engine corpus query system (Kilgarriff et al 2004) and make them available through the web service at http://www.sketchengine.co.uk. (Sign up for a free trial; all the corpora listed above, and more as the months proceed, will be available for you to explore.)

The Sketch Engine is a web-based Corpus Query System, which takes as its input a corpus of any language with an appropriate level of linguistic mark-up and offers a number of language-analysis functions like Concordance, word sketches, distributional thesaurus and sketch difference. Concordance is display of all occurrences from the corpus for a given query. This system accepts simple queries (lemma) as well as complex queries in CQL [4] format. A Word Sketch is a corpus-based summary of a word's grammatical and collocational behaviour. This also checks to see which words occur with the same collocates as other words, and on the basis of this data it generates a "distributional thesaurus". A distributional thesaurus is an automatically produced "thesaurus" which finds words that tend to occur in similar contexts as the target word. And finally Sketch Difference is a neat way of comparing two very similar words: it shows those patterns and combinations that the two items have in common, and also

C. Singh et al. (Eds.): ICISIL 2011, CCIS 139, pp. 312–313, 2011.
© Springer-Verlag Berlin Heidelberg 2011

those patterns and combinations that are more typical of, or unique to, one word rather than the other. Ideally, prior to loading into the Sketch Engine, we lemmatise and part-of-speech-tag the data and we can then prepare word sketches, distributional thesaurus and also sketch differences.

We are currently looking for collaborators with expertise in lemmatisers and taggers for one or more of the Indian languages, so we can jointly prepare world class resources for Indian languages to match those for European and East Asian ones.

References

1. Kilgarriff, A., Grefenstette, G.: Introduction to a Special Issue on Web as Corpus. Computational Linguistics 29(3) (2003)
2. Kilgarriff, A., Rychly, P., Smrz, P., Tugwell, D.: The Sketch Engine. In: Proc. EURALEX. Lorient, France (2004)
3. Kilgarriff, A., Reddy, S., Pomikalek, J., Avinesh, P.V.S.: A Corpus Factory for Many Languages. In: Proc. Language Resource and Evaluation Conference, Malta (May 2010)
4. Christ, O.: A modular and flexible architecture for an integrated corpus query system. In: COMPLEX 1994, Budapest (1994)

Localization of EHCPRs System in the Multilingual Domain: An Implementation

Sarika Jain[1], Deepa Chaudhary[2], and N.K. Jain[3]

[1] Department of Computer Application, CCS University, Meerut, Uttar Pradesh
jasarika@gmail.com
[2] Department of Computer Science and Engineering,
Indraprastha Engineering College, Ghaziabad
[3] Department of Electronics, Zakir Husain College, University of Delhi, New Delhi
drnkjain3365@gmail.com

Abstract. The increase of cross-cultural communication triggered by the Internet and the diverse language distribution of Internet users intensifies the needs for the globalization of the online intelligent system The Extended Hierarchical Censored Production Rules (EHCPRs) system might act as a generalized intelligent agent which takes care of context sensitivity in its reasoning. Efforts are to make the EHCPRs system online which can be localized as per the requirement of any specific multilingual domain of users.

Keywords: Localization, Internationalization, Context, EHCPRs System.

1 System Architecture

An Extended Hierarchical Censored Production Rule (EHCPR) is a unit of knowledge for representation in the EHCPRs based intelligent system. Any concept or object can be represented by employing the same uniform structure of an EHCPR. In an EHCPR, there are various operators to define different relations or dependencies of the objects with other objects. The operators are filled with fillers, which can be either atomic values or link to other EHCPRs. The EHCPRs System has two major components (figure 1): Declarative knowledge and procedural knowledge. The knowledge base consists of all the EHCPRs representing rules, definitions, or structures. The database stores all the instances of the concepts or objects that the system has come across till date in the form of EHCPRs.

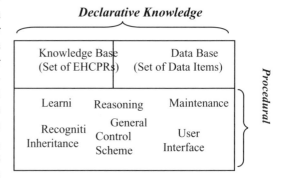

Fig. 1. Components of EHCPRs System

C. Singh et al. (Eds.): ICISIL 2011, CCIS 139, pp. 314–316, 2011.

Every EHCPR in the knowledge base is an instance of the class EHCPR defined in programming language Java as follows:

class EHCPR {String concept;
LinkedList preConditions;
EHCPR generality;
LinkedList specificity;
LinkedList censors;
LinkedList hasPart;
LinkedList hasProperty;
LinkedList hasInstance;   }

2   System Working

The EHCPRs System has been developed in India as a Ph.D. work without any financial help and is very widely published [ref. 1 - 3]. A prototype of the EHCPRs system is prepared to be deployed on the Internet for the online use by the multilingual global community of its users.

The EHCPRs system exhibit changes in precision of the decision with changes in constraint of amount of resources, i.e., more time, more precise answer. Also it acts with the different level of precision with change in context of time, place, or emotions (i.e., state) of the subject, system or both. The EHCPRs system is helpful for answering queries of the type: *"What is X doing"* [ref. 2], and is given in Table1.

1. Display: The EHCPRs System exhibits the characteristics of logic, Production Rule, Semantic Network, Frame, Neural Network etc., in representation, and refer the Figure 2, as a snapshot of the system generated Semantic Network.

2. Reasoning:
    Question Answering: If user asks the system "What is Dolly Doing". The system provides a context sensitive answer, Table 1.

    Recognition: If user looks at an flying object and asks the system about it, then the system may reply that either it is a Crow, or Kite.

Table 1. Variable Precise Reply with Change in Context

Context (Fuzzy Priority)	Output with different Specificity and Certainty
Very Low Priority User	Specificity Level is 0 *"Dolly is in city Delhi" with certainty 0.476*
Low Priority User	Specificity Level is 0*"Dolly is in city Delhi" with certainty 0.7917*
Medium Priority User	*"Dolly is at home" with certainty 0.624*
High Priority User	Specificity Level is 0 *"Dolly is in city Delhi" with certainty 0.7546* Specificity Level is 1 *"Dolly is outdoor" with certainty 0.6324000000000001*Specificity Level is 1. *"Dolly is working outdoor" with certainty 0.5565120000000001*
Very High Priority User	Specificity Level is 0 *"Dolly is in city Delhi" with certainty 0.8623999999999999* Specificity Level is 1 *"Dolly is outdoor" with certainty 0.7812* Specificity Level is 2 *"Dolly is working outdoor" with certainty 0.7343280000000001*

3. Search Engine to support the question answering   with inheritance and recognition
4. Learning: The system has a Knowledge base (KB) and a database (DB). KB contains knowledge in the form of EHCPRs. Lot many EHCPRs have been hard coded. More will be learned through the interaction of the EHCPRs System with the users [ref. 3].
5. User Interface employing the Natural language processing, Vision and so on.

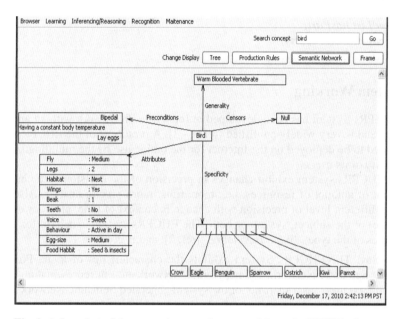

Fig. 2. A Snapshot of the semantic network extracted from the EHCPRs System

3   Conclusion

Efforts are to develop the EHCPRs system as a machine tool which can be localized as per the requirement of any specific domain of users. The first step is internationalization, i.e., separating the locale dependencies from the source code, and then comes localization, i.e., adapting to the needs of a given locale.

References

1. Jain, N.K., Bharadwaj, K.K., Marranghello, N.: Extended Hierarchical Censored Production Rules (EHCPRs) System: An Approach toward Generalized Knowledge Representation. Journal of Intelligent Systems 9, 259–295 (1999)
2. Jain, S., Jain, N.K.: Generalized Knowledge Representation System for Context Sensitive Reasoning: Generalized HCPRs System. In: Artificial Intelligence Review, vol. 30, pp. 39–42. Springer, Heidelberg (2009)
3. Jain, S., Jain, N.K.: Acquiring Knowledge in Extended Hierarchical Censored Production Rules (EHCPRs) System. International Journal of Artificial Life Research 1(4), 10–28 (2010)

Author Index

Printing: Mercedes-Druck, Berlin
Binding: Stein + Lehmann, Berlin